VOLUME SIX HUNDRED AND SEVENTY NINE

METHODS IN
ENZYMOLOGY

Integrated Methods in Protein
Biochemistry: Part B

METHODS IN ENZYMOLOGY

Editors-in-Chief

ANNA MARIE PYLE

Departments of Molecular, Cellular and Developmental Biology and Department of Chemistry
Investigator, Howard Hughes Medical Institute
Yale University

DAVID W. CHRISTIANSON

Roy and Diana Vagelos Laboratories
Department of Chemistry
University of Pennsylvania
Philadelphia, PA

Founding Editors

SIDNEY P. COLOWICK and NATHAN O. KAPLAN

VOLUME SIX HUNDRED AND SEVENTY NINE

METHODS IN
ENZYMOLOGY

Integrated Methods in Protein
Biochemistry: Part B

Edited by

ARUN K. SHUKLA
*Department of Biological Sciences and Bioengineering
Indian Institute of Technology,
Kanpur, India*

ACADEMIC PRESS

An imprint of Elsevier

ELSEVIER

Academic Press is an imprint of Elsevier
50 Hampshire Street, 5th Floor, Cambridge, MA 02139, United States
525 B Street, Suite 1650, San Diego, CA 92101, United States
The Boulevard, Langford Lane, Kidlington, Oxford OX5 1GB, United Kingdom
125 London Wall, London, EC2Y 5AS, United Kingdom

First edition 2023

ISBN: 978-0-323-99264-0
ISSN: 0076-6879

For information on all Academic Press publications
visit our website at https://www.elsevier.com/books-and-journals

Publisher: Zoe Kruze
Developmental Editor: Federico Paulo S. Mendoza
Production Project Manager: Sudharshini Renganathan
Cover Designer: Greg Harris

Typeset by STRAIVE, India
Transferred to Digital Printing 2023

Working together
to grow libraries in
developing countries

www.elsevier.com • www.bookaid.org

Contents

Contributors

Rei Abe-Yoshizumi
Department of Life Science and Applied Chemistry, Nagoya Institute of Technology, Nagoya, Japan

Tomoko Amimoto
Natural Science Center for Basic Research and Development, Hiroshima University, Higashi-Hiroshima, Japan

Tomoyuki Araki
Department of Biochemistry, Saitama Medical University, Saitama, Japan

Tom Casimir Bamberger
Department of Molecular Medicine, The Scripps Research Institute, La Jolla, CA, United States

Alexandra N. Barlow
Department of Chemistry, University of Massachusetts, Amherst, MA, United States

Misti Cartwright
Department of Chemistry and Biochemistry, University of Maryland Baltimore County, Baltimore, MD, United States

Vanessa Chaplin Momaney
Department of Chemistry, University of Massachusetts, Amherst, MA, United States

Shikha S. Chauhan
Department of Biochemistry and Molecular Biology, Pennsylvania State University, University Park, PA, United States

Shawn Chen
Global Health Drug Discovery Institute, Haidian, Beijing, China

Radim Chmelik
CEITEC—Central European Institute of Technology; Institute of Physical Engineering, Faculty of Mechanical Engineering, Brno University of Technology, Brno, Czech Republic

Florian Csarman
Department of Food Science and Technology, Institute of Food Technology, University of Natural Resources and Life Sciences, Vienna, Austria

Vincent G.H. Eijsink
Faculty of Chemistry, Biotechnology and Food Science, NMBU—Norwegian University of Life Sciences, Ås, Norway

Sharon N. Greenwood
Department of Molecular Biology, Rowan University School of Osteopathic Medicine, Stratford, NJ, United States

Koichi Honke
Department of Biochemistry, Kochi University Medical School, Nankoku, Japan

Shoko Hososhima
Department of Life Science and Applied Chemistry, Nagoya Institute of Technology, Nagoya, Japan

Isabella Jaen Maisonet
Department of Chemistry, University of Massachusetts, Amherst, MA, United States

Hideki Kandori
Department of Life Science and Applied Chemistry; OptoBioTechnology Research Center, Nagoya Institute of Technology, Nagoya, Japan

Takayasu Kawasaki
Accelerator Laboratory, High Energy Accelerator Research Organization, Tsukuba, Ibaraki, Japan

Michael J. Knapp
Department of Chemistry, University of Massachusetts, Amherst, MA, United States

Norihiro Kotani
Medical Research Center; Department of Biochemistry, Saitama Medical University, Saitama, Japan

Rashmi S. Kulkarni
Department of Molecular Biology, Rowan University School of Osteopathic Medicine, Stratford, NJ, United States

Roland Ludwig
Department of Food Science and Technology, Institute of Food Technology, University of Natural Resources and Life Sciences, Vienna, Austria

Lindsie Martin
Department of Chemistry and Biochemistry, Price Family Foundation Institute of Structural Biology, Stephenson Life Sciences Research Center, University of Oklahoma, Norman, OK, United States

Michael A. Mingroni
Department of Chemistry, University of Massachusetts, Amherst, MA, United States

Douglas A. Mitchell
Department of Chemistry; Carl R. Woese Institute for Genomic Biology; Department of Microbiology, University of Illinois at Urbana-Champaign, Urbana, IL, United States

Arisa Miyagawa-Yamaguchi
Department of Biochemistry, Kochi University Medical School, Nankoku, Japan

Kazuhiro Nakamura
Department of Laboratory Sciences, Gunma University Graduate School of Health Sciences, Maebashi, Gunma, Japan

Miyako Nakano
Graduate School of Integrated Sciences for Life, Hiroshima University, Higashi-Hiroshima, Hiroshima, Japan

Hisashi Okumura
Exploratory Research Center on Life and Living Systems (ExCELLS); Institute for Molecular Science, National Institutes of Natural Sciences; Department of Structural Molecular Science, SOKENDAI (The Graduate University for Advanced Studies), Okazaki, Aichi, Japan

Sandra Pankow
Department of Molecular Medicine, The Scripps Research Institute, La Jolla, CA, United States

Rakhi Rajan
Department of Chemistry and Biochemistry, Price Family Foundation Institute of Structural Biology, Stephenson Life Sciences Research Center, University of Oklahoma, Norman, OK, United States

Saadi Rostami
Department of Chemistry and Biochemistry, Price Family Foundation Institute of Structural Biology, Stephenson Life Sciences Research Center, University of Oklahoma, Norman, OK, United States

Safoura Salar
Department of Biological Sciences, Virginia Polytechnic Institute & State University, Blacksburg, VA, United States

Florian D. Schubot
Department of Biological Sciences, Virginia Polytechnic Institute & State University, Blacksburg, VA, United States

Lorenz Schwaiger
Department of Food Science and Technology, Institute of Food Technology, University of Natural Resources and Life Sciences, Vienna, Austria

Kyle E. Shelton
Department of Chemistry; Carl R. Woese Institute for Genomic Biology, University of Illinois at Urbana-Champaign, Urbana, IL, United States

Aaron T. Smith
Department of Chemistry and Biochemistry, University of Maryland Baltimore County, Baltimore, MD, United States

Ahrum Son
Department of Molecular Medicine, The Scripps Research Institute, La Jolla, CA, United States

Anton A. Stepnov
Faculty of Chemistry, Biotechnology and Food Science, NMBU—Norwegian University of Life Sciences, Ås, Norway

Verna Van
Department of Chemistry and Biochemistry, University of Maryland Baltimore County, Baltimore, MD, United States

Chu Wang
Synthetic and Functional Biomolecules Center, Beijing National Laboratory for Molecular Sciences, Key Laboratory of Bioorganic Chemistry and Molecular Engineering of Ministry of Education, College of Chemistry and Molecular Engineering; Peking-Tsinghua Center for Life Sciences, Academy for Advanced Interdisciplinary Studies, Peking University, Beijing, China

Heng Wang
Global Health Drug Discovery Institute, Haidian, Beijing, China

Emily E. Weinert
Department of Biochemistry and Molecular Biology; Department of Chemistry, Pennsylvania State University, University Park, PA, United States

Brian P. Weiser
Department of Molecular Biology, Rowan University School of Osteopathic Medicine, Stratford, NJ, United States

Fan Yang
Synthetic and Functional Biomolecules Center, Beijing National Laboratory for Molecular Sciences, Key Laboratory of Bioorganic Chemistry and Molecular Engineering of Ministry of Education, College of Chemistry and Molecular Engineering, Peking University, Beijing, China

John R. Yates III
Department of Molecular Medicine, The Scripps Research Institute, La Jolla, CA, United States

Alice Zenone
Department of Food Science and Technology, Institute of Food Technology, University of Natural Resources and Life Sciences, Vienna, Austria

Daniel Zicha
CEITEC—Central European Institute of Technology; Institute of Physical Engineering, Faculty of Mechanical Engineering, Brno University of Technology, Brno, Czech Republic

Preface

Probing the structure, function, and regulation of proteins is a key aspect of modern biology as it helps decipher the fundamental mechanisms of biological processes. This has direct relevance toward understanding the intricacies of cellular, physiological, pathophysiological, and disease conditions with potential implications for the design and development of novel therapeutics. Protein biochemistry in its broadest sense is an integral part of almost every life science laboratory across the world in one form or the other. The methodological advances in this area continue to emerge at incredible speed, and they play an instrumental role in making new discoveries, refining the existing mechanisms, and establishing new paradigms to push the field forward. In this backdrop, we present three volumes of *Methods in Enzymology* focused on "Integrated Methods in Protein Biochemistry" covering a broad range of topics including protein characterization, posttranslational modifications, protein-protein interactions, structural visualization, and computational analysis. Although protein biochemistry is arguably a vast topic, we have tried to touch upon as many aspects as possible in these three volumes. My sincere thanks to all the authors for taking time to write these chapters despite their busy schedules, and these volumes would not have been possible without their contributions. I acknowledge the wonderful production team of *Methods in Enzymology* for keeping the project on track and for all their efforts in handing the chapters in a timely manner. I very much look forward to your feedback and hope that you find this compendium useful in your research and teaching.

ARUN K. SHUKLA, PhD
Department of Biological Sciences and Bioengineering
Indian Institute of Technology, Kanpur, India

CHAPTER ONE

Biochemical analysis of protein–protein interfaces underlying the regulation of bacterial secretion systems

Safoura Salar and Florian D. Schubot*

Department of Biological Sciences, Virginia Polytechnic Institute & State University, Blacksburg, VA, United States
*Corresponding author: e-mail address: fschubot@vt.edu

Contents

Methods in Enzymology, Volume 679
ISSN 0076-6879
https://doi.org/10.1016/bs.mie.2022.07.030

1

Abstract

Bacterial pathogens such as *Pseudomonas aeruginosa* use complex regulatory networks to tailor gene expression patterns to meet complex environmental challenges. *P. aeruginosa* is capable of causing both acute and chronic persistent infections, each type being characterized by distinct symptoms brought about by distinct sets of virulence mechanisms. The GacS/GacA phosphorelay system sits at the heart of a complex regulatory network that reciprocally governs the expression of virulence factors associated with either acute or chronic infections. A second non-enzymatic signaling cascade involving four proteins, ExsA, ExsC, ExsD, and ExsE is a key player in regulating the expression of the type three secretion system, an essential facilitator of acute infections. Both signaling pathways involve a remarkable array of non-canonical interactions that we sought to characterize. In the following section, we will outline several strategies, we adapted to map protein–protein interfaces and quantify the strength of biomolecular interactions by pairing complex mutational analyses with FRET binding assays and Bacterial-Two-Hybrid assays with appropriate functional assays. In the process, protocols were developed for disrupting large hydrophobic interfaces, deleting entire domains within a protein, and for mapping protein–protein interfaces formed primarily through backbone interactions.

1. Introduction

The bacterium *Pseudomonas aeruginosa* is an opportunistic pathogen that may cause many distinct types of infections in a human host (Gellatly & Hancock, 2013; Sadikot, Blackwell, Christman, & Prince, 2005). While acute infection may occur in the heart, the eyes, an open wound and the bloodstream, chronic persistent biofilm infections by *Pseudomonas aeruginosa* are most prevalent in the lung tissues of predisposed individuals such as people suffering from cystic fibrosis (Depuydt, 2006; Gaynes & Edwards, 2005; Kadri, 2020; Mayhall, 2003; Sadikot et al., 2005; Wine, 1999). Beyond causing infections, *Pseudomonas aeruginosa* also displays a remarkable adaptability to survive in many environmental niches (Moradali, Ghods, & Rehm, 2017). Underlying this adaptability is a complex array of regulatory mechanisms that coordinates expression of diverse virulence factors and other adaptive responses. Compared with other bacterial species, *P. aeruginosa* has a large genome, containing approximately 6.3 million base pairs. *P. aeruginosa* uses over sixty two-component and phosphorelay systems to assimilate environmental cues into the appropriate responses at the gene expression level (Stover et al., 2000; Taylor, Zhang, & Mah, 2019). At the heart of these signaling systems is a sensor histidine kinase that either directly (two-component) or indirectly (phosphorelay) phosphorylate a response regulator (Francis et al., 2018; Liu, Sun, Zhu, & Liu, 2019; Zschiedrich, Keidel, & Szurmant, 2016). Response regulators are usually transcription factors, which, upon phosphorylation, regulate the expression of an associated regulon (Lohrmann & Harter, 2002). For many years the prevailing paradigm for these phosphor-transfer pathways postulated that crosstalk between different chains was selected against by evolution (Laub & Goulian, 2007). However, in recent years, this assumption has been refuted through the discovery of multiple webbed multi-kinase networks (MKNs) (Francis, Stevenson, & Porter, 2017). These MKNs use a variety of non-canonical interactions to coordinate gene expression (Francis et al., 2017; Francis & Porter, 2019). Many of the underlying mechanisms have yet to be characterized, although patterns are emerging. The signaling histidine kinase GacS and its cognate response regulator GacA are part of such a network encompassing at least seven linked phosphotransfer chains in *Pseudomonas aeruginosa* (Chambonnier et al., 2016; Goodman et al., 2004). In particular, our work focused on elucidating the complex interactions of GacS with the sensor histidine kinase family protein RetS. In the process, we applied several approaches to characterize their bimolecular interactions and map their interaction interfaces. In addition to sensor histidine kinases, *Pseudomonas aeruginosa* also employs the non-enzymatic ExsA-ExsD-ExsC-ExsE signaling cascade to link expression of genes associated with the type three secretion system, a major virulence mechanism of *Pseudomonas aeruginosa*, to the host-cell contact (Hauser, 2010). The interactions within this pathway are entirely based on an affinity sink, involving formation and dissociation of three distinct protein–protein complexes (Zheng et al., 2007). Through structural and biochemical studies, we examined the unusual and complex interactions between these four proteins, which are accompanied by shifting oligomeric states and distinct complex stoichiometries. Prior to host cell contact, the transcriptional activator ExsA is sequestered through binding to ExsD in a 1:1 complex, while the T3SS secretion chaperone ExsC forms a 2:1 complex with the small effector protein ExsE (Bernhards, Xing,

Vogelaar, Robinson, & Schubot, 2009; Thibault, Faudry, Ebel, Attree, & Elsen, 2009). Upon host cell contact, ExsE is secreted *via* the type three secretion system, thus liberating ExsC, which in turn releases ExsA by forming a 2:2 complex with ExsD (Galle, Carpentier, & Beyaert, 2013). ExsD alone is trimeric, a feature which may serve to add an element of thermoregulation to this cascade (Bernhards, Marsden, Esher, Yahr, & Schubot, 2013). Mapping the ExsD-ExsA interface proved particularly challenging because the complex was recalcitrant to crystallization and the bimolecular interface appears to be formed primarily through protein backbone interactions. Therefore, the traditional approach of replacing key side-chains through site-directed mutagenesis was not successful.

In the following sections, using the described signaling systems as models, we will outline effective and simple quantitative and qualitative approaches to biochemically characterize bimolecular interfaces. We will address particular challenges associated with obligate interfaces, the characterization of homodimeric interactions, interactions with high on and off rates, and interactions mediated primarily by backbone contacts. Along the way, we will describe our approach to introducing single site-specific mutations to actively disrupt putative protein–protein interfaces rather than performing the traditional alanine scanning analysis. Moreover, we will outline the step-by-step modeling process we applied on several occasions to delete entire domains within a protein sequence by bridging the adjoining regions with a stretch of multiple glycine residues. Not described are specific details of the *in vitro* autophosphorylation, *in vitro* transcription, and *in vivo* biofilm assays that have been used to functionally characterize the various protein variants. These assays have been published elsewhere in great experimental detail not only for the specific signaling systems but as routine microbiological approaches.

2. Case study 1. Quantifying homodimeric interactions using FRET-based *in vitro* titration

While isothermal titration calorimetry (ITC) remains the method of choice for measuring the dissociation constants of bimolecular interactions *in vitro*, we found that this technique did not work well for our systems. In general, ITC is well-suited for measuring interactions associated with significant changes in enthalpy (Duff, Grubbs, & Howell, 2011). However, many protein–protein interfaces are formed through largely hydrophobic interactions that involve small heat changes but are instead driven but by large gains in entropy. ITC is particularly unsuitable for measuring homodimerization because the sample in the cell and the titrant as the same. Therefore, dimerization will produce no net heat change. In such instances, we found Fluorescence Resonance Energy Transfer (FRET)-based measurements well-suited for quantifying homomeric and entropy-driven bimolecular interactions. We first applied this approach after the crystal structure of the periplasmic sensory domain of RetS (RetS$_{peri}$) suggested the formation of an asymmetric homodimer (Jing, Jaw, Robinson & Schubot, 2010). We sought to quantify this interaction using an *in vitro* FRET assay. A contact analysis of the asymmetric unit of the crystal guided the site-specific fluorescent labeling of RetS$_{peri}$. This analysis suggested that the ten amino-terminal amino acids of the RetS$_{peri}$ construct were solvent exposed and not involved in the dimerization. Therefore, in order to avoid interference of the fluorophores label with binding, a single cysteine was introduced by replacing serine residue 45 of the RetS$_{peri}$ construct to create the RetS$_{peri}$-S45C variant protein. Here we took advantage of the absence of cysteine residues elsewhere in the protein. If a target protein possesses more than one surface cysteine residue, replacement of the cysteines with serines often has little impact on protein stability because bacterial proteins rarely form disulfide bonds. Another, recently popularized, approach is the use of unnatural amino acids, which are incorporated during protein synthesis in order to introduce unique chemistries that may be used for site-specific labeling (Wals & Ovaa, 2014). We opted to use the thiol-reactive maleimides of the Alexa-Flour 488 and Alexa Flour 555 fluorophores to create our RetS$_{peri}$ FRET pair. The labeling reactions were highly efficient and reproducible. For our FRET experiments, we used a plate-based approach, wherein each well contained the exact same concentration of the FRET donor molecule (Alexa-488-RetS$_{peri}$) and an increasing concentration of the protein tagged with the fluorescent FRET acceptor (Alexa-555-RetS$_{peri}$). Universally, all experiments were performed in triplicate to ensure reproducibility and subjected to standard statistical analysis. To measure the binding affinity of the RetS$_{per}$ homodimer in a titration experiment, we had to consider several additional complexities. (1) The Alexa-488-RetS$_{peri}$ needed to be diluted to a concentration that was significantly below the projected dissociation constant to ensure that the protein is primarily present as a monomeric species. (2) Alexa-555-RetS$_{peri}$ was to be used as titrant at a concentration where a substantial proportion of the protein would be dimeric and thus not available for Alexa-488 RetS$_{peri}$

binding. Consequently, substantially more Alexa-555 RetS$_{peri}$ was required to saturate binding. (3) Because the FRET signal is comparatively small, bleed-through fluorescence arising from the direct excitation of the acceptor poses a significant challenge. To mitigate this, we chose 430 nm as excitation wavelength and measured fluorescence quenching of the Alexa-Fluor-488-RetS$_{peri}$ FRET donor at 522 nm rather than the FRET signal associated with Alexa-555 RetS$_{peri}$ excitation at ~600 nm. In order to remove background fluorescence stemming from the increasing concentration of the Alexa-555-RetS$_{peri}$, additional reference wells were set-up. These sample wells contained Alexa-555-RetS$_{peri}$ concentrations mirroring those in the actual titration experiment without the AlexaFlour-488-labeled-RetS$_{peri}$ counterpart. For the determination of the dissociation constant, we derived a distinct isotherm to account for the dimerization of the titrant. This isotherm is valid as long as the dissociation constant is not affected by the fluorescent tags, resulting in a dramatically improved curve fit compared to the standard equation (Fig. 1A). For this specific biological system, we performed cross-linking experiments with unlabeled RetS$_{peri}$ protein to confirm the specificity of the interaction. However, it is simpler and recommended to confirm the specificity of the measured interactions by conducting a FRET based competition experiment, wherein unlabeled titrant is titrated into a solution containing a fixed concentration ratio of the FRET pair. Here, the gradual recovery of the donor fluorescence corroborates that the observed interactions are indeed mediated by protein–protein contacts and not by interactions between the fluorophore molecules. Such control experiments are important because the ring systems of the fluorophores are prone to stacking interactions. We experienced this firsthand when we sought to measure ATP binding to the signaling histidine kinase RetS using the fluorescent nucleotide analog TNP-ATP. The initial titration experiment yielded an impressive sigmoidal curve (Fig. 1B). However, the nucleotide analog could not be dislodged from its binding site by ATP suggesting that the observed binding interactions between RetS histidine kinase domain and TNP-ATP were mediated by the fluorophore rather than the nucleotide moiety. Subsequent structural studies confirmed that residues critical for ATP binding are not conserved in RetS and the pocket, as a whole, has collapsed (Mancl, Ray, Helm, & Schubot, 2019).

Beyond the outlined approach, there are several traditional and newer alternative approaches to determine the oligomeric state of a protein and quantify the interaction *in vitro*. These include analytical ultracentrifugation

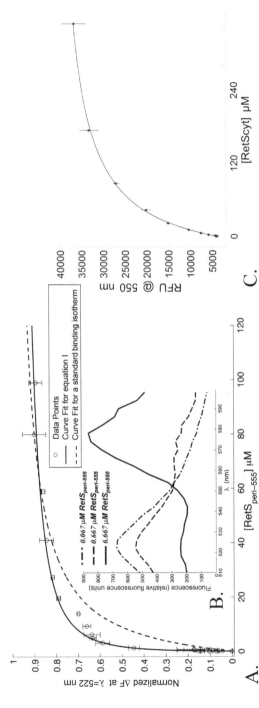

Fig. 1 (A) Shown are the curve fits of the FRET titration data using either a standard isotherm or eq. (I). The error bars signify the standard deviation of the ΔF calculated from the readings obtained from triplicate set-ups. (B) Sample plots of fluorescence spectra obtained during the FRET-based dimerization assay at three different concentrations of RetSperi-555. The spectra are generated by subtracting the background fluorescence produced by the same concentrations of RetSperi-555 in absence of RetSperi-488 from the raw data. (C) False positive results for the titration of RetS$_{cyt}$ into 25 μM TNP-ATP. Triplicate experiments were performed. Subsequent competition experiments demonstrated that binding was mediated by the fluorophore and not the nucleotide moiety. *Panel A and B were adapted from No author, 2008. Crystal structure and oligomeric state of the RetS signaling kinase sensory domain, Bone 23(1), 1–7.*

(AUC), multi-angle light scattering coupled with size-exclusion chromatography (SEC-MALS) and fluorescence correlation spectroscopy (FCS) (Ogawa & Hirokawa, 2018). While the latter has been adapted to both determine the size of an assembly as well as the strength of the interactions, AUC and SEC-MALS only provide insights into the oligomer size and come with the caveat that they are non-equilibriums techniques that may only be suitable for tightly formed assemblies. Microscale thermophoresis (MST) is a new capillary-based approach wherein the molecular motion of a fluorescently-labeled protein along a laser-induced temperature gradient is tracked to quantify bimolecular interactions (Jerabek-Willemsen, Wienken, Braun, Baaske, & Duhr, 2011). Although, MST has to date not be explicitly applied to characterize homo-oligomerization, the dependence of the diffusion coefficient on oligomer size should enable the determination of a dissociation constant by measuring the diffusion of a fluorescently labeled protein in the presence of increasing amount of unlabeled protein, similar to the protocols that were used to access this information through FCS.

2.1 Key equipment

- HiPrep 26/10 desalting column (GE Life Sciences)
- UV–visible spectrophotometer
- TECAN infinite M200 fluorescence intensity scanner (Tecan, inc.)

2.2 Key buffers, vectors, and reagents

- Buffer A: 20 mM Tris–HCl and 150 mM NaCl, pH 7.4
- Alexa fluor 488 maleimide, Alexa fluor 555 maleimide
- Vectors: pDONR201 and pDEST-HisMBP (Invitrogen)
- His_6-tobacco etch virus (TEV) protease
- QuikChange Mutagenesis kit (Agilent Technologies)

2.3 Procedure

1) The sequence and structure of the protein of interest are inspected to identify cysteine residues and their positioning in the protein. If no model is available, build a homology model using, for example, Swiss-Model (Waterhouse et al., 2018) or use Alphafold2 (Jumper et al., 2021) to generate a predicted structure without template.
2) If site-specific labeling is desired, replace all but one surface cysteine codons in the sequence with serine codons. This can be achieved *via*

site-directed mutagenesis or by purchasing a synthetic gene. There are several ways to identify suitable labelling sites: (1) packing analysis of known crystal structures may be used to select sites that avoid oligomerization interfaces. (2) If no structure is available, one may search the Protein Data Bank (PDB) (Berman et al., 2000) for structurally similar proteins and review the bimolecular interfaces in the related structure. (3) Generally, loop regions rich in charged amino acids are good targets as they are rarely involved in protein–protein interactions. The two termini are also often targeted for labeling

3) The protein of interest containing the modifications is expressed and purified. In the present example, the gene encoding retS$_{peri}$ was originally cloned into the pDEST-HisMBP expression vector using the two-step GATEWAY recombinational cloning approach (Reece-Hoyes & Walhout, 2018), where the pDONR201 vector was used as an intermediate for transferring the gene construct. To generate the RetS41–185-S45C protein the pDONR201-retS41-185 vector served as template during the site-directed mutagenesis reaction to ensure that no inadvertent mutations were introduced into the expression vector backbone. After introducing the cysteine codon using the Quikchange kit (Agilent Technologies), closely following the manufacturer's instructions, the construct was submitted to DNA sequencing. Subsequently, the LR reaction was performed to transfer the gene construct into the pDEST-HisMBP vector yielding the pDESTHMBP-retS41-185-S45C vector. The details of the construct design and purification are published elsewhere. It might be of interest that we almost always use His$_6$-MBP-fusion protein constructs with a TEV protease cleavage side introduced in the intervening regions. His$_6$-TEV protease is then used to cleave the His$_6$-MBP tag. In general, there are many different strategies for protein expression and purification. The choice of system very much depends on the particular protein of interest. The protocol will proceed assuming the purified protein is in hand

4) The RetS$_{peri}$-S45C variant was fluorescently labeled with thiol reactive Alexa Fluor 488 and Alexa Fluor 555 maleimides (Invitrogen) according to the manufacturer's instructions

5) Protein concentrations were determined using absorbance at 280 nm and verified *via* densitometry comparison to known standards following SDS-PAGE

6) Following overnight labeling the modified proteins were separated from the unincorporated fluorophore molecules through buffer exchange chromatography, using a HiPrep 26/10 desalting column

(GE Life Sciences) that had been equilibrated with buffer A. Key here is that the buffer does not contain a reducing agent such as DTT or TCEP that might interfere with the labeling reaction

7) The degree of labeling of both the Alexa Fluor 555-labeled form of the $RetS_{peri}$ variant (Alexa-555 $RetS_{peri}$) and the Alexa Fluor 488-labeled form of the $RetS_{peri}$ variant (Alexa-488 $RetS_{peri}$) were near 100% as assessed by UV-spectroscopy in conjunction with the estimated molar extinction coefficients of $RetS_{peri}$ and the respective fluorophores

8) The fluorescence resonance energy transfer measurements involved titrating a solution of 3 nM Alexa-488 $RetS_{peri}$ with Alexa-555 $RetS_{peri}$. Triplicate set-ups of 80 μL reactions containing 3 nM Alexa-488 $RetS_{peri}$ and 0 to 100 μM Alexa-555 $RetS_{peri}$ were transferred to a 96-well half area black polystyrene assay plate (Corning).

9) Background fluorescence produced by Alexa-555 $RetS_{peri}$ was accounted for by measuring and subtracting the signal of well solutions that did not contain Alexa-488 $RetS_{peri}$ but matching concentrations of Alexa-555 $RetS_{peri}$

10) The excitation wavelength was set to 430 nm, while the emission spectrum was recorded in between in the wavelength range between 510 and 646 nm using a 4 nm step-size. The integration time was set to 20 μs and all scans were carried out at a room temperature of 24°C

11) Dimer formation was monitored by recording the decrease in the peak fluorescence of RetSperi-488 at $\lambda = 522$ nm (Fig.1) and fitting the modified isotherm listed as Eq. (1) to the data using Matlab (The MathWorks) (Jing et al., 2010):

$$R_{5T} = \frac{k_d \ (FMax - Fmin)}{F - F_{min}} \left[\frac{2 \ (FMax - Fmin)}{F - F_{min}} - 3 \right] + R_{4T} \frac{Fmax - F}{Fmax - F_{min}} + K_D \qquad (1)$$

R_{5T} and R_{4T} are the total concentrations of $RetS_{peri-555}$ and $RetS_{peri-488}$, respectively, F is the fluorescence measured at $\lambda = 522$ nm, F_{max} is the fluorescence measured at $\lambda = 522$ nm in absence of $RetS_{peri-555}$, F_{min} the residual fluorescence of $RetS_{peri-488}$ due to incomplete quenching even when all $RetS_{peri-488}$ is bound to $RetS_{peri-555}$, and K_D is the dissociation constant of the $RetS_{peri}$ dimer (Jing et al., 2010).

3. Case study 2. Using *in vitro* FRET binding experiments to study competing biomolecular interactions

Building on the previous study, we also used FRET measurements to characterize competing homomeric and heteromeric interactions that characterize the RetS-GacS system. In this instance, we used a non-specific labelling approach. The RetS-GacS signaling system constitutes a stunning departure from the traditional paradigm of sensor histidine kinase-mediated signaling. The two histidine kinases interact directly, resulting in an inhibition of GacS-dependent signaling through RetS. The longstanding model posited dissociation of the respective homodimers and the formation of a RetS-GacS heterodimeric complex (Goodman et al., 2009). However, prior to our work this model had never been tested experimentally. Because previous studies had mapped the RetS-GacS interface to the cytosolic regions of the two proteins, we used constructs encompassing only those regions of the two proteins ($RetS_{cyt}$ and $GacS_{cyt}$). For the fluorescent labelling, we initially sought to take advantage of the emerging technical advances that enable the site-specific incorporation of unnatural amino acids using the amber stop codon (Wals & Ovaa, 2014). The unique chemistry of these unnatural amino acids (UAA) permits the subsequent fluorescent labeling of the protein of interest at a specific site. This approach constitutes an elegant alternative to the need for replacing multiple cysteines in the protein to ensure site-specific labeling. While the protocols for the incorporation of the UAA works very well, the subsequent labeling reactions did not yield consistent results in our hands. Therefore, we ultimately opted to perform non-specific labeling targeting surface exposed amines and used competition experiments with unlabeled proteins to corroborate the specificity of the observed binding interactions. Because both $RetS_{cyt}$ as well as $GacS_{cyt}$ contain multiple lysine residues, we had some concerns that proteins would be excessively labeled. However, we consistently only incorporated 1–2 fluorophores per molecule when only a small excess of the fluorophore over the proteins was used in the labeling reactions. Thus, the subsequent titration experiments proceeded smoothly and the results are shown in Fig. 2A–D. Using AlexaFluor 488 and Alexa Fluor 555-labeled samples, we were able to demonstrate that GacS forms a tight homodimer. AlexaFluor 488-labeled $GacS_{cyt}$ also readily bound to AlexaFluor 555-labeled $RetS_{cyt}$. The control experiments with unlabeled proteins confirmed the specificity of the respective interactions:

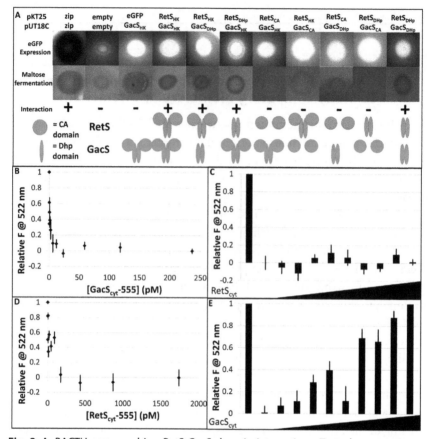

Fig. 2 A. BACTH assay probing RetS-GacS domain interactions. Tested constructs are listed above the panels. A GFP tag was added to the C terminus of all the RetS constructs to verify stable expression levels *in vivo* (top panel). RetS constructs and eGFP were expressed from pKT25 while GacS constructs were expressed from pUT18C. Strains were plated on MacConkey agar. Positive interactions give colonies a bright red color (bottom panel). Images show representative results of triplicate measurements. B. GacS homodimerization was verified by titrating $GacS_{cyt}$-555 into a solution of 250 fM $GacS_{cyt}$-488. Complete FRET quenching was observed near a ratio of 100:1 $GacS_{cyt}$-555:$GacS_{cyt}$-488. (C) $RetS_{cyt}$ binding does not disrupt the $GacS_{cyt}$-488:$GacS_{cyt}$-555 dimer, as addition of unlabeled $RetS_{cyt}$ has no measurable effect on the observed FRET signal. The first data bar represents fluorescence of a solution of 250 fM $GacS_{cyt}$-488. The second data bar represents the quenched fluorescence of a solution containing 250 fM $GacS_{cyt}$-488 and 25 pM $GacS_{cyt}$-555. The concentration of unlabeled $RetS_{cyt}$ runs from 0 to 8.7 nM. (D) FRET interactions between $RetS_{cyt}$-555 and $GacS_{cyt}$-488 demonstrate $RetS_{cyt}$-$GacS_{cyt}$ complex formation under the given experimental conditions. (E) Titration of unlabeled $GacS_{cyt}$ into a solution of 250 fM $GacS_{cyt}$-488 and 25 pM $GacS_{cyt}$-555 disrupts the dually labeled $GacS_{cyt}$-488:$GacS_{cyt}$-555 dimer. The first data bar represents the fluorescence of a solution of 250 fM $GacS_{cyt}$-488. The second data bar represents the quenched fluorescence of a solution containing 250 fM $GacS_{cyt}$-488 and 25 pM $GacS_{cyt}$-555. The concentration of unlabeled $GacS_{cyt}$ ranged from 0 to 125 pM. Experiments were performed in triplicate and error bars represent the standard error calculated from those triplicate experiments (Mancl et al., 2019). *This figure was taken and reproduced with permission from the publisher from J. M. Mancl, W. K., Ray, R. F., Helm, & F. D. 2019. Schubot, Helix cracking regulates the critical interaction between RetS and GacS in pseudomonas aeruginosa, Structure 27(5), 785–793.e5.*

The addition of unlabeled GacS$_{cyt}$ disrupted the GacS$_{cyt}$ homodimer, while unlabeled RetS$_{cyt}$ depleted the FRET signal obtained from the AlexaFluor 488-GacS$_{cyt}$: AlexaFluor 555-RetS$_{cyt}$ complex. To test the prevailing hypothesis that RetS inhibits GacS by interfering with GacS homodimerization, we also titrated unlabeled RetS$_{cyt}$ into wells containing a fixed ratio of the AlexaFluor 488-GacS$_{cyt}$: AlexaFluor 555-GacS$_{cyt}$ FRET pair. Remarkably, the unlabeled RetS protein failed to disrupt the FRET signal indicating that the GacS$_{cyt}$ dimer remains intact when RetS$_{cyt}$ binds (Mancl et al., 2019; Ryan Kaler, Nix, & Schubot, 2021). We subsequently discovered that RetS forms a domain-swapped dimer with GacS involving the proteins' DHp domains. This local interaction does not fully break the GacS$_{cyt}$ dimer because the GacS-HAMP domain maintains the homomeric interactions. With hindsight this result also highlights a limitation of the random fluorescent labeling approach. Because the FRET signal is distance dependent, fluorophores placed within the DHp domain of GacS and RetS for example would have been sensitive to RetS$_{cyt}$ binding during the competition experiment. Yet, the untargeted labeling approach did not permit detection of local structural changes.

3.1 Key equipment
- Superdex S200 26/60 column (GE Life Sciences)
- TECAN infinite M200 fluorescence intensity scanner (Tecan, inc.)
- UV–visible spectrophotometer

3.2 Buffers, vectors, and reagents
- Stabilization buffer (25 mM HEPES pH 7.5, 150 mM NaCl, 2 mM TCEP pH 8.0)
- Labeling buffer (150 mM sodium bicarbonate pH 8.3)
- Alexa fluor 488 NHS ester, Alexa fluor 555 NHS ester
- Vectors: pQE60RetScyt, pQE60GacScyt, pDONR201 and pDEST-HisMBP (Invitrogen)

3.3 Procedure
1) The expression vectors for C-terminally His-tagged RetS$_{cyt}$ and GacS$_{cyt}$ were a generous gift from the Porter lab at University of Exeter, UK. The protein purification protocols are published elsewhere and were closely followed. For the fluorescent labeling RetS$_{cyt}$ and GacS$_{cyt}$ were purified into labeling buffer using gel filtration

chromatography. The buffer composition is fairly flexible as long as buffer that contains free amine groups are avoided and the pH is kept slightly above neutral

2) Protein concentrations were determined using absorbance at 280 nm and verified *via* densitometry comparison to known standards following SDS-PAGE

3) Both protein samples were concentrated to 10 mg/mL and fluorescently labeled with amine-reactive NHS-esters of Alexa Fluor 488 or Alexa Fluor 555 (Invitrogen, Inc.) closely following the directions of the manufacturer

4) Unreacted fluorophores were removed *via* buffer exchange chromatography, using a HiPrep 26/10 desalting column (GE Life Sciences). At the same time the proteins were transferred into stabilization buffer

5) The degree of labeling was determined as directed by the manufacturer. In each case, labeled proteins were found to contain more than 1 but less than two molecules of fluorophore per protein molecule

6) The fluorescence resonance energy transfer measurements involved titrating a solution of fM Alexa-488 $GacS_{cyt}$ with either Alexa-555 $GacS_{cyt}$ or Alexa-555 $RetS_{cyt}$. Triplicate set-ups of 80 μL reactions containing 250 fM lAexa-488 $GacS_{cyt}$ and 0–100 μM of the respective titrant were transferred to a 96-well half area black polystyrene assay plate (Corning).

7) Controls for background fluorescence consisted of all experimental components present in the well minus the Alexa-488 labeled protein. Background-subtracted data was then normalized to the positive (F_{max}) and negative (F_{min}) controls (Mancl et al., 2019).

8) The excitation wavelength was set to 430 nm, while the emission spectrum was recorded in between in the wavelength range between 510 and 646 nm using a 4 nm step-size. The integration time was set to 20 μs and all scans were carried out at a room temperature of 24°C. FRET signals were monitored by recording the decrease in the peak fluorescence of Alexa-488 $GacS_{cyt}$ at $\lambda = 522$ nm

9) For the competition experiments, samples containing a fixed molar ratio of Alexa-488 $GacS_{cyt}$ and Alexa-555 $GacS_{cyt}$ were combined with increasing concentrations of either unlabeled $GacS_{cyt}$ or $RetS_{cyt}$

4. Case studies 3 to 6: Mapping novel protein-protein-interfaces through site-directed mutagenesis coupled with *in vitro* assays

4.1 Introduction

In order to understand the mechanism of action of a regulatory cascade, it is pivotal to not only determine the presence and quantify the strength of the underlying binding interactions but also to map the involved binding surfaces. Traditionally, alanine scanning mutagenesis was the method of choice for such an endeavor. However, due to the large number of candidate residues this approach is daunting and may well not be successful because a simple removal of one or even two interface residues may prove insufficient for disrupting an interaction. Here, we will describe alternative approaches that we applied effectively to map a number of novel protein–protein interfaces. As starting points for our analyses, we almost always use three-dimensional models of the protein. When no crystal structure is available, and even before the emergence of Alphafold2, numerous homology modeling programs have enabled us to produce a reasonable representation of most proteins' tertiary structure even in the absence of an experimentally determined atomic resolution structure. These structures constitute excellent starting points for the targeted experimental analysis of protein interfaces. The candidate surfaces may be further narrowed by attempting protein–protein docking and employing one of the many surface analysis algorithms aimed at predicting protein–protein interfaces. However, protein–protein interactions are often accompanied by substantial conformational changes. Some proteins, such as TyeA from Yesinia pestis, take this to the extreme, as they only assume stable tertiary structures when bound to another protein (Schubot et al., 2005). Below we will describe our efforts at pinpointing residues involved in the oligomerization of ExsD and binding of ExsD to one of its partners, the transcriptional activator ExsA. Remarkably, while ExsD is a relatively large 32 kDa protein with overall well-defined tertiary structure, it is the unstructured 23 residue amino-terminus that is critical for the interaction. To further complicate matters, the interactions between this region of ExsD and ExsA seem to be primarily mediated by the proteins' backbones. Whenever possible we validated our findings from binding studies through functional assays and strive to create gain-of-function protein variants. A loss-of-function mutation always raises concerns about the overall impact on the tertiary structure. Of course, such concerns may be mitigated through circular dichroism (CD) measurements or structural analysis. If performed inside the cell, as is the case for bacterial-two-hybrid (BACTH) studies, soluble expression may be monitored as a reasonable proxy for proper folding of a variant protein. In the following sections, we will describe our strategies for (1) disrupting large protein–protein interfaces, (2) deleting large sections of a protein by bridging adjoing regions through a glycine linker, (3) probing backbone contact-mediated interactions, and (4) monitoring protein-stability as a part of a BACTH-based mutational analysis of a bimolecular interface.

 ## 5. Case study 3. Examining the role of ExsD trimerization in the inhibition of ExsA-dependent transcription

The anti-activator protein ExsD has a predicted molecular mass of 32 KDa, however, the crystal structure suggested formation of a trimer, an observation also supported by the elution profile of ExsD from a gel filtration column (Bernhards et al., 2009). To determine the possible biological relevance of ExsD oligomerization, we introduced several point mutations at the putative interfaces. Because protein–protein interfaces tend to encompass large often hydrophobic surfaces, we generally do not favor the classic alanine scanning approach but instead opted to introduce arginine residues. The bulky charged side-chains of arginines are likely to be more disruptive when introduced at the heart of a hydrophobic interface and incorporation of a hydrophilic residue also reduces the risk of non-specific protein aggregation due to the exposure of a largely hydrophobic surface region. In the case of ExsD, the single M59R mutation converted ExsD into a monomeric protein as demonstrated by its delayed retention time on the GF column. Remarkably, the mutation made ExsD a more potent inhibitor of ExsA in our *in vitro* transcription assays. This fairly simple example also illustrates our strategy to build hypotheses that are testable through gain-of-function mutations, whenever feasible. The stepwise procedure for selecting and introducing the described mutation is outlined below.

5.1 Key equipment
• Cytiva Superdex™ 200 10/300 column
• PCR Thermocycler machine (T100 thermal Cycler, Bio-Rad)
• Shaker (Innova 44, New Brunswick Scientific)

5.2 Buffers, vectors, and reagents

- Vectors: pDONR201 and pDEST-HisMBP (Invitrogen)
- QuikChange Mutagenesis kit (Agilent Technologies)

5.3 Procedure

1) If a crystal structure is available, perform a packing analysis to identify potential protein–protein interfaces. The easiest approach for the non-expert is to use the PDBsum server, which provides a detailed analysis of all protein–protein interfaces in the asymmetric unit of the crystal

2) Examine the interfaces with a molecular structure analysis program such as Pymol and Chimera. Many convenient tutorials on the use of each software package are available. Here, we outline the steps involved in visualizing the structure in PYMOL

   (a) open the program by double-clicking on the icon.

   (b) download pdb file by selection FILE > Get PDB from the <FILE> tab. If a pdb code is known, simply enter the code and select <OK>. If it is a new structure is FILE > OPEN instead.

   (c) after the structure file has been opened, an object will be generated on the right-hand side of screen. Next to the object several letters will be shown: A(CTION), S(HOW), (H)IDE, (L)ABEL, and (C) OLOR. Click on SHOW > AS > CARTOON next to the object of. This should display the structure showing all molecules in the asymmetric unit. If the protein–protein interface involves crystallographic symmetry, the researcher can display symmetry-related molecules by selecting ACTION > GENERATE > SYMMETRY MATES next to the object. Selecting a radius will generate additional objects, each encompassing a symmetry related asymmetric unit within the given radius.

   (d) DISPLAY > SEQUENCE will provide a convenient means to select individual residues by simply highlighting them in the sequence. If the residues forming the putative interface are known, simply select them for each protein chain in the sequence. A new temporary object named SELE will appear on the right-hand panel. Selecting ACTION > COPY TO OBJECT > NEW creates a new permanent object containing only the selected residues.

   (e) SHOW > SIDE CHAIN > STICKS will display the side chains of the interface residues. Careful inspection should readily reveal the heart of the interface. Generally, an area containing multiple hydrophobic residues is well-suited. Ideally, select at least one

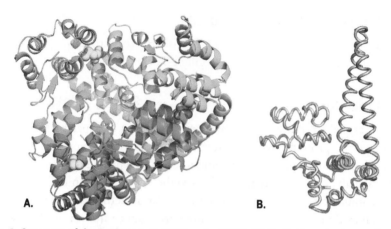

Fig. 3 Structure of the ExsD protein (PDB code: 3FD9). (A) ExsD trimer is shown. Atoms of the discussed M59 residue are shown as spheres. (B) Shown is a single ExsD molecule. The modeled four glycine linker is shown in red as an alternative loop.

residue from each chain that may be targeted. In the example of ExsD the hydrophobic patch included amino acids M59 (Fig. 3A) and L63 on one side and M217 and L221 on the other.

(**f**) A tentative idea of how replacing one of these residues might impact the interface may be obtained by selecting WIZARD > MUTAGENESIS and mutating this residue to an arginine. The arrow keys on the keyboard will subsequently navigate the different possible rotamer and highlight steric clashes as red discs. Ideally, all possible rotamers will produce significant clashes indicating that the envisioned mutation will very likely be disruptive.

(**g**) Any PYMOL session may be saved for future examination.

3) ExsA and ExsD were overexpressed in *E. coli* from a vector constructed by Gateway recombinational cloning (Invitrogen, Carlsbad, CA, USA). Initially, the nucleotide sequences of the ORFs were verified after being introduced into the pDONR201 vector, then recombined into the destination vector pDEST-HisMBP to create the expression vectors pFS-HMBPExsD and pFS-HMBPExsA. These vectors were designed to produce either ExsA or ExsD as a fusion to the C-terminus of an N-terminally His6-tagged *E. coli* maltose-binding protein (MBP). Design primers for introducing the desired mutation using the QuikChange Mutagenesis kit (Agilent Technologies). Closely follow manufacturer's instructions, including those for primer design. The protocols work well. The purification procedure for ExsA, ExsD and the ExsD-M59R variant are detailed elsewhere (Fessler, Michael, & Rudel, 2008). Both the

wild-type and the variant protein are required for the qualitative analysis. Because a hydrophobic interface might have been disrupted consider using chaotropic agents such as glycerol to enhance solubility and exercise caution when concentrating the protein, although that was not necessary for ExsD-M59R (Fessler et al., 2008).

4) For the qualitive analysis of protein assembly size, we performed analytical gel filtration chromatography using a series of protein standards to create a reference curve. Selection of the appropriate resin is critical here. In the present example a Cytiva Superdex™ 200 10/300 column was used, which covers a broad molecular weight range between 10 and 600 kDa. Because the elution volume of a protein is not only dependent on the molecular volume but also the molecule shape, absolute size determination *via* this approach alone is not possible. However, the most important reference is the native protein. In the case of ExsD, the crystal structure suggested trimer formation. Therefore, disrupting the putative interface should alter the molecular mass of the assembly from 96 kDa of the trimer to 32 kDa for the monomer. Indeed, ExsD-M59R variant displayed a dramatic shift towards a longer retention time compared to the wild-type protein, suggesting that the molecular assembly had become smaller.

5) Step 5 served to establish whether or not an observed or predicted protein–protein interface is indeed present *in vitro* and not simply the result of crystal packing. Ideally, this initial finding is followed up with a functional assay to determine the biological significance of a particular interface. ExsD is a transcriptional anti-activator protein inhibiting the function of the transcription factor ExsA. Therefore, an *in vitro* transcription assay was used to determine the impact of the M59R mutation. It turns out that ExsD trimerization has two opposing roles. On one hand, the trimer appears to stabilize ExsD, thus increasing its lifetime. On the other hand, monomeric ExsD-M59R proved to be a more potent inhibitor of ExsA-dependent *in vitro* transcription, suggesting that ExsD trimerization negatively impacts ExsD activity.

6. Case Study 4. Examining the functional role of the extensive coiled coil region in ExsD through domain deletion with a poly-glycine linker

The crystal structure of ExsD was found to contain an extensive coiled-coil region of unknown function. Because the ExsA-ExsD interface was entirely unknown at the time, we opted to first test the impact of deleting the entire domain, residues 138 to 202, on ExsD function. Simply deleting the involved residues could lead to structural strain and distortion in the adjoining regions. Therefore, we decided to bridge the gap using multiple a linker consisting of multiple glycines. We opted for glycine residues because these a more conformationally flexible presumably allowing the rest of the molecule to settle into a stable structure. However, we did not want to introduce too many glycines because this region might because susceptible to proteolysis or cause protein unfolding. Using the crystal structure as a template, we therefore conducted a simple modeling procedure to gauge the appropriate linker length. Since, we have applied this approach on a number of occasions successfully. The steps are outlined below. Because we are closely familiar with these software packages PYMOL and WINCOOT were used to perform the modeling.

6.1 Key equipment

• Cytiva Superdex™ 200 10/300 column

6.2 Buffers, vectors, and reagents

• The ExsDΔC-C variant was created by two sequential rounds of PCR. The first set of primers was:

  5′–GTGGAGAACCTGTACTTCCAGGGTATGGAGCAGGA AGAC–3′

  5′–GTGGAGAACCTGTACTTCCAGGGTGCGATCCCCG GCTGG–3′

  The second set was:

  5′–CGGGTCAACCTCGGAGGAGGAGGATCGGCACTGG CG–3′.

  5′–   GGGGACAACTTTGTACAAGAAAGTTGCTCATACT GGCAGAGCTGA–3′

• Vectors: pDONR201 and pDEST-HisMBP (Invitrogen)
• His$_6$-tobacco etch virus (TEV) protease
• QuikChange Mutagenesis kit (Agilent Technologies)

6.3 Procedure

1. Open the PYMOL program by double-clicking on the icon
2. download pdb file by selection FILE > Get PDB from the <FILE> tab. The PDB for ExsD is 3FD9. Enter the code and select <OK>. If it is a new structure is FILE > OPEN instead

3. After the structure file has been opened, an object will be generated on the right-hand side of screen. Next to the object several letters will be shown: A(CTION), S(HOW), (H)IDE, (L)ABEL, and (C)OLOR. Click on SHOW > AS > CARTOON next to the object of. This should display the structure showing all molecules in the asymmetric unit

4. DISPLAY > SEQUENCE will provide a convenient means to select individual residues by simply highlighting them in the sequence. To create a molecule that lacks residues 138 to 202 select all other residues for one of the three chains in the sequence. A new temporary object named SELE will appear on the righthand panel that contains all the selected residues. Selecting ACTION > COPY TO OBJECT > NEW creates a new permanent object containing only the selected residues

5. Under FILE > EXPORT MOLECULE > SELECTION select the newly created object and choose the PDB file format for the output file. This concludes the work with PYMOL

6. Open WINCOOT. Under FILE > OPEN COORDINATES select your newly created file. Also, go to FILE > FETCH PDB & MAP USING PDB REDO to download the original pdb file and map. The map is required by the program for the modeling, even though the glycine loop will obviously not fit the density. To avoid confusion, delete the original PDB file under DISPLAY MANAGER

7. In order to model a connecting loop between residues 137 and 203 of ExsD WINCOOT requires renumbering the amino acids to have the gap match the number of newly introduced glycines residues. As a rule of thumb, introduction of a single glycine will bridge a gap of about 6.4 Å. One or two additional residues should be added to prevent clashes and permit a low energy conformation for the loop residues. In this instance we opted to introduce four glycines. Therefore, under EDIT > RENUMBER RESIDUES selected 203 to C-terminus and entered an OFFSET of −61. Consequently residue 203 was now numbered 142

8. Open the MODEL/FIT/REFINE panel under CALCULATE. Go to DRAW > SEQUENCE VIEW to display the sequence and double click on residue 137

9. Select ADD TERMINAL RESIDUE on the MODEL/FIT/REFINE panel and click on the alpha carbon of 137. An alanine will be added. Repeat the steps until residue 141 has been added

10. Go to pulldown CALCULATE > MUTATE RESIDUE RANGE enter 138 to 141 into the appropriate boxes and four Gs into the sequence box

11. To close the gap first select REFINE/REGULARIZE CONTROL on the MODEL/FIT/REFINE panel and check the box for RAMACHANDRAN RESTRAINTS. Then select REGULARIZE ZONE and select residues 137 to 142 by simply clicking on the alpha-carbons. A refinement panel will appear, with a color coding that indicates the quality of the refinement. Green is desired for the RAMA PLOT as that indicates a realistic geometry for the introduced loop region. If the color is red, likely the introduction of additional glycines is required to bridge the gap properly. The newly built model is shown in Fig. 3B

12. Subsequent steps follow the usual protocol. The primer design is distinct because the primer pair contains codons for both boundary regions joined by four glycine codons. Therefore, for illustrative purposes, the primers are listed above. The remainder of the protocol follows the manufacturer's instructions of QuickChange Mutagenesis kit, which are excellent

13. The ExsD-ΔCC variant was expressed and purified according to previously established protocols that were used for wild-type ExsD and have been published elsewhere in great detail (Bernhards, 2013).

14. For qualitive analysis of protein assembly size, perform analytical gel filtration chromatography using a Cytiva Superdex™ 200 10/300 column. Remarkably, the ExsD-ΔCC protein appeared to be still trimeric, as its column retentions time suggested an apparent molecular weight of approximately 70 kDa

15. An *in vitro* transcription assay was used to determine the impact of the domain deletion. The details of the assay are published elsewhere (Bernhards, 2013). Remarkably, ExsD-ΔCC had a slightly lower IC_{50} than wild-type ExsD in the assay, demonstrating that the coiled coil domain is not required for ExsA binding

7. Case study 5. Using proline mutations to examine intermolecular backbone contacts between ExsD and ExsA

While we were able to determine the structures of ExsD alone and of the amino-terminal regulatory domain of the AraC-type transcription factor

ExsA, we were unable to crystallize an ExsD:ExsA complex. We had previously identified a conserved sequence motif in the structurally dynamic amino terminus of ExsD that is required for binding to the T3SS chaperone ExsC (Vogelaar, Jing, Robinson, & Schubot, 2010). Because the ExsD-ExsC interactions compete directly with the ExsD:ExsA complex formation (Thibault et al., 2009), we quickly established that the ExsD amino-terminus is also pivotal for inhibiting ExsA. The ExsA regulatory domain has the characteristic β-barrel structure encountered in canonical AraC-type transcription factors. The binding pocket for the usual small molecule regulatory ligand is located at the heart of the barrel structure. Therefore, our initial attempts to map the ExsD:ExsA interface focused on this area of ExsA. Altogether, we created more than 30 different ExsA variants. However, while most ExsA mutant proteins retained their ability to activate ExsA-dependent *in vitro* transcription, they also remained susceptible to inhibition by ExsD. Because the ExsD:ExsC interactions are partially mediated through the formation of an intermolecular beta-sheet, we hypothesized that ExsD and ExsA also form an intermolecular beta sheet. Because proline residues do not have free amide groups the introduction of prolines at select sites in ExsA should disrupt such interactions with ExsD. Using the crystal structure of ExsA-NTD as guide, we selected several residues in beta-strands β-1 and β-2 with outward facing amide groups for mutagenesis to avoid disrupting the intramolecular ExsA beta sheet (Fig. 4A). Using this approach, we were finally able to introduce two mutations strand β-2 of ExsA that abrogated inhibition of ExsA by ExsD. Subsequently, we applied a complementary approach to establish if the intermolecular backbone interactions are pivotal for complex formation independent of the amino acid sequence. We drew inspiration from a previous study wherein stretches of alanine residues were shown to assume different secondary structures depending on the sequence of the flanking regions (Shinchuk et al., 2005). According to these data we should be able to replace an entire stretch of ExsA residues without disrupting its secondary structure. Indeed, even a quintuple ExsA mutant protein wherein all five residues of strand β-1 had been replaced by alanines still interacted with ExsD (Fig. 2C). In summary, the selective incorporation of proline residues is an effective means to probe backbone-mediated intermolecular contacts, while the incorporation of stretches of alanine residues may be used to corroborate the sequence-independent binding mechanism. While our studies only focused on strand-strand interactions, there is no reason why a similar strategy could not be applied to protein loop regions.

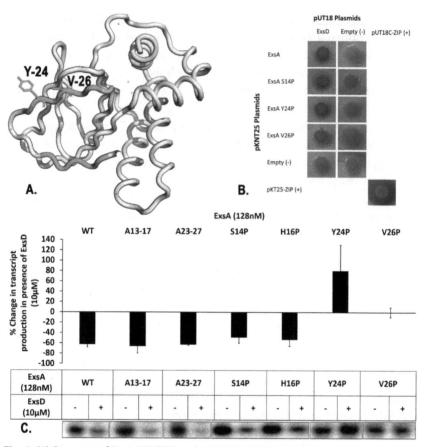

Fig. 4 (A) Structure of ExsA-NT (PDB code 4ZUA). Highlighted red are the two targeted beta strands. The side chains of Y24 and V26 are shown as sticks. (B) Results from a BACTH test demonstrate that, although ExsAY24P and ExsAV26P are insensitive to the regulation by ExsD, the two ExsA mutant still interact with ExsD, suggesting the presence of a larger ExsA-ExsD interface. Image shown is representative of the results we obtained from three biological replicates. (C) Residues in β-strands β-1 and β-2 of ExsA were replaced by stretches of alanines to determine if the side chains are important for ExsA binding. Both variants still respond to ExsD, demonstrating that the side chains of the substituted residues are not critical. However, proline substitutions of strand β-2 residues Y24 and the V26 impact ExsD dependent regulation. The bar diagrams show the averages obtained from three independent experiments with the error bars representing the standard deviations (Shrestha et al., 2020). *Figure 2B and C were taken and reproduced with permission from the publisher from reference.*

In the context of this particular study, it is important to highlight the difference between an indirect approach to probing ExsA:ExsD interactions and a direct measurement of binding. The two ExsA proline variants were no longer inhibited by ExsD in the *in vitro* transcription assay. However, we also probed the interactions of the two proteins directly using a BACTH approach (Fig. 2B). In the BACTH assay, we did still show binding between ExsD and the ExsA variant, leading us to conclude that the that strand-strand interactions are pivotal for ExsD function, but do not constitute the only interface between the two proteins.

7.1 Key equipment
- PCR Thermocycler machine (T100 thermal Cycler, Bio-Rad)
- Shaker (Innova 44, New Brunswick Scientific)
- HiPrep 26/10 desalting column (GE Life Sciences)

7.2 Buffers, vectors, and reagents
- Vectors: pDEST-His-MBP, pDONR221, pKNT25, pUT18, pUT18C-zip and pKT25-zip, pFS-HMBPExsD and pFS-HMBPExsA

7.3 Procedure
1) A three-dimensional structure is the best starting point for identifying residues suitable for the mutational analysis. If a crystal structure is available, the molecule may be inspected *via* molecular structure analysis programs such as Pymol and Chimera. Here, we outline the steps involved in visualizing the structure in PYMOL

    (a) Open the PYMOL program by double-clicking on the icon.

    (b) download pdb file by selection FILE > Get PDB from the <FILE> tab. If a pdb code is known, simply enter the code and select <OK>. If it is a new structure is FILE > OPEN instead. If no structure is available, use search the Alphafold database for a structural model. If no model is available yet, submit the sequence to the server and generate a model.

    (c) After the structure file has been opened, an object will be generated on the right-hand side of screen. Next to the object several letters will be shown: A(CTION), S(HOW),(H)IDE, (L)ABEL, and (C)OLOR. Click on SHOW > AS > RIBBON next to the object. This should display the backbone structure showing all molecules in the asymmetric unit. To identify residues suitable for replacement with proline, the backbone atoms have to be visualized. Click on SHOW > MAIN CHAIN > STICKS. Then select COLOR > BY ATOM to highlight backbone peptide bonds. Residues whose amide groups are not involved in intramolecular interactions may be suitable for

replacement with prolines. PYMOL may be used to introduce prolines into the structure using the WIZARD > MUTAGENESIS
option, to confirm that a proline side-chain would indeed not break
any intramolecular contacts. If a beta-strand is targeted, the amide
groups of consecutive peptide bounds should point in opposite
directions. Therefore, only every other residue should be suitable
for mutagenesis to proline. In the case of ExsA, the amide groups
Tyr-24 and Val-26 were selected because they point away from
the beta sheet.

2) Design primers for introducing the desired mutation using the
QuikChange Mutagenesis kit. Closely follow manufacturer's instructions, including those for primer design. The protocols work very well.
If a beta-strand is targeted two types of mutations should be introduced:
(1) to probe backbone contacts directly, proline should be placed at
strategically selected locations. (2) To determine if sidechains are
involved in binding as well or to corroborate that they are not involved,
one or several residues should be replaced by alanines. In the present
example Y-24 and V-26 were replaced by prolines and a third variant
was created that contained alanines from residue 23–27.

3) Variant proteins were expressed and purified using previously established
protocols.

4) An *in vitro* transcription assay using an ExsA-dependent promoter template served to establish whether or not an observed or predicted
protein–protein interface is indeed involved in intermolecular contacts
in vitro. In the case of ExsA, we were able to replace the residues 23 to
27 with alanine without disrupting ExsD-ExsA interactions. Yet, introduction of a single proline in position 24 or 26 made the corresponding
ExsA variant insensitive to the presence of ExsD (Fig. 3).

5) Because the *in vitro* transcription assay is an indirect approach for probing
ExsA-ExsD interactions, a BATH assay was also performed to directly
determine the impact of ExsA mutations on binding. The BACTH assay
measures the ability of adenylate cyclase to synthesize cAMP, which provides an output to assess binding between target proteins. Adenylate
cyclase has been split into two fragment peptides T18 and T25, which
must be in direct contact for cAMP synthesis to occur. The T18 and T25
DNA sequences are fused to the genes encoding target proteins in separate plasmids and co-expressed in *cya E. coli* strain BTH101. Activation
of cAMP synthesis, indicating physical interaction between the target
proteins, was visualized on MacConkey agar *via* the change in colony

color from clear to bright pink when the proteins of interest interact. The interaction of wtExsA and wtExsD served as a positive control. The T18 fragment is expressed from high copy number pUT18 vector, and the T25 fragment is expressed from the low copy number pKNT25 vector. Both *exsA* and *exsD* were fused to the 5′-ends of the fragment genes in the pKNT25 and the pUT18 vectors, respectively and co-transformed into chemically competent BTH101. BTH101 was also co-transformed with vectors (pUT18C-zip and pKT25-zip) to serve as an additional positive control for the assay. Co-transformation of the empty pKNT25 and pUT18 vectors into BTH101 served as a negative control. Further negative controls involved co-transformation of the empty pKNT25 vector with the pUT18-wtexsD plasmid, as well as co-transformation of all *exsA* expressing constructs with the empty pUT18 vector. After growing the cells overnight in LB under ampicillin (100 µg/mL) and kanamycin (50 µg/mL) selection at 37°C, the bacteria were plated on MacConkey agar plates and incubated at 30°C for 24 h. Transformants expressing interacting proteins cause increased production of β-galactosidase, which in turn leads to the development of a bright pink color. The lack of an interaction, on the other hand, results in colorless colonies on the indicator plates (Shrestha, Bernhards, Fu, Ryan, & Schubot, 2020).

8. Case study 6. Using BACTH assay to map protein-protein interfaces

The BACTH assay has emerged as a reliable tool for detecting protein–protein interactions. Apart from detecting an interaction this approach may also be adapted to map the location of an interface through mutational analysis. However, because you are looking for a loss-of-function mutation in order to corroborate the importance of a particular residue for an interaction, there is always the possibility that a loss of signal as the result of a particular mutation was not caused by a disruption of the interaction but because the mutation destabilized the protein structure. This problem becomes particularly vexing, when dealing with an obligate interface the proteins require the interactions for stability. Irrespectively of whether or not a predicted interface is expected to be expected for protein stability, the principal challenge arising from the fact that one is looking for a loss-of-function creates a need to independently verify that the modified protein is stable and soluble. Traditionally this may be achieved through western blotting. However, we found it more convenient to directly assess protein solubility by expressing GFP fusion construct of the target protein. This approach enabled us to observe and visualize soluble expression directly inside a colony. The BACTH assay is semi-quantitative, allowing for a relative assessment of complex affinity. Therefore, one possible solution to this conundrum is to only partially disrupt the protein–protein interface and observe weakened interactions. A fascinating relatively new alternative is to rationally design and engineer mutations that are predicted to increase complex stability (Pearce et al., 2020). In the RetS-GacS system, we first used the BACTH system to narrow the dimerization interface to the DHp domains. As can be seen in Fig. 2A, we were able to monitor soluble expression of the RetS-GFP fusions to confirm stability of the individual RetS subdomains.

8.1 Key equipment

- PCR Thermocycler machine (T100 thermal Cycler, Bio-Rad)
- Shaker (Innova 44, New Brunswick Scientific)
- Thin-walled PCR tubes (sterile)
- HiPrep 26/10 desalting column (GE Life Sciences)

8.2 Buffers, vectors, and reagents

- QuikChange Mutagenesis kit (Agilent Technologies)
- Primers used for creating BACTH assay constructs are listed below
    1. pKT25_RetS_XbaI_F 5'-GTCGACTCTAGAGCAGACCAAGG CCGAGTTC-3'
    2. pKT25_RetS_SacI_R 5'-CAGTGAAGAGCTCTTCTCCTTT ACTCATGAACTCG-3'
    3. pKT25_RetS_CA_Xba_F 5'-CAGGTCGACTCTAGAGGAGC TGGACGAAGTGCAGT TCGACCTCAAC-3'
    4. pUT18C_GacS_KpnI_R 5'-GAGCTCGGTACCTTACAGACT CAGG-3'
    5. pUT18C_GacS_CA_Xba_F 5'-CAGGTCGACTCTAGAGGTT CTGGAAAACCTCCCTTTC AATCTC-3'
    6. pKT25_GFP_XbaI_F 5'-CAGGGTCGACTCTAGAG ATGA GTAAAGGAGAAG-3'
    7. pKT25_GFP_KpnI_R 5'-GTTACTTAGGTACCCTATTTGT ATAGTTCATC-3

8. pUT18C_RetS_XbaI_F 5′-GTCGACTCTAGAGCAGACCAA GGCC-3′

9. pUT18C_RetS_KpnI_R 5′-GAGCTCGGTACCTCAGGCGG TGGGG-3

- Vectors: pKT25_eGFP, pKT25_RetSHK, pKT25_RetSDHp, pKT25_RetSCA, pKT25zip, pUTC18GacSHK, pUTC18GacSDHp, pUTC18GacSCA

8.3 Procedure

1. The RetS and GacS BACTH constructs were synthesized by GENEWIZ. The C-terminus of all the RetS constructs are tagged by GFP to verify stable expression levels *in vivo* (Mancl et al., 2019). Towards this end, the different *retS* coding sequences were fused to the 5′ end of the eGFP coding region through insertion of additional codons encoding a linker region (GSAGSAAGSGEF) (Waldo, Standish, Berendzen, & Terwilliger, 1999) and inserted into the XbaI and KpnI sites of pKT25.

2. The RetS and GacS DHp constructs were created *via* site-directed mutagenesis using the Quikchange kit (Agilent Technologies).

3. RetS and GacS CA domain constructs, as well as the eGFP control, were generated *via* standard cloning procedures using the primers listed above.

4. The BACTH assay was performed as described in the previous section. The eGFP fluorescence was monitored by imaging the plate using the Alexa-fluor 488 setting of a Bio-Rad Chemidoc MP imaging system (Mancl et al., 2019).

References

Berman, H. M., et al. (2000). The protein data bank. *Nucleic Acids Research*, *28*(1), 235–242.

Bernhards, R. C. (2013). Molecular interactions of type III secretion system transcriptional regulators in Pseudomonas aeruginosa. *ExsA and ExsD.*

Bernhards, R. C., Marsden, A. E., Esher, S. K., Yahr, T. L., & Schubot, F. D. (2013). Self-trimerization of E xs D limits inhibition of the P seudomonas aeruginosa transcriptional activator E xs A in vitro. *The FEBS Journal*, *280*(4), 1084–1094.

Bernhards, R. C., Xing, J., Vogelaar, N. J., Robinson, H., & Schubot, F. D. (2009). Structural evidence suggests that antiactivator ExsD from Pseudomonas aeruginosa is a DNA binding protein. *Protein Science*, *18*(3), 503–513. https://doi.org/10.1002/pro.48.

Chambonnier, G., et al. (2016). The hybrid histidine kinase LadS forms a multicomponent signal transduction system with the GacS/GacA two-component system in Pseudomonas aeruginosa. *PLoS Genetics*, *12*(5), 1–30. https://doi.org/10.1371/journal.pgen.1006032.

Depuydt, P., et al. (2006). Outcome in bacteremia associated with nosocomial pneumonia and the impact of pathogen prediction by tracheal surveillance cultures. *Intensive Care Medicine*, *32*, 1773–1781. https://doi.org/10.1007/s00134-006-0354-8.

Duff, M. R., Grubbs, J., & Howell, E. E. (2011). Isothermal titration calorimetry for measuring macromolecule-ligand affinity. *Journal of Visualized Experiments*, *55*, 2–5. https://doi.org/10.3791/2796.

Fessler, M., Michael, B., & Rudel, L. L. (2008). Brown and Sheean, self-trimerization of ExsD limits inhibition of the Pseudomonas aeruginosa transcriptional activator ExsA in vitro. *Bone*, *23*(1), 1–7. https://doi.org/10.1111/febs.12103.Self-trimerization.

Francis, V. I., & Porter, S. L. (2019). Multikinase networks: Two-component signaling networks integrating multiple stimuli. *Annual Review of Microbiology*, *73*, 199–223. https://doi.org/10.1146/annurev-micro-020518-115846.

Francis, V. I., Stevenson, E. C., & Porter, S. L. (2017). Two-component systems required for virulence in Pseudomonas aeruginosa. *FEMS Microbiology Letters*, *364*(11), 1–22. https://doi.org/10.1093/femsle/fnx104.

Francis, V. I., et al. (2018). Multiple communication mechanisms between sensor kinases are crucial for virulence in Pseudomonas aeruginosa. *Nature Communications*, *9*(1). https://doi.org/10.1038/s41467-018-04640-8.

Galle, M., Carpentier, I., & Beyaert, R. (2013). Structure and function of the type III secretion system of Pseudomonas aeruginosa. *Current Protein & Peptide Science*, *13*(8), 831–842. https://doi.org/10.2174/138920312804871210.

Gaynes, R., & Edwards, J. R. (2005). Overview of nosocomial infections caused by gram-negative bacilli. *Clinical Infectious Diseases*, *41*, 848–854. https://doi.org/10.1086/432803.

Gellatly, S. L., & Hancock, R. E. W. (2013). Pseudomonas aeruginosa: New insights into pathogenesis and host defenses. *Pathogens and Disease*, *67*, 159–173. https://doi.org/10.1111/2049-632X.12033.

Goodman, A. L., Kulasekara, B., Rietsch, A., Boyd, D., Smith, R. S., & Lory, S. (2004). A signaling network reciprocally regulates genes associated with acute infection and chronic persistence in Pseudomonas aeruginosa. *Developmental Cell*, 7(5), 745–754. https://doi.org/10.1016/j.devcel.2004.08.020.

Goodman, A. L., Merighi, M., Hyodo, M., Ventre, I., Filloux, A., & Lory, S. (2009). Direct interaction between sensor kinase proteins mediates acute and chronic disease phenotypes in a bacterial pathogen. *Genes & Development*, *23*(2), 249–259. https://doi.org/10.1101/gad.1739009.

Hauser, A. R. (2010). The type III secretion system of Pseudomonas aeruginosa: Infection by injection. *Nature Reviews Microbiology*, *7*(9), 654–665. https://doi.org/10.1038/nrmicro2199.The.

Jerabek-Willemsen, M., Wienken, C. J., Braun, D., Baaske, P., & Duhr, S. (2011). Molecular interaction studies using microscale thermophoresis. *Assay and Drug Development Technologies*, *9*(4), 342–353. https://doi.org/10.1089/adt.2011.0380.

Jing, X., Jaw, J., Robinson, H. H., & Schubot, F. D. (2010). Crystal structure and oligomeric state of the RetS signaling kinase sensory domain. *Proteins*, *78*(7), 1631–1640. https://doi.org/10.1002/prot.22679. PMID: 20112417; PMCID: PMC3621116.

Jumper, J., et al. (2021). Highly accurate protein structure prediction with AlphaFold. *Nature*, *596*(7873), 583–589. https://doi.org/10.1038/s41586-021-03819-2.

Kadri, S. S. (2020). Key takeaways from the U.S. CDC's 2019 antibiotic resistance threats report for frontline providers. *Critical Care Medicine*, *48*, 939–945. https://doi.org/10.1097/CCM.0000000000004371.

Laub, M. T., & Goulian, M. (2007). Specificity in two-component signal transduction pathways. *Annual Review of Genetics*, *41*, 121–145. https://doi.org/10.1146/annurev.genet.41.042007.170548.

Liu, C., Sun, D., Zhu, J., & Liu, W. (2019). Two-component signal transduction systems: A major strategy for connecting input stimuli to biofilm formation. *Frontiers in Microbiology*, *10*. https://doi.org/10.3389/fmicb.2018.03279.

Lohrmann, J., & Harter, K. (2002). Plant two-component signaling systems and the role of response regulators. *Plant Physiology*, *128*(2), 363–369. https://doi.org/10.1104/pp.010907.

Mancl, J. M., Ray, W. K., Helm, R. F., & Schubot, F. D. (2019). Helix cracking regulates the critical interaction between RetS and GacS in pseudomonas aeruginosa. *Structure*, *27*(5), 785–793.e5. https://doi.org/10.1016/j.str.2019.02.006.

Mayhall, C. G. (2003). The epidemiology of burn wound infections: Then and now. *Clinical Infectious Diseases*, *37*, 543–550. https://doi.org/10.1086/376993.

Moradali, M. F., Ghods, S., & Rehm, B. H. A. (2017). Pseudomonas aeruginosa lifestyle: A paradigm for adaptation, survival, and persistence. *Frontiers in Cellular and Infection Microbiology*, *7*. https://doi.org/10.3389/fcimb.2017.00039.

Ogawa, T., & Hirokawa, N. (2018). Multiple analyses of protein dynamics in solution. *Biophysical Reviews*, *10*(2), 299–306. https://doi.org/10.1007/s12551-017-0354-7.

Pearce, R., Huang, X., Setiawan, D., Zhang, Y., Arbor, A., & A. (2020). Arbor, EvoDesign: Designing protein-protein binding interactions using evolutionary Interface profiles in conjunction with an. *Optimized Physical Energy Function*, *431*(13), 2467–2476. https://doi.org/10.1016/j.jmb.2019.02.028.EvoDesign.

Reece-Hoyes, J. S., & Walhout, A. J. M. (2018). Gateway recombinational cloning. *Cold Spring Harbor Protocols*, *2018*(1), 1–6. https://doi.org/10.1101/pdb.top094912.

Ryan Kaler, K. M., Nix, J. C., & Schubot, F. D. (2021). RetS inhibits Pseudomonas aeruginosa biofilm formation by disrupting the canonical histidine kinase dimerization interface of GacS. *The Journal of Biological Chemistry*, *297*(4), 101193. https://doi.org/10.1016/j.jbc.2021.101193.

Sadikot, R., Blackwell, T., Christman, J., & Prince, A. (2005). Pathogen-host interactions in Pseudomonas aeruginosa pneumonia. *American Journal of Respiratory and Critical Care Medicine*, *171*, 1209–1223. https://doi.org/10.1164/rccm.200408-1044SO.

Schubot, F. D., et al. (2005). Three-dimensional structure of a macromolecular assembly that regulates type III secretion in Yersinia pestis. *Journal of Molecular Biology*, *346*(4), 1147–1161. https://doi.org/10.1016/j.jmb.2004.12.036.

Shinchuk, L. M., Sharma, D., Blondelle, S. E., Reixach, N., Inouye, H., & Kirschner, D. A. (2005). Poly-(L-alanine) expansions form core β-sheets that nucleate amyloid assembly. *Proteins: Structure, Function, and Genetics*, *61*(3), 579–589. https://doi.org/10.1002/prot.20536.

Shrestha, M., Bernhards, R. C., Fu, Y., Ryan, K., & Schubot, F. D. (2020). Backbone interactions between transcriptional activator ExsA and anti-activator ExsD facilitate regulation of the type III secretion system in Pseudomonas aeruginosa. *Scientific Reports*, *10*(1), 1–9. https://doi.org/10.1038/s41598-020-66555-z.

Stover, C. K., et al. (2000). Complete genome sequence of Pseudomonas aeruginosa PAO1, an opportunistic pathogen. *Nature*, *406*(6799), 959–964. https://doi.org/10.1038/35023079.

Taylor, P. K., Zhang, L., & Mah, T.-F. (2019). Loss of the two-component system TctD-TctE in Pseudomonas aeruginosa affects biofilm formation and aminoglycoside susceptibility in response to citric acid. *mSphere*, *4*(2), 1–15. https://doi.org/10.1128/msphere.00102-19.

Thibault, J., Faudry, E., Ebel, C., Attree, I., & Elsen, S. (2009). Anti-activator ExsD forms a 1:1 complex with ExsA to inhibit transcription of type III secretion operons. *The Journal of Biological Chemistry*, *284*(23), 15762–15770. https://doi.org/10.1074/jbc.M109.003533.

Vogelaar, N. J., Jing, X., Robinson, H. H., & Schubot, F. D. (2010). Analysis of the crystal structure of the ExsCExsE complex reveals distinctive binding interactions of the pseudomonas aeruginosa type III secretion chaperone ExsC with ExsE and ExsD. *Biochemistry*, *49*(28), 5870–5879. https://doi.org/10.1021/bi100432e.

Waldo, G. S., Standish, B. M., Berendzen, J., & Terwilliger, T. C. (1999). Rapid protein-folding assay using green fluorescent protein. *Nature Biotechnology*, *17*(7), 691–695. https://doi.org/10.1038/10904.

Wals, K., & Ovaa, H. (2014). Unnatural amino acid incorporation in E. coli: Current and future applications in the design of therapeutic proteins. *Frontiers in Chemistry*, *2*, 1–12. https://doi.org/10.3389/fchem.2014.00015.

Waterhouse, A., et al. (2018). SWISS-MODEL: Homology modelling of protein structures and complexes. *Nucleic Acids Research*, *46*(W1), W296–W303. https://doi.org/10.1093/nar/gky427.

Wine, J. J. (1999). The genesis of cystic fibrosis lung disease. *Journal of Clinical Investigation*, *103*, 309–312. https://doi.org/10.1172/JCI6222.

Zheng, Z., Chen, G., Joshi, S., Brutinel, E. D., Yahr, T. L., & Chen, L. (2007). Biochemical characterization of a regulatory cascade controlling transcription of the Pseudomonas aeruginosa type III secretion system. *The Journal of Biological Chemistry*, *282*(9), 6136–6142. https://doi.org/10.1074/jbc.M611664200.

Zschiedrich, C. P., Keidel, V., & Szurmant, H. (2016). Molecular mechanisms of two-component signal transduction. *Journal of Molecular Biology*, *428*(19), 3752–3775. https://doi.org/10.1016/j.jmb.2016.08.003.

CHAPTER TWO

Quantitative structural proteomics in living cells by covalent protein painting

Ahrum Son, Sandra Pankow, Tom Casimir Bamberger, and John R. Yates III*

Department of Molecular Medicine, The Scripps Research Institute, La Jolla, CA, United States
*Corresponding author: e-mail address: jyates@scripps.edu

Contents

Methods in Enzymology, Volume 679
ISSN 0076-6879
https://doi.org/10.1016/bs.mie.2022.08.046

Abstract

The fold and conformation of proteins are key to successful cellular function, but all techniques for protein structure determination are performed in an artificial environment with highly purified proteins. While protein conformations have been solved to atomic resolution and modern protein structure prediction tools rapidly generate near accurate models of proteins, there is an unmet need to uncover the conformations of proteins in living cells. Here, we describe Covalent Protein Painting (CPP), a simple and fast method to infer structural information on protein conformation in cells with a quantitative protein footprinting technology. CPP monitors the conformational landscape of the 3D proteome in cells with high sensitivity and throughput. A key advantage of CPP is its' ability to quantitatively compare the 3D proteomes between different experimental conditions and to discover significant changes in the protein conformations. We detail how to perform a successful CPP experiment, the factors to consider before performing the experiment, and how to interpret the results.

Abbreviations

ACN	acetonitrile
BCA	bicinchoninic acid
CAA	chloroacetamide
CH$_2$O	formaldehyde
CID	collisional induced dissociation
CIL	Cambridge Isotope Laboratories
CPP	covalent protein painting
3D	three-dimensional
DDA	data dependent acquisition
DIA	data independent acquisition
EDTA	2,2′,2″,2‴-(ethane-1,2-diyldinitrilo)tetraacetic acid
FDR	false discovery rate
HEPES	2-[4-(2-hydroxyethyl)piperazin-1-yl]ethanesulfonic acid
HPLC	high pressure liquid chromatography
IAA	iodoacetamide
MS	mass spectrum
MS/MS	fragment ion mass spectrum

MudPIT	multidimensional protein identification technology
NaBD$_3$CN	sodium cyanoborodeuteride
PBS	phosphate buffered saline
PTM	post-translational modification
SDC	sodium deoxycholate
TCEP	tris(2-carboxyethyl)phosphine

1. Introduction

Proteins determine cellular viability, purpose, and fate, and the function of a protein strictly follows its structure. Hence, it is important to know the conformational state in which a protein executes its function. Conformational changes such as the allosteric activation of proteins by small molecules or the regulation of protein-protein interactions in the assembly and disassembly of larger protein complexes determine the biological outcome of the protein's activity. Somatic mutation or chemical modification of proteins can directly impede protein folding and thus, function, which can cause diseases like cancer. Age-correlated proteopathies constitute a specific class of disease that are caused by a misfolding of proteins into 3D conformations that are detrimental to cell viability. The newly adopted 3D conformations, which are observed in misfolded tau protein or misfolded amyloid-β peptide, for example, are toxic because the misfold self-templates, which means that a misfolded conformer of the protein can initiate the misfolding of another copy of the same protein, yielding an amplification cascade that is solely based on the misfolded protein conformation. Because of their self-amplifying nature, this class of proteopathies initially progresses slowly but increases in speed and cellular destruction with time and age. Propagating the misfold in cells only requires a close spatial proximity of replicate protein molecules in a cellular sub-compartment such as in endosomes and lysosomes. While most misfolded proteins are recognized by the proteostasis network and are rapidly degraded, a few select, misfolded proteins can accumulate on the outside or inside of cells, impeding the integrity of the extra- or intracellular space which eventually culminates in cell death.

Chemical footprinting technologies have been used to elucidate misfolded proteins for a century. However, these analyses were limited to a single, purified protein. The advent of modern bottom-up mass spectrometry increased the scope of analysis from a single protein to a complete proteome and later revolutionized protein conformational analysis. Additional technologies broadened the range of bottom-up proteomic applications, and isotope-resolved analysis of individual peptides enabled quantification of proteins in a proteome. Labeling peptides with isotope-defined atoms also made it possible to directly compare the abundance of proteins and proteoforms in proteomes for relative quantification (Bamberger et al., 2018), either between different samples or within a sample using isotope-defined peptides spiked in as external standards.

In general, the access to an amino acid for chemical modification in a protein is determined by the amino acid's immediate environment, and chemical reactivity is reduced or absent when access is sterically hindered. Chemical footprinting reveals whether an amino acid at a specific site in a protein is accessible for chemical modification or not. Here, we describe Covalent Protein Painting (CPP) which is a chemical footprinting method to measure the chemical reactivity of the amino acid lysine in proteins. First, CPP quantifies the chemical reactivity of the lysine ε-amine using isotope defined labeling reagents (Bamberger et al., 2021). In contrast to modern structure determination tools with high spatial resolution like protein X-ray crystallography, CPP provides spatial information across a whole proteome in living cells. Then and in a second step, CPP results from two different experimental conditions are compared to find differences in the accessibility of lysine sites for chemical modification. This step identifies changes in the 3D proteome between different experimental conditions.

Different chemical labeling strategies for chemical footprinting exist and can help elucidate protein structure. These techniques include hydrogen-deuterium exchange, which is now established as a routine application (Pascal et al., 2007). Carboxy and hydroxy group footprinting in proteins is performed with several different small reactants that are characterized by a highly reactive ester group as in diethylpyrocarbonate (Zhou & Vachet, 2012). Lysine sites have also been used in different chemical footprinting approaches to infer protein structure information. For example, lysine sites can be labeled with succimidylanhydride (Kahsai et al., 2011) and *N*-acetylimidazole (Xiang et al., 2015). The recent advent of isotope mass balanced and gas-phase cleavable "Tandem Mass Tags" (TMT) enabled highly elaborate protein footprinting labeling schemes on purified protein complexes (Zhou & Vachet, 2013). TMT facilitates comparison of samples under different conditions and makes techniques like thermal shift assays accessible (Mateus et al., 2020). The use of TMT with up to 16 different analytical reporter ions increased the amount of conformational information that can be obtained for a protein complex under different experimental conditions.

CPP and a few other chemical footprinting techniques have been developed to analyze the structural proteome in living cells. For example, living cells and organisms were analyzed with the hydroxyl radical labeling strategy Fast Photochemical Oxidation of Proteins (FPOP) (Johnson, Di Stefano, & Jones, 2019). Hydroxyl radical footprinting (Espino, Mali, & Jones, 2015) is a method that produces a vast quantity of hydroxyl radicals in direct proximity of the protein to rapidly label its solvent accessible surface. In contrast, CPP combines isotope defined labeling with the facility of chemical dimethylation of the ε-amine of lysine to quantify the accessibility of lysine sites for chemical labeling.

2. Experiment layout and general considerations

CPP measures the relative fraction of protein molecules in which an individual lysine site was accessible for chemical modification. Accessible lysine sites are chemically modified with two methyl groups at their side chain ε-amines, which are primary amines. Each methyl group is introduced by a formaldehyde molecule that forms a Schiff's base (hydroxymethylamine) with a free primary amine on lysine (Kallen & Jencks, 1966). The Schiff's base is reduced with cyanoborohydride, a mild reducing reagent. Due to the kinetics of the reaction (Kallen & Jencks, 1966), primary amines are rapidly methylated in living cells (Bamberger et al., 2021). A quenching reaction inactivates the labeling reagents and suppresses unwanted chemical modification of lysine sites that were initially inaccessible. A second methyl group is added to the now-secondary amine to convert the ε-amine to a dimethylated, tertiary amine with two methyl groups. The primary amines of N-termini of proteins and peptides are also modified with dimethylation and must be considered during the analysis of mass spectrometry data.

The basic strategy of CPP is to quantify the relative ratio of amino acid sites in cells that are accessible versus inaccessible for chemical labeling (Fig. 1). Using CPP, any lysine site that is sufficiently accessible is chemically modified, whereas lysine sites that are not accessible remain unmodified. After quenching the labeling reaction, proteins are extracted from the labeled cells, purified, denatured, and disulfide bonds are reduced prior to endoproteolytic digestion of the proteins. Endoproteolytic digestion is a critical step in the CPP protocol because it allows us to digest proteins into peptides that can then be detected, identified, and quantified with bottom-up mass spectrometry. The CPP protocol utilizes Chymotrypsin, which cleaves

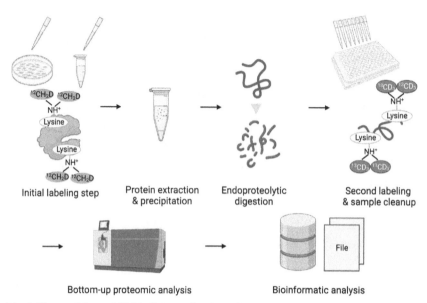

Fig. 1 The workflow of CPP in living cells. The solvent-accessible lysine sites are initially modified with dimethylation ($^{12}CH_2D$). After the initial labeling reaction in living cells, proteins are extracted, purified, and digested into peptides. The newly exposed lysine sites were modified with the different isotope dimethylation ($^{13}CD_3$). Based on the results from proteomic analysis, the relative ratio of solvent-accessible to solvent-inaccessible lysine sites are quantified.

C-terminal of phenylalanine, tyrosine and tryptophan but does not cleave at lysine, and is therefore agnostic towards the dimethylation status of lysine residues.

The endoproteolytic digestion of proteins into peptides reveals all non-modified lysine sites. These now accessible ε-amines of lysine are subsequently labeled with dimethyl moieties that are generated with a second, distinct set of isotope-defined labeling reagents. After the second labeling step, all lysine sites are chemically modified with dimethyl groups. Because the endoproteolytic digestion is insensitive to the modification status of the lysine site, the chemical properties of the endoproteolytic peptides are identical and independent of whether the lysine site was labeled in the initial step or the second labeling step. A lysine site that was initially accessible for dimethylation is now chemically indistinguishable from the same lysine site that was labeled following digestion of the proteome. The only difference in the peptides is the isotope composition of the dimethyl groups at the lysine ε-amine of the lysine site, and this mass difference is only revealed following mass spectrometric analysis. The advantage of the CPP labeling strategy is that both groups of the same peptide with differentially labeled ε-amines at the same lysine site are measured with identical sensitivity in a standard bottom–up proteomic experiment.

All peptides are then subjected to bottom–up proteomic analysis. A multidimensional protein identification technology (MudPIT)-based bottom–up proteomic experiment is recommended for highest sensitivity and coverage of a proteome (Washburn, Ulaszek, & Yates, 2003). MudPIT resolves peptides two-dimensionally based on the peptide's hydrophobicity and cationic strength. Acidified peptides are first loaded onto a reversed phase C18 trap column, then moved from the trap column onto a strong cation exchange column. Based on the cationic character of the peptides, peptides are released from the strong cation exchange resin with an ionic buffer that increases in ionic strength with each elution step. Released peptides are then bound to an analytical reversed phase C18 column that is in line with the mass spectrometer and that separates peptides based on hydrophobicity. The chromatographically separated peptides are then introduced into the mass spectrometer by electrospray ionization at a finely extruded glass tip with a sub micrometer orifice.

The mass spectrometer measures the mass of the peptide ions. Lysine-containing peptides appear with a charge $z \geq +2$ due to the contribution of the charge from the N-terminus and lysine (or arginine) residues in the peptide sequence. Thus, following an initial mass scan, the charge of the peptide ions is determined and mass peaks with charge $z \geq +2$ are selected for collision induced fragmentation (CID). CID breaks peptide precursor ions into smaller fragments that reveal the amino acid sequence of the peptide. The fragment ions of the CID-fragmented peptide precursor ion are detected by the mass spectrometer and recorded in a MS/MS spectrum. Following mass spectrometric data acquisition, MS/MS spectra are searched against a database of predicted peptide fragment ions and the best match peptide for each MS/MS spectrum is reported (Eng, McCormack, & Yates, 1994). Bottom–up mass spectrometry analysis of proteolytic peptides is a well-established proteomics method and has been extensively used to rapidly analyze and identify the proteome of many organisms (Zhang, Fonslow, Shan, Baek, & Yates, 2013).

The individual lysine site's surface accessibility is retrieved after identification of the peptides that resulted from the endoproteolytic digestion of proteins in the proteome. All precursor ions of a peptide with a lysine site are selected, and the chromatographic elution peak of the peptide is reconstructed based on the recurrent presence of the precursor ions in the MS survey scans, and chromatographic peaks are recorded for each peptide that eluted from the analytical reversed phase column and that was identified. Because dimethylation of surface accessible lysine sites was initially performed with one set of isotope-defined reagents (light) and initially inaccessible lysine sites that were exposed after digestion with chymotrypsin were then labeled with a second, distinct set of isotope-defined reagents (heavy), the chromatogram of the precursor ions of labeled peptides is split into two peaks of different mass based on the isotope composition of the dimethyl label. The area under the chromatographic peak for each peptide precursor ion is determined, and a relative ratio of the light-labeled to the heavy-labeled lysine molecules per lysine site is calculated. This ratio reflects the ratio of peptide molecules and thus proteins (proteoforms) in which a specific lysine site was initially accessible or inaccessible for chemical labeling. In this way, many lysine sites per proteome can be interrogated in a single experiment. Because the relative accessibility for chemical modification is quantified for each of the lysine sites that have been identified, a large amount of 3D structure-specific information is generated.

CPP is a powerful tool for the chemical footprinting of proteins and proteomes, but there are some limitations. CPP does not cover the entire amino acid sequence of a single protein to obtain structural information because CPP monitors only one amino acid (lysine) for chemical modification. Second, it is possible that none of the lysine sites in a protein change in chemical reactivity due to alterations in protein conformation or due to the assembly of the protein into a protein complex. In such a case, analyzing a lysine site's chemical reactivity may not reveal any structural alterations of the protein even if all lysine sites of a protein are detected in a proteomic experiment. At least partial inhibition of chemical reactivity at one of the lysine sites in a protein is required to detect a change in protein conformation with CPP. This prerequisite can be assessed in a preliminary CPP experiment.

The anionic character of the ε-amine typically excludes lysine from the hydrophobic core of protein domains and favors their position on the solvent accessible surface of proteins. A CPP analysis of a 3D proteome reveals ionic interactions between the anionic ε-amine of lysine and a counter cation that is sufficiently strong to withstand disruption during the initial labeling reaction. However, many ionic interactions of proteins with small molecules or other proteins are transient. Because non-covalent affinities depend on different environmental factors such as the ionic strength of a solution or concentration of the interaction partners, it is important to understand which conformational constraints CPP can capture, and which are beyond the reach of the CPP approach. In a CPP experiment that analyzes cell or tissue samples, the proteome is labeled for 15 min on ice (4 °C). Although cooling cells to 4 °C limits the diffusion of molecules, especially in cellular membranes, any molecular dissociation of a protein complex during the initial labeling will yield access to dimethylation of a lysine site. Molecular interactions that are not stable for up to 15 min on ice might not be detected with CPP. Thus, CPP detects lysine sites as inaccessible for chemical modification (dimethylation) only if protein conformations and interactions remain unaltered during the initial labeling step.

CPP has two major advantages over other chemical footprinting technologies. First, it can survey many lysine sites in a proteome in a single experiment. There are an estimated 25,000 different lysine sites in a standard proteome and a well-designed CPP experiment can detect up to 5,000 of them. Thus, it is possible to cover a large part of the structural proteome with CPP. Second, the fraction of protein molecules in which a lysine site was inaccessible for chemical labeling can be quantified with high sensitivity. This is important because it allows us to accurately determine the fraction of molecules in which a lysine site is not accessible in a cell. For example, the fraction of protein molecules that is incorporated into a protein complex or the fraction of protein molecules that are enzymatically active might be quantified.

The ability to measure the relative proportions of protein molecules that are accessible or inaccessible is especially useful when comparing different experimental conditions to uncover changes in the structural proteome. Because CPP measures conformational changes indirectly by assessing the chemical modification at lysine, it enables us to screen a cellular proteome quickly for many different changes in protein structure and conformation. Rather than focusing on a single conformational change in living cells, CPP can be used to scan and quantitatively compare two or more proteomes for structural changes (Bamberger, Diedrich, Martìnez-Bartholomé, & Yates, 2022). For example, a single point mutation can transform a protooncogene to an oncogene, which can dysregulate cell growth and influence the 3D proteome of cells. In this case, CPP measured the lysine site accessibility in protooncogenic cells that are non-tumorigenic. The CPP results of the non-tumorigenic cells were compared to the CPP results that were measured in tumorigenic cells that harbor the oncogenic mutation. The comparison revealed any quantitative structural change across the 3D proteome. These changes are quantified with a measurement of error, so statistical analysis can be applied to assess the significance of the structural changes observed between two proteomes. This statistical analysis includes the use of Benjamini-Hochberg based corrections for multiple hypothesis testing. Thus, CPP is very well suited to discover changes in the structural proteome that are due to a perturbation or any other experimental condition, unless a structural change does not influence the accessibility of a lysine site for chemical modification either directly or indirectly.

It is important to note that CPP results can be influenced by posttranslational modifications (PTMs) on lysine. Lysine sites in proteins can be naturally mono- and dimethylated, but natural methylation groups have a natural isotope distribution. Therefore, users are advised to utilize non-natural isotope combinations of formaldehyde and cyanoborohydride during the initial and the second labeling step in CPP. The use of non-natural, isotope defined reagents allows us to differentiate naturally occurring dimethylation from chemical modifications that are introduced during CPP. CPP is agnostic towards any other PTMs of lysine (i.e., acetylation, ubiquitinoylation) because these modifications are not chemically identical to the peptides that were chemically modified in CPP. Thus, changes in the abundance of peptides modified with different PTMs at the very same lysine site remain unaccounted for. However, the presence of additional PTMs can be revealed by a standard search for PTM modified peptides. The PTM search results are not incorporated in a CPP result, but individual lysine sites can be flagged as being post-translationally modified.

Finally, it is important to realize that CPP measures the ratio of inaccessible over accessible molecules per lysine site and does not measure or compare the absolute abundance of proteins. In fact, all chemical footprinting measurements in CPP are performed independently of protein abundance. However, the protein abundance can influence the accuracy of measurement. A smaller area under the precursor ion peak due to a low abundance of a protein yields to results that are less accurate because of a diminished signal to noise ratio.

3. General equipment and chemical supplies
3.1 General equipment

- Centrifuge (Microfuge 20R, Beckman Coulter)
- Bath sonicator (M1800, Branson)
- Thermomixer (ThermoMixer® F1.5, Eppendorf) or 30 °C incubator

- C-18 tips, 100 μL bed (87784, Thermo Scientific)
- Vacuum dry evaporator (Labconco)
- Multi-pipette
- 96-well plate

3.2 General chemical supplies

- 1× phosphate buffered saline (PBS), pH 7.4
- Formaldehyde, $^{12}CH_2O$ (ULM-9498, Cambridge Isotope Laboratories (CIL))
- Formaldehyde, $^{13}CD_2O$ (CDLM-4599, CIL)
- Sodium cyanoborodeuteride, $NaBD_3CN$ (DLM-7364, CIL)
- 1 M 2-[4-(2-hydroxyethyl) piperazin-1-yl] ethane sulfonic acid (HEPES) buffer, pH 7.4
- 10% Ammonium bicarbonate
- 0.25% Trypsin-EDTA (1×), optional
- Methanol
- Chloroform
- Sodium deoxycholate (SDC), optional
- BCA Protein Assay Kit (23225, Thermo Scientific)
- Tris(2-carboxyethyl) phosphine hydrochloride (TCEP), pH 7.0
- Iodoacetamide (IAA) or chloroacetamide (CAA)
- Chymotrypsin, sequencing grade
- Formic acid
- 100% Acetonitrile (ACN)
- 100% Methanol

4. Initial labeling step

This step labels lysine sites that are accessible for chemical labeling with dimethyl groups. We present alternative procedures for two distinct types of samples: the first is for adherent cells grown in a cell dish and the second is for cryopreserved cells that were collected as a pellet at the bottom of a tube. These procedures differ based on the initial state of cells, but the mechanism and reagents for dimethylation are the same. In both procedures, the primary ε-amine of the accessible lysine reacts with formaldehyde to generate a Schiff base that is reduced by sodium cyanoborodeuteride.

Caution: Formaldehyde and sodium cyanoborodeuteride must be carefully handled. Formaldehyde is a carcinogenic, volatile organic compound, so the work area for should be well-ventilated. Working in a chemical hood is recommended. Sodium cyanoborodeuteride is destructive to the mucous membranes and respiratory tract and may be fatal if inhaled or absorbed through skin. The initial labeling step requires approximately 40 min of effort.

4.1 Equipment

- Centrifuge (Microfuge 20R, Beckman coulter)

4.2 Materials and stock solutions

- 1× PBS, pH 7.4
- Formaldehyde, $^{12}CH_2O$ (ULM-9498, Cambridge Isotope Laboratories (CIL))
- Sodium cyanoborodeuteride, $NaBD_3CN$ (DLM-7364, CIL)
- 1 M HEPES buffer, pH 7.4
- 10% Ammonium bicarbonate
- 0.25% Trypsin-EDTA (1×)
- Stock solution 1: 1% formaldehyde (CH_2O), 0.3 mM sodium cyanoborodeuteride ($NaBD_3CN$) in 20 mM HEPES, pH 7.4
- Stock solution 2: 10% ammonium bicarbonate, freshly prepared at the day of use.

4.3 Procedure

Adherent cells are allowed to reach > 80% or less confluency in a 100 mm cell culture dish, depending on the cell culture experiment that is performed. Please note that the supernatant above the cell pellet should be removed completely prior to flash freezing a cell pellet (50–100 μL volume) at −80 °C. The number of cells can be adjusted depending on the cell type or goal of study.

(A) Adherent cells

1. Wash the cells twice with 1× PBS, pH 7.4 and discard PBS.
2. Add 2 mL of Solution 1 (1% formaldehyde, 0.3 mM sodium cyanoborodeuteride in 20 mM HEPES, pH 7.4) to the cell culture dish.
3. Incubate the cells in Solution 1 for 15 min at 4 °C.
4. Add 200 μL of Solution 2 to a final concentration of 1% ammonium bicarbonate to quench the reaction.
5. Incubate the samples at room temperature (RT) for 10 min on a horizontal laboratory shaker.
6. A small plastic scraper is recommended to detach cells from the surface of the cell culture dish. Optionally, 0.25% Trypsin-EDTA can be added to

the cell dish and incubate for 5 min at RT prior to dislodging the cells. Aspirate the cells and supernatant.

7. Collect the aspirate into 1.5 mL reaction vials.

(B) Flash frozen cell pellet

1. Add 500 μL of Solution 1 (1% formaldehyde, 0.3 mM sodium cyano-borodeuteride in 20 mM HEPES, pH 7.4, solution at 4 °C) to the flash frozen cell pellet before it is thawed for the initial labeling. Make sure only the cells were frozen without additional supernatant.

2. Incubate for 15 min at 4 °C. Gently aspirate the frozen cell pellet to mix cells with the labeling reagent continuously. Once the cell pellet is completely dissolved, gently mix the tube content twice with 5 min intermediate incubation.

3. Add 50 μL of Solution 2 to a final concentration of 1% ammonium bicarbonate to quench the reaction. Mix sample vigorously.

4. Incubate the samples at RT for 10 min.

4.4 Notes

The reagent volumes are adjusted for one cell culture plate with a 10 cm diameter or a flash frozen cell pellet of ≤100 μL volume. Please scale all reagent volumes linearly either with the area of a cell culture plate or the volume of a cell pellet.

Cells might undergo lysis in variable amounts depending on the cell type during the initial labeling step. Therefore, cells and supernatant are collected, and proteins are extracted rapidly to remove reactants. Immediate processing of samples in the next step is important because excess reactants remain in the reaction solution that may continue to label lysine sites as the proteome gradually denatures.

5. Protein extraction

This section describes the steps to homogenize cells and extract proteins. During extraction, the sample must be kept at 4 °C (on ice) to avoid protein degradation. The protein extraction steps require approximately 180 min of effort.

5.1 Equipment

- Bath sonicator (M1800, Branson)
- Thermomixer (ThermoMixer® F1.5, Eppendorf)
- Centrifuge (Microfuge 20R, Beckman Coulter)

5.2 Materials and stock solutions

- HEPES, pH 7.4
- Methanol
- Chloroform
- Sodium deoxycholate (SDC)
- Solution 3: 1% SDC in 20 mM HEPES

5.3 Procedure

1. Partition 100 μL aliquots of the supernatant into new tubes, if necessary. The volume per tube needs to be 100 μL, so adjust smaller volumes to 100 μL with 1× PBS.
2. Add 400 μL of methanol to the tube and vortex vigorously to precipitate the proteins in the sample.
3. Add 100 μL of chloroform to the tube and vortex vigorously.
4. Centrifuge at 15,000 × g at 4 °C for 30 min. The proteins will pellet at the bottom of the tube.
5. Discard the supernatant. Be sure not to disturb the pellet. The aqueous layer does not need to be completely removed at this step since additional steps to wash the protein pellet will be performed.
6. Add 800 μL of methanol to the microcentrifuge tube.
7. Centrifuge the sample at 15,000 × g at 4 °C for 10 min, carefully aspirate and discard the supernatant. Make sure not to disturb the pellet with the micropipette tip.
8. Repeat steps 6 and 7 twice.
9. Remove as much of the supernatant as possible. Be sure not to disturb the pellet.
10. Dissolve the precipitated proteins by adding 100 μL of Solution 3 (1% SDC in 20 mM HEPES) and mixing vigorously for 1 min.
11. Sonicate the sample in a water bath sonicator to resolubilize all proteins with the help of the detergent for 1 h.

5.4 Notes

- If Solution 3 is too sticky to be aspirated with the pipette, equilibrate the solution to 37 °C for a brief time, i.e., in a 37 °C incubator. Solution 3 dissolves the protein pellet and a volume of 100 μL of solution is required per sample.
- We recommend vortexing the tube vigorously between each addition of methanol, chloroform, and water to ensure complete mixing.

These solutions should not be added all at once. When water is added, the sample becomes cloudy white, which indicates protein precipitation.

- Please note that when preparing a large sample set, we recommend removing only a few samples from the centrifuge at a time at step 9 and 11 to avoid dislodging the precipitate from the tube wall while handling additional samples.
- The protein pellet can be frozen and stored at −80 °C to pause the procedure but avoid repeated freezing and thawing.
- An MS compatible detergent such as Rapigest (Waters) can be used as an alternative to SDC. Please ensure to incorporate all additional steps that are necessary to inactivate the detergent prior to mass spectrometric analysis of the sample.

6. Endoproteolytic digestion

This section describes the critical step of endoproteolytic digestion of the proteome with the endoprotease Chymotrypsin. The amount of protein in the sample is measured prior to digestion (for example, with a Bicinchoninic acid (BCA) assay), and disulfide bonds are reduced and alkylated to unfold proteins for an efficient digestion with Chymotrypsin. Reduction and alkylation of the proteome requires approximately 120 min, and digestion with Chymotrypsin is performed overnight. Following endoproteolytic digestion, Chymotrypsin is inactivated, and detergent (SDC) is removed by precipitation. Acidification and precipitation require approximately 90 min of labor.

6.1 Equipment

- Thermomixer (ThermoMixer® F1.5, Eppendorf)
- Centrifuge (Microfuge 20R, Beckman Coulter)

6.2 Materials and stock solutions

- BCA Protein Assay Kit (23225, Thermo Scientific)
- HEPES, pH 7.4
- Sodium deoxycholate (SDC)
- Tris(2-carboxyethyl) phosphine hydrochloride (TCEP), pH 7.0
- Iodoacetamide (IAA)
- Chymotrypsin, sequencing grade
- Formic acid

- Solution 4: 1% SDC in 20 mM HEPES, pH 7.4
- Solution 5: 50 mM TCEP, 5% SDC in 20 mM HEPES, pH 7.4
- Solution 6: 120 mM in 20 mM HEPES, pH 7.4
- Solution 7: 0.1 μg/μL Chymotrypsin (endoprotease:protein = 1:100 (w:w)) with 20 mM HEPES, pH 7.4
- Solution 8: 15% formic acid in H_2O.

6.3 Procedure

1. Protein concentration should be determined by BCA assay with BCA Protein Assay kit. Follow the instructions from Thermo Fisher Scientific.

2. Adjust the volume of 200 μg of protein to 80 μL (2.5 μg/μL) with Solution 4 (1% SDC in 20 mM HEPES, pH 7.4). The amount of protein is dependent on the aim of experiment or type of sample. If the volume of 200 μg of protein is more than 80 μL, increase the volume of all solutions proportionally except Solution 7 (0.1 μg/μL chymotrypsin) since the concentration of Chymotrypsin in Solution 7 is based on a weight-to-weight (w:w) ratio of endoprotease to protein.

3. Denature the proteins and reduce the disulfide bonds in the sample by adding 20 μL of Solution 5 to final concentration of 10 mM TCEP, 1% SDC in 20 mM pH 7.4, and then incubate the sample for 1 h at 60 °C while shaking at 500 rpm. Alternatively, if other detergents such as Rapigest are used, protein can be heat denatured at 95 °C for 10 min.

4. Alkylate the reduced disulfide bonds by adding 20 μL of Solution 6 to a final concentration of 20 mM IAA in 20 mM HEPES pH 7.4. Incubate the sample for 30 min at room temperature in the dark.

5. Add 20 μL of Solution 7 and incubate overnight at 30 °C with shaking at 500 rpm.

6. Following digestion, quench the reaction by acidifying with 10 μL of Solution 8 (15% formic acid) to a final concentration of 1% formic acid. The reaction becomes cloudy-white when acidified but clears with vigorous mixing. Incubate them at 37 °C for 30 min while shaking at 500 rpm. Precipitated SDC will appear at the bottom of tube after incubation.

7. Centrifuge the samples at 15,000 × g at 4 °C for 30 min. Transfer the clear samples to a new tube. Collect as much of the sample as possible.

8. Centrifuge the samples at 15,000 × g at 4 °C for 30 min and transfer the supernatant to a new tube to remove any remaining precipitated SDC. Make sure that no precipitated SDC is collected with the sample.

6.4 Notes

- Solution 4 is used to adjust the volume of samples to 80 µL before reduction and alkylation of the disulfide bonds. Solution 4 can be viscous and should be prepared immediately before use.
- Solution 5 reduces disulfide bonds to free sulfhydryl groups, and 20 µL of Solution 5 is required per sample. Solution 5 may get turbid once TCEP is added, but it turns translucent upon extensive mixing. Solution 5 can be viscous and should be prepared immediately before use.
- Solution 6 alkylates free sulfhydryl groups. Solution 6 is light sensitive and must be shielded from direct exposure to light.
- When preparing a large sample set, we recommend removing only a few samples from the centrifuge at a time at step 8 to prevent dislodging the precipitate from the tube walls while handling other samples.

7. Second labeling, sample cleanup

This section describes the procedure for labeling newly exposed lysine sites after digestion. The samples are desalted and prepared for subsequent mass spectrometric analysis. The procedure for the second dimethyl-labeling step is like the initial labeling, but the combination of isotopomers of formaldehyde and cyanoborohydride is different. The second labeling and sample cleanup takes approximately 30 min of labor once reagents are prepared. The samples can be stored frozen at 80 °C prior to mass spectrometric analysis, however repeated freezing and thawing of the samples should be avoided. Two protocol variants are presented for either ≤ 6 or > 6 samples

7.1 Equipment

- C18 tip, 10 or 100 µL bed (for example, 87784, Thermo Fisher Scientific)
- Centrifuge (Microfuge 20R, Beckman Coulter)
- Multi-pipette (for multiple samples)
- Vacuum dry evaporator (Labconco)

7.2 Materials and stock solutions

- 96-well plate
- 100% Acetonitrile (ACN)
- 100% Methanol
- Formic acid
- HEPES, pH 7.4

- Formaldehyde, $^{13}CD_2O$ (CDLM-4599, CIL)
- Sodium cyanoborodeuteride, $NaBD_3CN$ (DLM-7364, CIL)
- Ammonium Bicarbonate
- Sample vial
- 0.1% Trifluoroacetic acid
- Solution 9: 0.1% formic acid in water
- Solution 10: 20 mM HEPES, pH 7.4
- Solution 11: 1% Formaldehyde ($^{13}CD_2O$), 0.3 mM Sodium cyanoboro-deuteride ($NaBD_3CN$), 20 mM HEPES, pH 7.4
- Solution 12: 1% Ammonium Bicarbonate
- Solution 13: 40% ACN in 0.1% formic acid
- Solution 14: 60% ACN in 0.1% formic acid
- 5% ACN in 0.1% formic acid

7.3 Procedure

This protocol is used when ≤6 samples are prepared simultaneously. Follow the manufacturer's recommendation for activating the C18 tip and loading the sample to the tip.

1. If using Thermo Fisher Scientific C18 tips (87784), wet the tip by aspirating twice 100 μL of 70% ACN.
2. Equilibrate the tip by aspirating twice 100 μL of 0.1% TFA.
3. Aspirate ten times 100 μL of sample.
4. Rinse the tip by aspirating twice 100 μL of 0.1% TFA.
5. Perform the second dimethylation labeling step by aspirating five times 100 μL of Solution 11. Immerse the tips in Solution 11 and incubate the tips with the solution for 15 min at room temperature.
6. Quench the reaction by aspirating five times 100 μL of Solution 12.
7. Wash the C18 tips by aspirating once 100 μL of Solution 10.
8. Wash C18 tips by aspirating once 100 μL of Solution 9.
9. Elute the peptides by aspirating five times 100 μL Solution 13.
10. Elute the peptides by aspirating five times 100 μL of Solution 14.
11. Collect the peptides by combining the solutions from step 9 and 10.
12. Lyophilize the desalted samples until dry or dilute samples to a final concentration of 5% ACN.

This procedure is used when >6 samples are prepared simultaneously.

1. Add sample and solutions to appropriate wells of 96 well plates or solvent reservoirs (Fig. 2). Add 150 μL of solution to each well to prevent bubbles from being aspirated into the tip. A multi-pipettor can be used

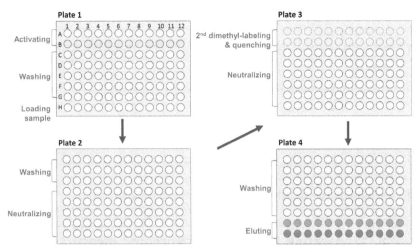

Fig. 2 Suggested design and use of the four 96 well plates for the simultaneous preparation of up to 12 CPP samples. The dimethylation of the peptides and the desalting are performed sequentially but within one integrated workflow. The color of the wells indicates different solutions used during the procedure. Arrows indicate the workflow.

to repeatedly aspirate the solution in each well and to advance up to 12 samples simultaneously through the protocol.

2. Activate a C18 tip by aspirating five times 100 μL of 100% methanol in row A of plate 1.

3. Activate a C18 tip by aspirating five times 100 μL of 100% ACN in row B of plate 1.

4. Wash the C18 tip by aspirating once 100 μL of Solution 9 in each row C to G of plate 1.

5. Load peptides onto the C18 tip by aspirating ten times the sample in row H of plate 1.

6. Wash C18 tips by aspirating once 100 μL of Solution 9 in each row A to C of plate 2.

7. Adjust the pH of the C18 tips by aspirating once 100 μL of Solution 10 in each row D to H of plate 2.

8. Perform the second dimethylation labeling step by aspirating five times 100 μL of Solution 11 in row A of plate 3. Immerse the tips in Solution 11 and incubate for 15 min at room temperature.

9. Quench the reaction by aspirating five times 100 μL of Solution 12 in row B of plate 3. Immerse the tips in Solution 12 and incubate for 10 min at room temperature.

10. Wash the C18 tips by aspirating once 100 μL of Solution 10 in row C to H of plate 3.

11. Wash C18 tips by aspirating once 100 μL of Solution 9 in row A to F of plate 4.

12. Elute the peptides by aspirating five times 100 μL of Solution 13 in row G of plate 4.

13. Elute the peptides by aspirating five times 100 μL of Solution 14 in row H of plate 4.

14. Collect the peptides by combining the solutions from rows G and H of plate 4.

15. Lyophilize the desalted samples to near dryness.

8. Preparation for mass spectrometric measurement and bottom-up proteomic analysis of sample

After labeling, digestion and clean-up, the sample is ready for mass spectrometric analysis. Depending on the chromatographic separation and sensitivity of the mass spectrometer, between 1 and 10 μg peptide per sample is loaded on a reversed phase column equipped with an electrospray ionization tip. Peptides are separated by reversed phase high pressure liquid chromatography (HPLC). Eluting peptides are electrospray ionized at the nanospray tip and introduced into the mass spectrometer. Two key features are recorded during mass spectrometry analysis. First, a survey mass spectrum of the eluted peptides is repetitively recorded so that the chromatographic elution peaks of peptides are resolved. At least five points of measurement across a chromatographic elution peak of a peptide are necessary for a robust quantification. Second, precursor ions are selected for CID, and MS/MS or fragment ion mass spectra are recorded. Many different mass spectrometers can be used in a bottom-up proteomic experiment. CPP does not require the use of mass spectrometers from specific vendors because CPP analysis can be performed on data that is acquired by any of the available instruments that allow for an isotope based proteomic quantitation experiment.

Bottom-up proteomic experiments come with inherent limitations. One limitation is a well-known trade-off between chromatographic separation of peptides that determines coverage of a proteome versus the speed of sample measurement and thus, sample throughput. Highest coverage of a structural proteome requires excellent chromatographic separation of peptides which typically requires extensive prefractionation of peptides as in a MudPIT experiment. The shallow chromatographic elution buffer gradients that are employed to improve the resolution of peptide peaks come at the cost of prolonged measurement time. The multi-step MudPIT protocol delivers the best proteomic coverage for the least amount of sample, but if sample throughput is more important than depth of proteomic coverage, there are good HPLC solutions. For instance, using the EvoSep system, peptides can be conveniently loaded onto pipet tips that can trap and desalt peptides prior to loading onto a reversed phase column. EvoSep tips conveniently merge peptide elution with direct loading onto the analytical reversed phase HPLC chromatographic column.

Mass spectrometers have been developed for fast and efficient peptide sample measurement. The mass spectrometer must have a sufficiently high mass resolution of the survey mass spectra to capture isotopic peak patterns of peptide precursor ions and sufficiently high repetition rate of MS survey spectra to resolve individual peptide peaks along the time axis. An automatic charge determination or ion mobility selection of higher charged ($z \geq +2$) peptides is required for an efficient coverage of lysine sites in the proteome if a data dependent acquisition (DDA) mode is chosen because endo-proteolytic digestion with Chymotrypsin generates charge $z = +1$ peptides. Mass spectrometric data can also be acquired in data independent acquisition (DIA) mode; however, as described below, additional deconvolution of fragment ion spectra and extensive analysis of raw data is required prior to a standard data analysis. Whether data was acquired in DDA or DIA mode, results are the same.

8.1 Equipment

- Centrifuge (Microfuge 20R, Beckman Coulter)
- Bath sonicator (M1800, Branson)
- Easy-nLC (Thermo Fisher Scientific)
- Orbitrap Lumos (Thermo Fisher Scientific)

8.2 Buffers and materials

- Snap top polypropylene vial
- 0.1% formic acid in water
- Reverse-phase chromatography C18 column (100 μm × 250 mm, 1.7 μm inner diameter)
- Mobile phase A: 5% ACN, 0.1% formic acid in water
- Mobile phase B: 20% water, 0.1% formic acid in ACN
- Solution 15: 3% ACN in 0.1% formic acid

8.3 Procedure

1. Dissolve the dried peptides in Solution 15 to a final concentration of 0.5 µg/µL.
2. Vortex the sample vigorously for 1 min and sonicate the sample in a bath sonicator for 1 min.
3. Centrifuge the samples at $15,000 \times g$ at $4\,°C$ for 30 min and transfer the samples into the sample vial.
4. Separate the peptides by reverse-phase separation using a C18 column using a gradient of Solvent A (5% ACN, 0.1% formic acid in water) and Solvent B (20% water, 0.1% formic acid in ACN). Inject 2 µg of peptides, into the peptide Nanocapillary Easy-nLC 1200 system coupled to an Orbitrap Lumos mass spectrometer (Thermo Fisher Scientific).
5. Run a gradient of Solvent B from 0% to 25% over 90 min, from 25% to 40% over 30 min and from 40% to 100% over 10 min and then maintain 100% Solvent B for 10 min at a flow rate of 300 nL/min.
6. Use the following parameters to collect data on the mass spectrometer:
   (a) Set the polarity of the mass spectrometer to positive in the tune file. On the Orbitrap mass spectrometer (Thermo Fisher Scientific) set the spray voltage to 2.5 kV and heat the entrance capillary of the mass spectrometer to $325\,°C$.
   (b) Select full MS/DD-MS2 as scan type of the mass spectrometer and set method duration to 120 min and default charge state to $z \geq +2$.
   (c) Perform a MS survey scan in profile mode from 375 to 1,500 m/z at a resolution of $R = 120,000$ with a maximum ion injection time of 50 ms and an automatic gain control target value of 4×10^5 counts.
   (d) Select the ions with the top 20 highest intensities (>5,000 counts) for directed MS/MS fragmentation while excluding ions with an unassigned charge state or a charge state of $z < +2$ and $z > +8$.
   (e) Collect MS/MS scans at a resolution of $R = 30,000$ with normalized collision energy of 28, and with a maximal ion injection time of 100 ms or with a maximal automatic gain control value of 3×10^3 counts. Set the dynamic time exclusion window for precursor ions to 10 s and an isolation window of 0.7 m/z.

8.4 Notes

- Please check for complete digestion with an appropriate peptide quantification assay and adjust the final size of the peptide sample that is analyzed with mass spectrometry.

9. Quantification of lysine site accessibility and interpretation of results

Once mass spectrometric data is acquired, bioinformatic data analysis can begin. Peptides are identified first, and then the relative fraction of accessible over inaccessible lysine molecules per site is determined. The relative accessibility per individual lysine site can then be compared between different experiments or experimental conditions. A statistical analysis of the data will reveal significant differences in lysine site accessibility, which helps identify lysine sites that are worth considering for further analysis.

9.1 Equipment

A personal computer with a large random–access memory (RAM) and long-term data storage (hard drive) is sufficient for most parts of the data analysis. Many novel peptide search engines make use of a vector graphics card for fast processing of data. A computer cluster can speed data analysis and allows multiple users to access the software tools but is not required.

9.2 Software examples

Several different free software tools are available at GitHub, others are provided by the vendor of the mass spectrometers for data dependent and data independent sample analysis (Codrea & Nahnsen, 2016). For example, Sequest (Xu et al., 2015) is a database search engine that was originally developed for data dependent analysis of fragment ion spectra. Sequest evaluates MS/MS spectra based on a cross correlation analysis of the acquired fragment ion spectra with a predicted fragment ion spectrum generated "on the fly" from a protein database that includes all annotated protein sequences of a species. Typically, this protein database comprises a complete human proteome that is updated monthly and released by Uniprot (The, 2018). Newer versions of Sequest-like search engines include Prolucid (Xu et al., 2015) and Blazmass (Chatterjee et al., 2016). Additionally, the peptide search engine Morpheus can be used. Morpheus is fast and efficient but requires high mass resolution and high-quality mass spectra (Wenger & Coon, 2013). Several of the commonly used software tools are conveniently integrated into the Integrated Proteomics pipeline (IP2) which is now available as a ready-to-use solution (the PASER box) sold by the vendor Bruker.

9.3 Procedure

The basic steps in data analysis are listed here.

9.3.1 Extract mass spectrometric data from the vendor-specific raw data format into a data format that is compatible with a database search engine

For example, RawConverter (Wenger & Coon, 2013) extracts monoisotopic precursor peaks from raw files that are generated by Orbitrap mass spectrometers (Thermo Fisher Scientific). RawConverter writes survey ion (MS1) and fragment ion (MS2) mass spectra into separate files in a flat file format that is easy to manually inspect and that is compatible with several different downstream data analysis software solutions. Both file types (MS1 and MS2) serve as input for diverse subsequent data analysis steps. Set the data conversion parameters in RawConverter to:

- Extract monoisotopic peaks.
- Deconvolute precursor mass spectra for ion charge to ion charge $z = +1$.
- Store fragment ion mass spectra in the MS2 file format.

9.3.2 Search fragment ion mass spectra for matches to the protein database that is provided by the user

Proteome databases in fasta file format are compatible with most fragment ion search engines. We recommend downloading the latest version of a curated human proteome database from Uniprot.

Set up the search parameters for the database search. Each search engine includes many different and unrelated search parameter options and specific parameters need to be adjusted by each user. The search parameters listed here are a recommended starting point for analyzing CPP experiments. The database search should

- Consider only precursor ion peaks with a charge $z \geq 2$.
- Consider only peptides that are cleaved C-terminal to the endoprotease Chymotrypsin specific sites F, Y, and W.
- Consider only peptides that are cleaved with Chymotrypsin at both the N- and C-terminus.
- Include a static modification for dimethylation at the N-terminus of the peptides. Set the mass shift for the N-terminal mass according to the isotope defined labeling reagents that were used in the second chemical labeling step.
- Add the mass of the light isotope defined chemical label to the mass of lysine as static mode.

- Include a dynamic modification at lysine that reflects the mass difference between heavy and light isotope defined masses
- Filter the search engine results for peptides of interest.

The result of the database search lists all peptides that matched the fragment ion mass spectra. For example, the search engine ProLuCID lists up to 10 different peptide sequences per fragment ion mass spectrum. The output sqt file lists peptide matches rank ordered with the best matching peptide at the first position. Different scores are reported for each peptide, which allows the dataset to be filtered for the best matching peptides. Scores like the cross-correlation score or the e-value (expectation value) for each peptide can reflect how well a theoretical fragment ion spectrum matched the existing fragment ion spectrum. Based on the peptide scores, the list is filtered either in part by hard cutoff filters or by dynamic cutoff filters using Bayesian inference.

The initial database search also comprises a copy of the protein sequence database in which all protein sequences were reversed. This reversed protein database is searched in parallel to the standard protein database. Because reversed protein sequences do not exist, any matches of a reversed peptide sequence to a fragment ion spectrum can be considered a peptide identification that is false. The number of false positive identifications in the reversed protein database provides an estimate of how many false positive identifications may be included in the set of peptide identifications that matched to the normal protein database. The fraction of peptide identifications in the reversed database relative to the peptide identifications in the protein database determines an estimated false discovery rate (FDR). The FDR estimate allows parameters to be adjusted to ensure that a small number of false positive peptide identifications is included in the result while a maximal number of true positive peptide identifications is listed.

The parameters that are used to filter results depend on the search engine. We utilize the most recent release of DTASelect to filter the search results that are reported in sqt file format. DTASelect allows for a flexible adjustment of parameters, is efficient, and includes a variety of elaborate filtering tools (Tabb, McDonald, & Yates, 2002). Adjust the FDR to 1% on fragment ion spectrum level. Following this initial filtering step, the data then needs to be filtered to include only sequences that harbor a single lysine site. A single lysine site per peptide is required to unambiguously quantify its accessibility for chemical modification. Peptide sequences without a lysine site are not informative, whereas a peptide sequence with more than one lysine site yields an ambiguous result because it is unclear how much each lysine site in the peptide contributed to the final quantification result.

9.3.3 Quantify the relative chemical surface accessibility per lysine site

Each peptide with a lysine residue will be dimethylated "light" or "heavy" depending on the accessibility for chemical labeling in the first labeling step. Heavy and light peptides elute simultaneously from the chromatographic reversed phase column and the precursor ion spectrum is recorded for both. The elution peaks of the light and heavy peptide elute simultaneously but differ in mass. This difference in mass is due to the difference between light and heavy dimethyl tags. The light and heavy elution peaks also differ in intensity and in area under the chromatographic peak. A ratio of the areas under the chromatographic peaks for the light and heavy peptide ion precursors represents the ratio of the heavy to the light labeled lysine sites.

There are many different software solutions available to calculate the intensity of peptide elution peaks. Here, we use Census (Park, Venable, Xu, & Yates, 2008), which was explicitly developed to quantify heavy to light peptide elution peaks of isotope labeled peptides. Essential parameters for a successful quantification of lysine site accessibility for chemical modification with CPP are:

- Set the range of reported values for the heavy to light ratio to 1000:1 to 1:1000.
- Report peptides that are heavy or light only labeled.

In this way, the result also includes all peptides for which only a light or heavy peptide peak was detected and the area under the curve quantified. These peptide quantifications are sometimes called singletons. The output file lists all peptides and the ratio of the heavy to light areas under the chromatographic elution peaks. The values indicate in which fraction of protein molecules one specific lysine site was inaccessible for chemical modification.

9.3.4 Compare experiments to identify lysine sites that are significantly altered in accessibility for chemical labeling

A single CPP experiment can identify the relative inaccessibility of lysine sites for chemical modification in a proteome, thus uncovering the 3D proteome in a single experimental condition. Multiple CPP experiments can be used to compare the results between different experimental conditions. Several different software solutions are available to compare isotope labeled proteomes and can be easily adopted for analyzing and comparing CPP experiments.

Here, we employ ProteinClusterQuant (and its extension tool) to accurately compare two or more different CPP experiments and report lysine sites that are significantly altered in accessibility for chemical labeling.

Importantly, an average ratio value is derived from technical replicate measurements per experimental condition, which allows the calculation of a random error of measurement (typically expressed as standard deviation). Given that the accessibility for chemical modification at each lysine is quantified, it can be efficiently compared between experiments.

The results report lysine sites that are significantly different in accessibility for chemical modification between two or several experimental conditions. We recommend correcting the results for the accuracy with which the individual lysine sites were quantified. Therefore, we recommend correcting for multiple hypothesis testing and reporting significant differences along with a corresponding false discovery rate. For example, a Bonferroni-Hochberg based post-hoc correction of discoveries can be used and is implemented in ProteinClusterQuant.

9.4 Notes

Database search: Depending on the time required to search the proteome database, a semi-Chymotryptic search can be performed. In this case only one end of the peptides matches the requirement to be cleaved by Chymotrypsin.

Labeling of peptide N-termini: All proteins are endoproteolytically digested prior to the second chemical labeling step to reveal all lysine sites that were not labeled with dimethyl groups during the initial labeling step. The endoproteolytic digestion introduces a primary amine at the peptide N-terminus that is eventually dimethylated during the second labeling step.

Choice of endoprotease: It is important to note that because Chymotrypsin cleaves with different efficiencies C-terminally to several hydrophobic amino acids with aromatic side chains, different peptides can include the same lysine site in a protein. Depending on the distribution of the hydrophobic amino acids in a protein sequence, different peptides may be generated that all cover the same lysine site. Therefore, it is important to consider all ratio values that were measured for a given lysine site. All peptides covering the same lysine site should be grouped to calculate the average and standard deviation of measurement for the relative accessibility for chemical modification. This step has been integrated into ProteinClusterQuant (Bamberger et al., 2018). ProteinClusterQuant displays a proteome in a bipartite network in which peptides and proteins are connected based on sequence identity. The bipartite network represents the proteome that was identified and quantified in a bottom-up proteomic experiment.

Interpretation of relative lysine site accessibility: The relative surface accessibility of each quantified lysine site is expressed as a ratio of protein molecules in which the lysine site was inaccessible over the protein molecules in which the lysine site was accessible for initial labeling. This ratio value can be converted to a $\log_2$ or $\log_{10}$ scale for a simplified calculation of measurement of error and error propagation. However, results are more difficult to interpret on a logarithmic scale. Ratio values can be converted into "percent inaccessible" protein molecules per lysine site for a quick comprehension of the results.

Conformational alterations in even a small percentage of protein molecules might have profound biological effects, like increased enzymatic activity. While we have observed high reproducibility of results across different experiments, smaller changes in lysine site accessibility might be more difficult to reconcile in follow up experiments. It is usually easier to follow up larger changes in lysine site accessibility that alter the majority of the respective protein molecules. Other selection criteria might include analyzing further protein conformations in which at least 20% or every fifth protein molecule has changed in surface accessibility.

Choosing the right isotope combinations: Some of the mass shifts that are introduced by isotope-defined labeling reagents are within the low Da range. This can cause a significant overlap of peptide isotope peaks between naturally dimethylated peptides (PTM methylation) and chemically induced dimethylation, even though naturally dimethylated peptides usually represent only an exceptionally small fraction of all peptides. Natural dimethylation can influence the results significantly only in exceedingly rare cases. To reduce any inference between natural dimethylation and CPP, we recommend using cyanoborodeuteride (the heavy version of sodium cyanoborohydride) as the isotope defined labeling reagent during the initial labeling step. The initial labeling step requires copious amounts of labeling reagent, and the cost of isotope defined formaldehyde is high, whereas sodium cyanoborodeuteride is inexpensive. Any overlap of isotope-defined peptide precursor peaks between the naturally occurring dimethylation of a lysine site and the initially accessible proportion of the lysine site may slightly influence the measurement of accessible lysine. Because the fraction of lysine accessible protein molecules is likely to be much larger than the fraction of protein molecules with an inaccessible lysine site, it is better to use cyanoborodeuteride in the initial labeling step. Any error in measurement influences the larger measurement value less, and the larger value is typically associated with the accessible fraction of a lysine site. In addition, the heavy labeled peptide precursor peak is influenced extraordinarily little by naturally occurring dimethylation at the same lysine site. Because the mass difference between the heavy isotope dimethyl group and a naturally occurring dimethyl group is large, the area of the chromato-graphic peak of the heavy labeled peptide precursor ion remains unaffected and thus less influenced by the naturally occurring dimethylation.

10. Summary and conclusions

This chapter describes the fundamental steps for Covalent Protein Painting (CPP), a chemical footprinting experiment at lysine sites. CPP was established in living human cells in cell culture but should provide a fast and comprehensive quantitation of the accessibility of lysine sites in the pro-teome of any cells. The technique has been efficiently adopted for snap-frozen human brain tissue and cell culture samples, and we expect a rapid adoption of CPP to other biological samples. Quantification of lysine site accessibility is based on an isotope measurement with mass spectrometry, a method that is well established in many different proteomic data analysis approaches. The quantification results in CPP are readily available, and the relative accessibility of lysine sites can be compared directly between differ-ent experiments on a large scale. For example, we used CPP to rapidly quan-tify and compare lysine site accessibility in all 60 cell lines in the NCI60 cell line panel (Shoemaker, 2006). Lysine sites that are altered in surface acces-sibility indicate where additional experiments are necessary to determine whether protein-protein interactions or protein conformations have chan-ged between different experimental conditions.

Lysine site accessibility and labeling efficiency in proteins may also pro-vide evidence to guide *ab initio* protein folding and protein structure predic-tion. The high efficiency of quantification and the precision of measurement of CPP may be used to model the local environment of a lysine site in an isolated protein. CPP results can be combined with results obtained from other methods of protein structure prediction or compared to existing protein structure data to infer the homeostasis of a protein in different conformational and protein-protein interactions in a cell. Changes in this homeostasis can be quickly revealed by comparing different experimental conditions and can be tracked across further experiments.

Covalent Protein Painting is an easy and efficient protocol for per-forming chemical footprinting experiments on lysine sites. The method can be adopted to additional amino acids if quantitative isotope labeling of the amino acid can be achieved. The extensive coverage of lysine sites in a proteome makes CPP an exceptionally efficient tool for the discovery of conformational or structural changes on a large scale.

Acknowledgments

We are indebted to Claire Delahunty for carefully reading the manuscript. We thank the Yates laboratory for continuous support and many helpful discussions.

Funding

This work was funded by NIH grant R03 AG047957-02, R33 CA212973-01 and R01 HL131697 awarded to John R. Yates III.

References

Bamberger, C., Diedrich, J., Martìnez-Bartholomé, S., & Yates, J. R. (2022). Cancer conformational landscape shapes tumorigenesis. *Journal of Proteome Research*, *21*(4), 1017–1028. https://doi.org/10.1021/acs.jproteome.1c00906.

Bamberger, C., Martinez-Bartolome, S., Montgomery, M., Pankow, S., Hulleman, J. D., Kelly, J. W., et al. (2018). Deducing the presence of proteins and proteoforms in quantitative proteomics. *Nature Communications*, *9*(1), 2320. https://doi.org/10.1038/s41467-018-04411-5. PubMed PMID: 29899466; PMCID: PMC5998138.

Bamberger, C., Pankow, S., Martínez-Bartolomé, S., Ma, M., Diedrich, J., Rissman, R. A., et al. (2021). Protein footprinting via covalent protein painting reveals structural changes of the proteome in Alzheimer's disease. *Journal of Proteome Research*, *20*(5), 2762–2771. https://doi.org/10.1021/acs.jproteome.0c00912.

Chatterjee, S., Stupp, G. S., Park, S. K., Ducom, J. C., Yates, J. R., 3rd, Su, A. I., et al. (2016). A comprehensive and scalable database search system for metaproteomics. *BMC Genomics*, *17*(1), 642. https://doi.org/10.1186/s12864-016-2855-3. PMID: 27528457; PMCID: PMC4986259.

Codrea, M. C., & Nahnsen, S. (2016). Platforms and pipelines for proteomics data analysis and management. *Advances in Experimental Medicine and Biology*, *919*, 203–215. https://doi.org/10.1007/978-3-319-41448-5_9. PubMed PMID: 27975218.

Eng, J. K., McCormack, A. L., & Yates, J. R. (1994). An approach to correlate tandem mass spectral data of peptides with amino acid sequences in a protein database. *Journal of the American Society for Mass Spectrometry*, *5*(11), 976–989. https://doi.org/10.1016/1044-0305(94)80016-2. PubMed PMID: 24226387.

Espino, J. A., Mali, V. S., & Jones, L. M. (2015). In cell footprinting coupled with mass spectrometry for the structural analysis of proteins in live cells. *Analytical Chemistry*, *87*(15), 7971–7978. https://doi.org/10.1021/acs.analchem.5b01888. PubMed PMID: 26146849.

Johnson, D. T., Di Stefano, L. H., & Jones, L. M. (2019). Fast photochemical oxidation of proteins (FPOP): A powerful mass spectrometry-based structural proteomics tool. *The Journal of Biological Chemistry*, *294*(32), 11969–11979. https://doi.org/10.1074/jbc.REV119.006218. PMID 31262727.

Kahsai, A. W., Xiao, K., Rajagopal, S., Ahn, S., Shukla, A. K., Sun, J., et al. (2011). Multiple ligand-specific conformations of the beta2-adrenergic receptor. *Nature Chemical Biology*, *7*(10), 692–700. https://doi.org/10.1038/nchembio.634. PubMed PMID: 21857662; PMCID: PMC3404607.

Kallen, R. G., & Jencks, W. P. (1966). Equilibria for the reaction of amines with formaldehyde and protons in aqueous solution. A re-examination of the formol titration. *The Journal of Biological Chemistry*, *241*(24), 5864–5878. Epub 1966/12/25. PubMed PMID: 5954364.

Mateus, A., Kurzawa, N., Becher, I., Sridharan, S., Helm, D., Stein, F., et al. (2020). Thermal proteome profiling for interrogating protein interactions. *Molecular Systems Biology*, *16*(3), e9232. https://doi.org/10.15252/msb.20199232. PMID: 32133759.

Park, S. K., Venable, J. D., Xu, T., & Yates, J. R., 3rd. (2008). A quantitative analysis software tool for mass spectrometry-based proteomics. *Nature Methods*, *5*(4), 319–322. https://doi.org/10.1038/nmeth.1195. PMID: 18345006; PMCID: PMC3509211.

Pascal, B. D., Chalmers, M. J., Busby, S. A., Mader, C. C., Southern, M. R., Tsinoremas, N. F., et al. (2007). The deuterator: Software for the determination of backbone amide deuterium levels from H/D exchange MS data. *BMC Bioinformatics*, *8*, 156. https://doi.org/10.1186/1471-2105-8-156. PubMed PMID: 17506883.

Shoemaker, R. H. (2006). The NCI60 human tumour cell line anticancer drug screen. *Nature Reviews. Cancer*, *6*(10), 813–823. https://doi.org/10.1038/nrc1951. PMID: 16990858.

Tabb, D. L., McDonald, W. H., & Yates, J. R., 3rd. (2002). DTASelect and contrast: Tools for assembling and comparing protein identifications from shotgun proteomics. *Journal of Proteome Research*, *1*(1), 21–26. PMID: 12643522; PMCID: PMC2811961.

The, U. P. C. (2018). UniProt: A worldwide hub of protein knowledge. *Nucleic Acids Research*, *47*(D1), D506–D515. https://doi.org/10.1093/nar/gky1049.

Washburn, M. P., Ulaszek, R. R., & Yates, J. R., 3rd. (2003). Reproducibility of quantitative proteomic analyses of complex biological mixtures by multidimensional protein identification technology. *Analytical Chemistry*, *75*(19), 5054–5061. Epub 2004/01/08. PubMed PMID: 14708778.

Wenger, C. D., & Coon, J. J. (2013). A proteomics search algorithm specifically designed for high-resolution tandem mass spectra. *Journal of Proteome Research*, *12*(3), 1377–1386. https://doi.org/10.1021/pr301024c. PubMed PMID: 23323968; PMCID: PMC3586292.

Xiang, S., Kato, M., Wu, L. C., Lin, Y., Ding, M., Zhang, Y., et al. (2015). The LC domain of hnRNPA2 adopts similar conformations in hydrogel polymers, liquid-like droplets, and nuclei. *Cell*, *163*(4), 829–839. https://doi.org/10.1016/j.cell.2015.10.040. PMID: 26544936; PMCID: PMC4879888.

Xu, T., Park, S. K., Venable, J. D., Wohlschlegel, J. A., Diedrich, J. K., Cociorva, D., et al. (2015). ProLuCID: An improved SEQUEST-like algorithm with enhanced sensitivity and specificity. *Journal of Proteomics*, *129*, 16–24. https://doi.org/10.1016/j.jprot.2015.07.001. PubMed PMID: 26171723; PMCID: PMC4630125.

Zhang, Y., Fonslow, B. R., Shan, B., Baek, M.-C., & Yates, J. R. (2013). Protein analysis by shotgun/bottom-up proteomics. *Chemical Reviews*, *113*(4), 2343–2394. https://doi.org/10.1021/cr3003533.

Zhou, Y., & Vachet, R. W. (2012). Diethylpyrocarbonate labeling for the structural analysis of proteins: Label scrambling in solution and how to avoid it. *Journal of the American Society for Mass Spectrometry*, *23*(5), 899–907. https://doi.org/10.1007/s13361-012-0349-3. PubMed PMID: 22351293; PMCID: PMC3324597.

Zhou, Y., & Vachet, R. W. (2013). Covalent labeling with isotopically encoded reagents for faster structural analysis of proteins by mass spectrometry. *Analytical Chemistry*, *85*(20), 9664–9670. https://doi.org/10.1021/ac401978w. PMID: 24010814; PMCID: PMC3819941.

CHAPTER THREE

Probing protein misfolding and dissociation with an infrared free-electron laser

Hisashi Okumura[a,b,c], Takayasu Kawasaki[d], and Kazuhiro Nakamura[e,*]

[a]Exploratory Research Center on Life and Living Systems (ExCELLS), National Institutes of Natural Sciences, Okazaki, Aichi, Japan
[b]Institute for Molecular Science, National Institutes of Natural Sciences, Okazaki, Aichi, Japan
[c]Department of Structural Molecular Science, SOKENDAI (The Graduate University for Advanced Studies), Okazaki, Aichi, Japan
[d]Accelerator Laboratory, High Energy Accelerator Research Organization, Tsukuba, Ibaraki, Japan
[e]Department of Laboratory Sciences, Gunma University Graduate School of Health Sciences, Maebashi, Gunma, Japan
*Corresponding author: e-mail address: knakamur@gunma-u.ac.jp

Contents

Methods in Enzymology, Volume 679
ISSN 0076-6879
https://doi.org/10.1016/bs.mie.2022.08.047

Abstract

Misfolding is observed in the mutant proteins that are causative for neurodegenerative disorders such as polyglutamine diseases. These proteins are prone to aggregate in the cytoplasm and nucleus of cells. To reproduce cells with the aggregated proteins, gene expression system is usually applied, in which the expression construct having the mutated DNA sequence of the interest is transfected into cells. The transfected DNA is finally converted into the mutant protein, which is gradually aggregated in the cells. In addition, a simple method to prepare the cells having aggregates inside has been recently applied. Peptides were first aggregated by incubating them in water. The aggregates are spontaneously taken up by cells because aggregated proteins generally transfer between cells. Peptides with different degrees of aggregation can be made by changing the incubation times and temperatures, which enables to examine contribution of aggregation to the toxicity to the recipient cells. Moreover, such cells can be used for therapeutic researches of diseases in which aggregates are involved. In this chapter, we show methods to induce aggregation of peptides. The functional analyses of the cells with aggregates are also described. Then, experimental dissociation of the aggregates produced using this method by mid infrared free electron laser irradiation and its theoretical support by molecular dynamics simulation are introduced as the therapeutic research for neurodegenerative disorders.

1. Introduction

Deposits of insoluble proteinaceous materials in multiple organs and tissues are called amyloid (Landrieu, Dupre, Sinnaeve, El Hajjar, & Smet-Nocca, 2022). Amyloidosis is categorized as diseases having deposition of amyloid. There are 2 types of amyloidosis; systemic amyloidosis and localized amyloidosis. The cause of amyloid formation is believed to be conformational changes from native functional structures to misfolded structures.

Amyloid fibrils are aggregated proteins with filamentous appearance. These are structurally stable because the polypeptide backbone is tightly packed into stacked β-sheet structures and residue side chains are intertwined (Landrieu et al., 2022). Amyloid formation gradually proceeds through nucleation, growth and maturation. Initially, aggregation–prone sequences are exposed and soluble oligomers or amorphous aggregates are generated. These aggregates can act as nuclei of the fibrillization process. During the growth step, new monomers are added to the nuclei and eventually fibrils rapidly grow. The maturation step includes conversion from the protofilament to high-ordered fibrillary structure through protein-protein interactions at the protofilament interfaces (Landrieu et al., 2022). Indeed, polyglutamine (polyQ) aggregates are classified into various species such as oligomer, amorphous aggregate, amyloid fibril, annular aggregate and inclusion body (Adegbuyiro, Sedighi, Pilkington, Groover, & Legleiter, 2017; Jayaraman, Kodali, et al., 2012; Jayaraman, Mishra, et al., 2012; Legleiter et al., 2010; Poirier et al., 2002; Wacker, Zareie, Fong, Sarikaya, & Muchowski, 2004).

Cells with aggregates work as tools for therapeutic researches of aggregates-related diseases. Induction of aggregation of proteins in cultured cells is generally achieved by overexpression of the genes encoding aggregation-prone mutant proteins of interest in cells. Recently, a simple procedure to obtain cultured cells having aggregates inside has been reported. The procedure is based on an interesting phenomenon "transfer between cells" of aggregated proteins that are causative for neurodegenerative disorders (Rodriguez, Marano, & Tandon, 2018). Although the mechanisms by which aggregates are taken up by cells have not been fully understood, accumulating body of evidences suggests involvement of the interaction of aggregates with proteins on cell membrane (Pearce & Kopito, 2018) and subsequent clathrin-dependent endocytosis (Ruiz-Arlandis, Pieri, Bousset, & Melki, 2016). In addition, direct connection between cells might enable the transfer. PolyQ (Costanzo et al., 2013), α-synuclein (Dieriks et al., 2017), tau (Tardivel et al., 2016), PrPSc (Gousset et al., 2009) and amyloid β (Aβ) (Dilna et al., 2021) transfer from a cell to other cells using the tunneling nanotubes. Subsequent analysis revealed that Aβ oligomers-induced damage of plasma membrane and the repair system followed by PAK1 dependent endocytosis and actin remodeling led to tunneling nanotube-like conduits (Dilna et al., 2021).

Therefore, cultured cells can spontaneously incorporate aggregated peptides added in the culture medium. Using the cells having aggregates, toxicity of the aggregates to cultured cells can be functionally studied. Peptides need to be aggregated before addition to cultured cells, which enables to prepare peptides with various degrees of aggregation, thereby making it possible to compare the toxicity among these aggregates.

In this chapter, we first show how to induce aggregation of peptides that are associated with diseases such as polyQ diseases and also how to introduce the aggregates into cultured cells. Examples of functional analysis of the cells with aggregates are also shown. Then, dissociation of the aggregates produced by the method above is described as a therapeutic research.

The dissociation was done by mid infrared (IR) free electron laser (FEL) irradiation. Finally, methodology of molecular dynamics (MD) simulation that theoretically supports the dissociation process of aggregates by the IR-FEL is introduced.

2. Preparation of cells having aggregates

The peptides used were polyQ, polyleucine (polyL), polyserine (polyS) and polyalanine (polyA). We first describe relations of these repeated sequences to diseases.

2.1 PolyQ-containing aggregates

Trinucleotide repeats such as CTG, CUG, CAG, CGG, CCG, GAA and GCN are found in a significant number of genes (Verma, Khan, Bhagwat, & Kumar, 2019). The repeats are located in coding regions as well as non-coding 5′ and 3′ untranslated regions of genes. When the numbers of the repeats in genes are longer and the resultant amino acid repeats in the corresponding proteins are beyond the pathological threshold, the proteins tend to form aggregates and often lead to developmental and neurological disorders (Richard, 2020). The aggregated proteins are suggested to cause pathological states such as increased autophagy, impaired axonal transport and mitochondrial dysfunction.

One of the common triplet repeat in genes is CAG repeats that are associated with polyQ diseases. Dentatorubropallidoluysian atrophy, spinal and bulbar muscular atrophy, six autosomal dominant forms of spinocerebellar ataxia (SCA1, SCA2, SCA3, SCA6, SCA7, SCA17) and Huntington's disease (HD) are classified into polyQ diseases (Gonzalez-Alegre, 2019). PolyQ diseases show threshold phenomena with high penetrance that occurs when the repeat numbers exceed the disease-specific limits (Lieberman, Shakkottai, & Albin, 2019). The expansion of glutamine repeats in the causative proteins leads to age-related and progressive neuronal degeneration. The repeat number determines the severity and time of onset of diseases (Verma et al., 2019). For instance, the normal repeat number of glutamine in Huntingtin (HTT), the causative protein of HD, is 6-35, whereas, that of pathogenic polyQ tract is 36-121 (Lieberman et al., 2019).

Multiple regions in the nervous system such as the striatum, brainstem, cerebellum, dentate nucleus, substantia nigra, thalamus and spinal cord are affected in polyQ diseases (Lieberman et al., 2019). In these regions, morphological abnormalities such as atrophy of dendritic arborization, axonal swellings and loss of synapses are seen (Lieberman et al., 2019). Therefore, symptoms of polyQ diseases are also highly-diversified. Patients with HD are, for example, suffered from impaired gait, incoordination, chorea, speech difficulties, dysphagia, cognitive impairment, personality changes and mood disorders (Lieberman et al., 2019).

PolyQ-containing aggregated proteins are often found in the affected neurons. In addition, molecular chaperones, ubiquitin-binding proteins and proteasome are occasionally included in the aggregates. Pathogenesis of polyQ diseases is diverse (Crotti et al., 2014; Dong et al., 2011; Harding, 1983; Lieberman et al., 2019; Miller et al., 2011; Nagai et al., 2007; Olejniczak, Urbanek, & Krzyzosiak, 2015; Pandey & Rajamma, 2018; Paulson, Shakkottai, Clark, & Orr, 2017; Trottier et al., 1995; Velazquez-Perez, Rodriguez-Labrada, & Fernandez-Ruiz, 2017; Wu et al., 2005; Yang, Dunlap, Andrews, & Wetzel, 2002). Analysis of SCA1-model mice revealed that the function of multiprotein complex is changed in the presence of the expanded polyQ domain leading to neuronal dysfunctions (Jafar-Nejad, Ward, Richman, Orr, & Zoghbi, 2011). Likewise, changes in gene expression, axonal transport, mitochondrial function and metabolism are implicated in HD pathogenesis (Lieberman et al., 2019). Although it is still controversial if highly-aggregated polyQ-containing inclusions are neuroprotective or neurotoxic, it was reported that mutant HTT inclusions sequester the small toxic species and thereby mitigating apoptosis at early phase, and then, lead to necrotic cell death by sequestration of other proteins (Ramdzan et al., 2017). These findings justify dissociation of polyQ aggregates as a therapeutic strategy for polyQ diseases.

Regarding transfer between cells of expanded polyQ-containing proteins, mutant HTT can propagate between cells (Pearce & Kopito, 2018). Importantly, internalized polyQ aggregates in a cell have the ability to seed aggregation of polyQ proteins in the recipient cell even below the pathogenic threshold through the prion-like mechanism (Ren et al., 2009). Seeding was also observed using aggregates from the model animals and HD patients (Gupta, Jie, & Colby, 2012; Tan et al., 2015). Thus, aggregated polyQ peptides are likely incorporated by cells.

2.2 RAN products from polyQ diseases-causative genes

Recently, accumulating body of evidences has shown repeat-associated non-AUG (RAN) translation from polyQ diseases-responsible genes containing CAG repeat (Pearson, 2011). RAN translation is not specific to polyQ diseases but is shared by genes that are causative for other neurode-generative disorders such as frontotemporal dementia and amyotrophic lateral sclerosis (Hutten & Dormann, 2019). In the case of RAN translation seen in genes having expanded CAG repeats, the sense transcripts theoretically generate proteins containing homopolymeric polyalanine (polyA) and polyserine (polyS) tracts in addition to polyQ, while the antisense transcripts generate polyleucine (polyL), polycystein (polyC) and polyA (Banez-Coronel et al., 2015). The RAN proteins from the genes for polyQ diseases have been found using cell models, animal models and human tissues. The animal studies included SCA8 and myotonic dystrophy type 1 (DM1) (Zu et al., 2011), in which accumulation of SCA8 polyA and DM1 polyQ were detected using previously established mouse models (Zu et al., 2011).

Functionally, RAN proteins from ataxin 3, the causative gene of SCA3, included polyQ and polyA and impaired nuclear integrity and induced apoptosis (Jazurek-Ciesiolka et al., 2020). More importantly, RAN translation of mutant HTT transcript generated proteins containing the 4 homopolymeric repeats other than polyQ. These proteins were found in white matter, the caudate putamen and the cerebellum of the brain from HD patients (Banez-Coronel et al., 2015). In these brain regions, microglial activation, apoptosis and neuronal loss were seen (Banez-Coronel et al., 2015). These results suggest that the RAN products might contribute to the dysfunctions. Importantly, not only expanded polyQ-containing proteins but also the RAN products seem to be toxic because polyL encoded by mixed DNA repeats was reported to be significantly more toxic than that of polyQ in mammalian cells (Dorsman et al., 2002). Thus, functions of cells having polyL and polyS inside were investigated. The other RAN product, polyC could not be synthesized.

2.3 PolyA aggregates

In polyA disease, the other triplet expansion disease, polyA expansions are seen in transcription factors that are pivotal for development. Therefore, polyA diseases are generally congenital disorders in contrast to polyQ diseases that affect aged brains. The exception is a late-onset polyA disease, oculopharyngeal muscular dystrophy (OPMD). The causative molecules of polyA diseases are, for instance, PHOX2B (congenital central hypoventilation syndrome), RUNX2 (cleidocranial dysplasia), HOXA13 (hand–foot–genital syndrome), SOX3 (X linked Hypopituitarism), FOXL2 (blepharophimosis, ptosis and epicanthus inversus syndrome), ZIC2 (holoprosencephaly), ARX (mental retardation) and HOXD13 (synpolydactyly).

Regarding pathogenesis of polyA expansions in polyA diseases, loss-of-normal functions of the causative molecules by polyA expansions can be raised by a reduction in the protein level in mice with expanded alanine (Hughes et al., 2013). This notion was supported by the observation that the phenotypes of mice having expanded polyA in the protein were similar in other mutant mice having reduced levels of the same protein (Vest et al., 2017). However, toxic gain-of-function likely underlies the mechanism based on presence of intranuclear aggregates and apoptotic skeletal muscle by transgenic expression of PABPN1 having expanded alanine (Davies et al., 2005; Davies, Rose, Sarkar, & Rubinsztein, 2010; Davies, Sarkar, & Rubinsztein, 2006, 2008). Thus, cells having polyA aggregates work as tools to test the causal relationship between polyA aggregation and cellular dysfunctions.

2.4 Equipment
- Scanning electron microscopy
- Transmission electron microscopy
- Spectropolarimeter for circular dichroism (CD) analysis

2.5 Reagents
- Peptides
- Trifluoroacetic acid
- Hexafluoroisopropanol
- DMSO

2.6 Procedure
General procedure is outlined here. However, the experimental conditions are different among peptides. Therefore, specific conditions for each peptide appears below.
1. The peptides were first dissolved in DMSO or a 1:1 mixture of trifluoroacetic acid and hexafluoroisopropanol to prepare the stock solution (10 mg/mL).
2. Peptides (10–1000 µg/mL) are incubated in water or PBS at room temperature (RT) (approximately 20 °C) or 37 °C for 10 min–7 days with or without shaking to induce aggregation.

3. Analysis of ultrastructure of resultant aggregates is carried out using scanning and transmission electron microscopy.
4. The secondary structure is determined by CD analysis.
5. Introduction of the aggregates into cells is done by adding the aggregates to culture medium.
6. The functional analyses of the recipient cells differ among peptides and cells. The analysis of each cell with each aggregate is described in the latter section.

2.6.1 PolyQ

1. PolyQ diseases-causative proteins contain expanded polyQ repeats and the flanking specific sequences. However, length of synthesized peptides is limited and proteins having polyQ repeat with only short flanking sequences are prone to aggregate. Therefore, the polyQ peptide used in our previous works essentially had only glutamine. The repeat number of glutamine was 69, which is beyond the pathogenic threshold. Two lysine were added to the both sides of the glutamine repeat. To enable detection of the peptide in cells, a fluorophore TAMRA was conjugated to the peptide. Accordingly, the sequence of the pure polyQ peptide was TAMRA-KK(Q_{69})KK-NH$_2$ (69Q).
2. 69Q in the stock solution was diluted in PBS to be a final concentration of 100 μg/mL, which was then incubated for 2 days at 37 °C to induce aggregation.
3. For functional analysis of cultured cells having 69Q, the peptide was added to the cells at a concentration of 7.5 μg per bottom area (cm^2) (approximately 15 μg/well) of the well in culture plates.

2.6.2 PolyL and polyS

1. The following observations were informative to design the RAN peptides. The ataxin 3 sequences surrounding the repeat region proved to determine the RAN translation initiation site and the efficiency. In addition, the RAN translation with polyQ repeats started at non-cognate codons located upstream of the CAG repeats, whereas, the RAN polyA proteins were likely translated within the repeats (Jazurek-Ciesiolka et al., 2020). Based on the observations, the polyL and polyS peptides did not essentially have amino acids before the repeats. Although the 15Q with short glutamine repeat showed weak aggregation (Mohara et al., 2018), a short repeat of leucine was predicted to form strong aggregates because of its hydrophobic property. Therefore, the repeat numbers

of polyL and polyS were set at 13. KKW and KK were added before and after these repeats, respectively. Again, a fluorophore TAMRA was added to the peptides to enable detection of the peptides in cells. The sequences of polyS and polyL peptides were TAMRA-KKW(S_{13})KK-NH$_2$ (13S) and TAMRA-KKW(L_{13})KK-NH$_2$ (13L), respectively.

2. To induce internalization of the aggregates into cultured cells, the aggregated 13S (1 mg/mL) after incubation for 48 h at 37 °C with shaking was applied to the culture medium to be a final concentration of 10 μg/mL. As for 13L, the aggregates produced in a concentration of 1 mg/mL was too large to be taken up by cultured cells. Thus, the stock solution of 13L peptide was directly added to the culture medium to be a final concentration of 10 μg/mL in the medium.

2.6.3 PolyA

To compare the toxicity of aggregates among those with different degrees of aggregation, polyA peptides with different degrees of aggregation were introduced in cells and toxicity of each aggregate to cells was examined.

1. Based on a previous literature (Bernacki & Murphy, 2011), the sequence of the polyA peptide used was TAMRA-KKW(A_{13})KK-NH$_2$ (13A) (Iizuka et al., 2021).
2. Secondary structure of 13A incubated for 4 h at RT or 37 °C was checked using CD analysis and BestSel software.
3. 13A incubated for 4 h at RT and that for 7 days at 37 °C were prepared for weak and strong aggregates, respectively.
4. These polyA aggregates were applied to neuron-like PC12 cells to see neurite outgrowth.

2.7 Notes

1. Regarding mechanisms by which aggregates pass through the cell membrane bilayer, exogenous HTT Exon1 aggregates were reported to be taken up by clathrin-dependent endocytosis in cultured cells, which was followed by transfer to lysosomal compartments (Ruiz-Arlandis et al., 2016). Dynamin is involved in clathrin-mediated endocytosis to form a ring on the neck of invaginated pits. Then, the vesicles are pinched off from the plasma membrane (Sever, 2002). Importantly, electron microscopic analysis detected the pits that were invaginated from the cell membrane and contain 13L and 13S aggregates (Fig. 1). In addition, 13S and 13L aggregates were found in closed vesicle-like structure and the membrane of the vesicle resembled cell membrane

13S   13L

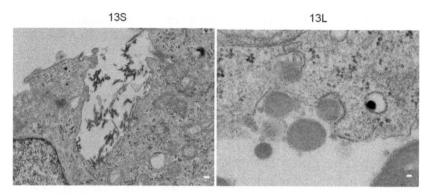

Fig. 1 Transmission electron microscopy images of 13S and 13L. Scale bars, 1 μm (left); 200 nm (right). *From Owada, R., Mitsui, S., & Nakamura, K. (2022). Exogenous polyserine and polyleucine are toxic to recipient cells. Scientific Reports, 12(1), 1685. doi:10.1038/s41598-022-05720-y is reused.*

having bilayer structure. Thus, entry of the polyS and polyL aggregates in cultured cells might use endocytosis.

2.8 Expected outcomes

The protocol induces aggregation of peptides and the aggregates spontaneously enter cells. Using the cells, toxicity of the aggregates can be studied. We show the examples below.

2.8.1 PolyQ
1. Incubation of 69Q for 2 days at 37 °C resulted in fibril-like structure, as proved by scanning electron microscopy (Kawasaki, Ohori, Chiba, Tsukiyama, & Nakamura, 2016).
2. Addition of aggregated 69Q peptide to neuron-like SH-SY5Y cells led to an internalization with a frequency of 50% (Mohara et al., 2018).
3. A peptide with a 15 glutamine repeat and 4 lysine (TAMRA-KK(Q$_{15}$)KK-NH$_2$, 15Q) was used as the control. Total length of neurites was shorter and number of neurites was fewer in SH-SY5Y cells with 69Q than those with 15Q (Mohara et al., 2018). The result indicates a toxicity of long polyQ aggregate to cells.

2.8.2 PolyL and polyS
1. Incubation of 13L in water for 10 min at even RT produced large clusters of aggregates like spherical agglomerates (Owada, Mitsui, & Nakamura, 2022). In contrast, the same condition (incubation for 10 min at RT) resulted in only small clusters of 13S aggregates. 13S incubated in water for 48 h at 37 °C with shaking induced rod-like big cluster of 13S

aggregates. 13L and 13S showed different ultrastructure as evidenced by transmission electron microscopy (Owada et al., 2022). The 13S appeared fibrillar aggregates. In contrast, the internal structure of 13L aggregates was complex with many grains and fibers (Fig. 1).

2. Incorporation of the two aggregates in cultured cells was verified by the ortho images along vertical axis produced from z-stack confocal images taken at every 1 μm (Owada et al., 2022). The internalization was further proved by transmission electron microscopy analysis.

3. Toxicity of aggregated 13L and 13S was first examined using neuron-like PC12 cells having the aggregates inside. To mimic the stage of neuronal degeneration in vitro, the cells were fully differentiated with NGF, then, the aggregates were incorporated in the cells and retraction of neurites was assessed. Total length of neurites was shorter and number of branch point was fewer in cells having 13S, indicating that 13S led to degeneration of neuron-like cells. The viability of differentiated PC12 cells was not changed in the presence of the aggregates compared to those treated with TAMRA alone (Owada et al., 2022).

4. When the toxicity was examined using KT-5 astrocytic cells, marked hypertrophy of the cells as demonstrated by the larger area of cells was seen in the presence of 13S. The hypertrophy resembles hypertrophic reactive astrocytes observed after injuries (Gaudet & Fonken, 2018). 13L led to longer processes of the cells than TAMRA alone (Owada et al., 2022). Thus, the two aggregates differentially change the morphology of astrocyte. Inspection of KT-5 cells with transmission electron microscopy revealed extensive cell death by 13S and 13L, as evidenced by destroyed intracellular organelles. Consistently, lower viability was seen in KT-5 cells with 13L and 13S.

2.8.3 PolyA

1. Aggregation of peptides is partly parallel to its secondary structure. CD analysis showed that anti-parallel β-sheet content was significantly higher in the 13A incubated at 37 °C than that in RT. Conversely, α-helix was lower in the 13A incubated at 37 °C than that in RT (Iizuka et al., 2021).

2. When the size of 13A aggregates of confocal images was measured, 13A incubated for 7 days at 37 °C proved to form larger aggregates than those for 4 h at RT at concentrations of 100 μg/mL and 1 mg/mL. Furthermore, scanning electron microscopy images clarified that the surface of the aggregates prepared by incubation for 7 days at 37 °C was more complex than those for 4 h at RT (Iizuka et al., 2021).

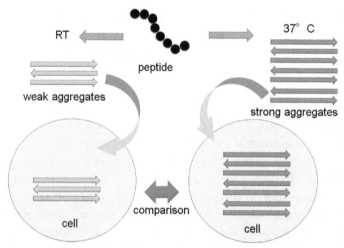

Fig. 2 Schema showing preparation of cells having peptides with different degrees of aggregation.

3. The 13A aggregates were spontaneously taken up by PC12 cells, as assessed by z-stack serial confocal images and transmission electron microscopy images.
4. Neurite outgrowth of PC12 cells were disturbed in cells with strong 13A aggregates but not in cells with weak 13A aggregates in vitro (Fig. 2) (Iizuka et al., 2021).

2.9 Advantages

This procedure does not require gene expression system and special reagents but uses only peptides. Entry of aggregates into cells occurs spontaneously. Therefore, this method is not time-consuming and does not cost so much. Moreover, the method enables to compare the toxicity among aggregates with different degrees of aggregation.

2.10 Limitations

The peptides used have only homopolymeric polyQ, polyL, polyS and polyA, However, polyQ-containing proteins and their RAN products contain flanking specific sequences. Furthermore, although the numbers of the homopolymeric repeats in the peptides were 13 or 69, those in the endogenous proteins are likely longer than those. These points are limitations of the current simple procedure to mimic cells having polyQ and the RAN products.

3. Dissociation of aggregates by FEL

The peptides used were polyQ and Aβ. We already introduced polyQ diseases. Therefore, we describe the association of aggregated Aβ with Alzheimer's disease (AD).

AD is a neurodegenerative disorders affecting 47 million people in 2015 in the world. Furthermore, the number was predicted to be 75 million in 2030 and 132 million by 2050 (Elonheimo, Andersen, Katsonouri, & Tolonen, 2021; Oumata et al., 2022). The major symptom of AD is dementia (Scheltens et al., 2016). Aβ results from the sequential proteolysis of amyloid precursor protein and accumulation of Aβ was hypothesized as a major cause of AD (Hardy, Duff, Hardy, Perez-Tur, & Hutton, 1998; Hardy & Higgins, 1992; Selkoe & Hardy, 2016). The amyloid hypothesis suggests that the deposition of oligomerized Aβ leads to multiple downstream events such as deposition of neurofibrillary tangles, neuroinflammation and vascular damage. To explore the therapeutic approaches for AD, FEL was applied to Aβ fibril as was done to polyQ.

3.1 Equipment

• IR-FEL instrument

The oscillation system of the IR-FEL is briefly described. The size of IR-FEL instrument installed in a synchrotron radiation facility is comparatively small; about 10 m and 5 m in length and width, respectively. The apparatus mainly consists of three parts (Glotin, Chaput, Jaroszynski, Prazeres, & Ortega, 1993; Knippels, van de Pol, Pellemans, Planken, & van der Meer, 1998; Lamb et al., 1999): (1) linear accelerator containing high radio frequency (RF) electronic gun and α-magnet, (2) undulator (periodic magnetic field), and (3) resonator mirrors (Fig. 3). The electron beam is oscillated from the RF gun at 2,856 MHz and is accelerated near to the photon rate in the accelerator tube. The electron bunches are successively introduced into the undulator, and the synchrotron radiation emitted from the wiggler motion of electron beam is captured in the resonator and is amplified through the interaction with successive electron bunches. A part of the amplified infrared ray is emitted from a coupling hole in one of the two resonator mirrors and is sent to a user station. The time structure of the IR-FEL is composed of two pulses, macro-pulse and micro-pulse: a macro-pulse has a duration of approximately 2 μs, and the repetition rate is 5 Hz. The macro-pulse consists of a train of micro-pulse with a duration of 1–2 ps, and the interval of the

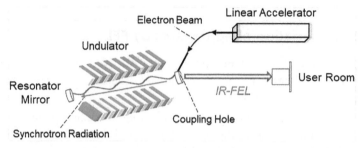

Fig. 3 An illustration of IR-FEL oscillation system. The laser beam is transported from a coupling hole on a resonator mirror to a user room.

micro-pulse is 350 ps. The laser energy reaches 40–50 mJ/cm^2 per macro-pulse at maximum as measured using a thermal energy meter. The oscillation wavelength is tunable within the mid-infrared region (5–10 μm), and the half-width at the half maximum of the IR-FEL spectrum is about 0.1–0.2 μm.

• Infrared microscope

The secondary structure of peptides was examined using the system with an infrared microscope combined with a Fourier transform infrared spectrometer. The absorption spectra were obtained within the mid-IR range of 700–4000 cm^{-1}.

3.2 Procedure

1. Preparation of aggregated peptides by incubating at 37 °C.
2. FEL irradiation to aggregated peptides and to cells having aggregates inside is described in the Section 3.4. Expected outcomes. The wavelengths applied ranged from 5.0 to 7.2 μm.
3. The functional analysis of the irradiated cells having aggregates is also described in the latter section.

3.3 Notes

1. In the mid-infrared region, various vibrational modes of molecules are contained. A remarkable feature of the IR-FEL is that intense and monochromatic radiation from IR-FEL enables us to excite various vibrational modes state-selectively. In recent years, several facilities of IR-FEL all over the world are available to users, and various experimental studies have been conducted in biomedical (Edwards et al., 1994) and molecular science (Andersson et al., 2020; Elferink et al., 2018) fields.

3.4 Expected outcomes

The FEL dissociates aggregation of peptides and restores the dysfunctions of cells caused by the aggregates. The examples are shown below.

3.4.1 PolyQ aggregates

1. Fibril-like structure of 69Q aggregates disappeared after the irradiation tuned at 6.1 μm but not at 5.5 μm as assessed by scanning electron microscopy (Kawasaki et al., 2016) (Fig. 4).
2. Percentage of β-sheet structure of naked 69Q peptide decreased after irradiation at 6.1 μm (Kawasaki et al., 2016).
3. Irradiation at 6.1 μm to cultured cells having aggregated 69Q inside decreased the β-sheet content (Kawasaki et al., 2016).
4. Functional analyses were done as follows (Mohara et al., 2018). SH-SY5Y and PC12 cells were fully differentiated into neuron-like cells with retinoic acid and NGF, respectively. Then, 69Q was incorporated in these cells to see degeneration of neurons. The cells were given irradiation at 6.1 μm the next day. Three days later, morphological analysis was carried out. The shorter lengths of individual neurite and total length of neurites as well as fewer number of neurites in the presence of 69Q were rescued by the irradiation. Likewise, to mimic differentiation stage of SH-SY5Y cells, addition of 69Q and FEL irradiation to undifferentiated SH-SY5Y cells were done on the same day and then, differentiation was induced using retinoic acid. Three days later, the morphological examination was done. Again, the three parameters above were corrected by the FEL irradiation.

Irradiation at 5.5 μm   Irradiation at 6.1 μm

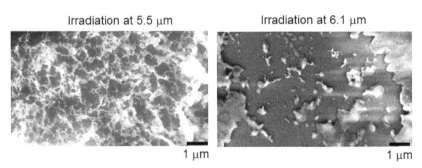

1 μm   1 μm

Fig. 4 Scanning electron microscopy images of 69Q aggregates after FEL irradiation at 5.5 and 6.1 μm. *From Kawasaki, T., Ohori, G., Chiba, T., Tsukiyama, K., & Nakamura, K. (2016). Picosecond pulsed infrared laser tuned to amide I band dissociates polyglutamine fibrils in cells. Lasers in Medical Science, 31(7), 1425–1431. doi:10.1007/s10103-016-2004-x are reused with permission from Springer Nature.*

3.4.2 Aβ fibril

1. Aβ peptide in PBS incubated at 37 °C yielded the fibril structures (Kawasaki, Yaji, Ohta, Tsukiyama, & Nakamura, 2018).
2. Energy absorbed by Aβ was high at the wavelength of 6.1 μm, which justified application of this wavelength to Aβ fibril (Kawasaki et al., 2018).
3. FEL at 6.1 μm but not 5.0 and 7.2 μm reduced percentage of β-sheet structure of Aβ (Kawasaki et al., 2018).
4. Congo-red signal was reduced after the irradiation (Kawasaki et al., 2018).
5. However, effect of the FEL irradiation on Aβ fibril was not so apparent under dry condition (Kawasaki et al., 2018). Thus, water molecule seems to maintain the dissociation effect by FEL, which was theoretically supported by MD simulation (Okumura, Itoh, Nakamura, & Kawasaki, 2021).

3.5 Advantages

The IR-FEL irradiation system can supply intense vibrational excitation energy to the corresponding functional groups mode-selectively compared with the commercially available infrared lasers. The vibration excitation of the chemical bonds can induce multi-photon absorption and dissociation reaction on the various biomolecules. In addition, the wavelength variability is also the feature overcoming the commercially available laser instruments that often emit laser with a fixed wavelength.

3.6 Limitations

The IR-FEL irradiation system is involved in a synchrotron-radiation facility, and the operation and the maintenance for the quantum beam oscillation are usually conducted by special technicians who are familiar with the radiation works. In addition, enormous costs are required for the laser operations. Therefore, a negative point in the usage of the IR-FEL is supposed to be an inconvenience for the general users.

4. MD simulation for the dissociation of protein aggregates with an IR-FEL

Process of dissociation of protein aggregates by IR-FEL has been studied not only using the experimental techniques (Kawasaki et al., 2016, 2018, 2020; Kawasaki, Fujioka, Imai, Torigoe, & Tsukiyama, 2014; Kawasaki, Fujioka, Imai, & Tsukiyama, 2012; Kawasaki, Tsukiyama, & Irizawa, 2019) but also using theoretical techniques (Hoang Viet et al., 2015; Okumura et al., 2021). The theoretical technique to study the dissociation process by laser irradiation is MD simulation. This section describes how MD simulations are performed for amyloid-fibril dissociation with an IR-FEL.

The MD simulation study for the laser-induced dissociation of protein aggregates consists of two steps: (1) equilibrium MD simulation and (2) nonequilibrium MD simulation for the laser irradiation. The first step, equilibrium MD simulation, is performed to determine the appropriate wavenumber of the laser for the system. In IR-FEL experiments, a sample is irradiated with an infrared laser that corresponds to the amide I band. The amide I band is associated with the stretching vibration of backbone C=O that can form an intermolecular hydrogen bond, as shown in Fig. 5. Due to the limitations of the force field used in the simulation, the C=O stretching vibration of the model system may differ slightly from the experimental value, so it is necessary to determine first the resonance wavenumber of the C=O stretching vibration for the target system. For this purpose, equilibrium MD simulations are performed in the first step. Then, in the second step, nonequilibrium MD simulations are performed with a time-varying electric field that mimics a laser with this wavenumber. These simulations make it possible to reproduce the dissociation process of the protein aggregates on a computer. This section explains this simulation method, taking up the dissociation process of the amyloid fibril of Aβ peptides (Luhrs et al., 2005) as an example.

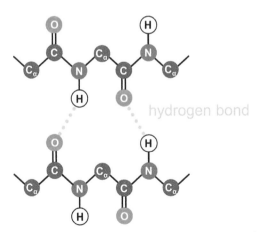

Fig. 5 Schematic illustration of hydrogen bonds between two peptides. *Reprinted with permission from Okumura, H., Itoh, S. G., Nakamura, K., & Kawasaki, T. (2021). Role of water molecules and helix structure stabilization in the laser-induced disruption of amyloid fibrils observed by nonequilibrium molecular dynamics simulations.* The Journal of Physical Chemistry B, *125(19), 4964–4976. doi:10.1021/acs.jpcb.0c11491. Copyright 2021 American Chemical Society.*

4.1 Equipment

- Computer, or supercomputer if possible, where an MD simulation can be performed.
- MD simulation software.
- Compiler to compile the MD simulation program.
- Modeling software, if necessary.

4.2 Procedure

1. Prepare the initial molecular structure of the target protein system. The initial structure is usually taken from Protein Data Bank (PDB). If missing structures or atoms exist, complement them using modeling software. Place the target protein system in an explicit solvent, such as water. Note that the total electric charge of the whole system must be zero, adjusting the number of counter ions. For example, the initial conformation of the amyloid fibril of Aβ peptides is shown in Fig. 6A.
2. To determine the C=O stretching mode of the model system, the equilibrium MD simulations of the target system are performed in advance of the nonequilibrium MD simulations. The MD simulations are often performed at a constant temperature, such as room temperature.
3. Calculate the infrared absorption spectrum of the C=O stretching vibration. To calculate this spectrum, the time derivative $\dot{r}_{C=O}$ of the distances of the C=O double bonds that form hydrogen bonds in the β-sheet structure is saved in the equilibrium MD simulations. Calculate the autocorrelation function $A(t)$ of $\dot{r}_{C=O}$ given by

$$A(t) = \frac{1}{N_{C=O\,bonds}} \sum_{C=O\,bonds} \langle \dot{r}_{C=O}(t)\dot{r}_{C=O}(0)\rangle,$$

4. where the bracket $\langle \cdots \rangle$ means an ensemble average, and $N_{C=O\,bonds}$ is the number of the C=O double bonds in the β-sheet structure (Okumura et al., 2021). The infrared absorption spectrum $I(\omega)$ is then calculated by the Fourier transform of $A(t)$

$$I(\omega) = \int_{-\infty}^{\infty} A(t)\exp(-\omega t)\,dt.$$

5. Note that the angular frequency ω is related to the wavenumber ν by $\omega = 2\pi c\nu$, where c is the speed of light. Determine the resonance wavenumber ν_{res} of the β-sheet structure as the peak position of the absorption spectrum. As an example, the infrared absorption spectrum of the Aβ amyloid fibril is shown in Fig. 7. In this case, ν_{res} was determined as $1676\,cm^{-1}$.

(A) Before irradiation   (B) After 100 pulses

(C) After 500 pulses   (D) After 1000 pulses

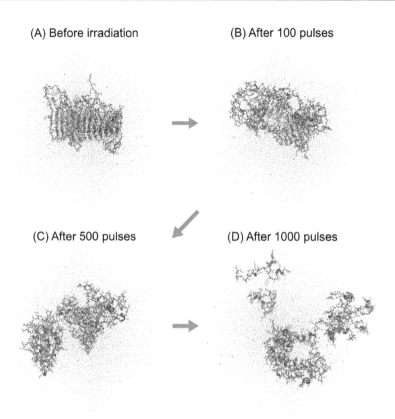

Fig. 6 Snapshots during the dissociation process of the Aβ amyloid fibril in the nonequilibrium MD simulation with an IR-FEL (A) before laser irradiation, (B) after 100 pulses, (C) after 500 pulses, and (D) after 1000 pulses. *The figure images were created using PyMOL (The PyMOL Molecular Graphics System, V.S., LLC.).*

6. The nonequilibrium MD simulations are performed in a time-varying electric field corresponding to the IR-FEL. In experiments, the IR-FEL consists of a series of macro-pulses, and each macro-pulse consists of about 6,000 consecutive micro-pulses (Kawasaki et al., 2012, 2014). To mimic such a laser irradiation, the electric field was applied as a series of Gaussian-distributed pulses, as shown in Fig. 8A. Each pulse corresponds to the micro-pulse in the IR-FEL experiments (Kawasaki et al., 2012, 2014) and is expressed as

$$E = E_0 \exp\left\{-\frac{(t - t_0)^2}{2\sigma^2}\right\} \cos(\omega_{\text{res}}(t - t_0)),$$

where E_0 is the maximum intensity of the electric field, t_0 is the time at $E = E_0$, σ is the standard deviation of the Gaussian distribution, and ω_{res}

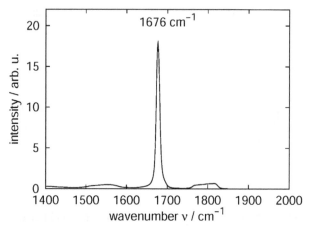

Fig. 7 The infrared absorption spectrum of the stretching vibration of the backbone C=O double bonds that form hydrogen bonds in the β-sheet structure of the Aβ amyloid fibril. *Reprinted with permission from Okumura, H., Itoh, S. G., Nakamura, K., & Kawasaki, T. (2021). Role of water molecules and helix structure stabilization in the laser-induced disruption of amyloid fibrils observed by nonequilibrium molecular dynamics simulations.* The Journal of Physical Chemistry B, 125(19), 4964–4976. doi:10.1021/acs. jpcb.0c11491. *Copyright 2021 American Chemical Society.*

is the resonance angular frequency related to the resonance wavenumber ν_{res} such that $\omega_{res} = 2\pi c \nu_{res}$. This electric field oscillates with the resonance wavenumber ν_{res}, as shown in the enlarged view of the electric field in Fig. 8B.

7. The MD simulation program needs to be modified to realize the MD simulation with this electric field. The electric force corresponding to the electric field pulse must be added to the program. The electric force $\boldsymbol{F}_i$ acting on atom i, which has an electric charge q_i, is given by

$$\boldsymbol{F}_i = q_i E_0 \widehat{\boldsymbol{z}},$$

when the electric force is applied along the z-axis. Here, $\widehat{\boldsymbol{z}}$ is the unit vector along the z-axis. Once the forces are modified, compile the program.

8. Perform the nonequilibrium MD simulations in the electric field using the modified MD simulation program. The temperature is usually controlled, for example, at a room temperature, to exhaust heat caused by the external electric field. Observe the secondary structure of the protein aggregates. The electric-field pulses are applied until the β-sheet structure is destroyed. MD simulations from several different initial conformations and/or velocities are desirable for statistical analysis.

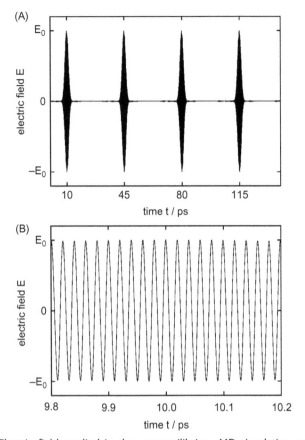

Fig. 8 (A) Electric field applied in the nonequilibrium MD simulation as a series of Gaussian-distributed pulses. (B) Enlarged view of the electric field pulse shown in panel (A). *Reprinted with permission from Okumura, H., Itoh, S. G., Nakamura, K., & Kawasaki, T. (2021). Role of water molecules and helix structure stabilization in the laser-induced disruption of amyloid fibrils observed by nonequilibrium molecular dynamics simulations.* The Journal of Physical Chemistry B, *125(19), 4964–4976. doi:10.1021/acs. jpcb.0c11491. Copyright 2021 American Chemical Society.*

4.3 Notes

1. There is no commercial MD simulation program that can handle such a time-varying electric field. Those who need it have to modify an MD program by themselves. In the case of the Aβ amyloid fibril exemplified here, one of the authors (H. O.) developed and used an original MD program, Generalized-Ensemble Molecular Biophysics (GEMB). This program has been applied to several protein and peptide systems (Itoh, Tanimoto, &

Okumura, 2021; Okumura, 2012; Okumura & Itoh, 2014, 2016; Okumura, Itoh, & Okamoto, 2007; Yamauchi & Okumura, 2017).

2. In the nonequilibrium MD simulations with the laser pulse, the pressure should not be controlled, although the temperature can be controlled. If one performs a nonequilibrium MD simulation with constant pressure in such a time-varying electric field, not only the C=O double bond but also the volume oscillates so fast that the simulation can stop unexpectedly.

3. The interval of the electric-field pulses can be set to be an appropriate value, such as an experimental value of 350 ps, or much shorter, for example, 35 ps, to save the computation time (Okumura et al., 2021). The maximum intensity of the electric field E_0 should usually be set to a value significantly greater than the experimental value to observe the dissociation process of the protein aggregates in a reasonable simulation time. For example, E_0 was set to be 1×10^8 V/cm in Reference (Okumura et al., 2021), whereas a typical experimental value is the order of 10^6 V/cm.

4.4 Expected outcomes

We can observe the dissociation of the protein aggregates by the non-equilibrium MD simulation. For example, the dissociation process of the Aβ amyloid fibril is shown in Fig. 6 (Okumura et al., 2021). The amyloid fibril was gradually destroyed by the laser pulses. To discuss this result quantitatively, the ratio of the amino acid residues that formed the intermolecular parallel β-sheet structure was calculated according to the DSSP criteria (Kabsch & Sander, 1983), as shown in Fig. 9. Almost all the intermolecular β-sheet structures were destroyed after 1000 pulses in this case.

4.5 Advantages

The advantage of the nonequilibrium MD simulation to investigate the dissociation process with an IR-FEL is that the detailed behavior of molecular dynamics can be observed at an atomic level. For example, the role of water molecules in the disruption process of the Aβ amyloid fibril was clarified by the nonequilibrium MD simulation (Okumura et al., 2021). Fig. 10 shows enlarged snapshots of the Aβ amyloid fibril in a typical MD trajectory and the electric field intensity at that time (red circles). In Fig. 10A, six intermolecular hydrogen bonds are formed between the two β-strands (inside the purple dashed line). These intermolecular hydrogen bonds are broken by the electric field pulse as in Fig. 10B. However, these intermolecular

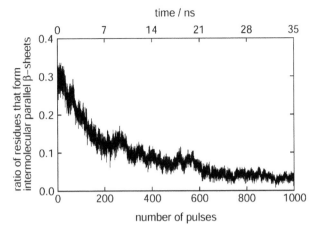

Fig. 9 The ratio of the residues that form the intermolecular β-sheet structure in the nonequilibrium MD simulation of the Aβ amyloid fibril. *Reprinted with permission from Okumura, H., Itoh, S. G., Nakamura, K., & Kawasaki, T. (2021). Role of water molecules and helix structure stabilization in the laser-induced disruption of amyloid fibrils observed by nonequilibrium molecular dynamics simulations.* The Journal of Physical Chemistry B, 125(19), 4964–4976. doi:10.1021/acs.jpcb.0c11491. *Copyright 2021 American Chemical Society.*

hydrogen bonds reformed simultaneously, and the two β-strands were completely repaired (Fig. 10C). Before this pulse, the intermolecular hydrogen bonds between the Aβ peptides were repeatedly broken and repaired after each pulse in the same way. However, immediately after the hydrogen bonds between the Aβ peptides were broken during the next pulse, a water molecule nearby (the pink-highlighted water molecule) entered the space between C=O and N—H, where the intermolecular hydrogen bond had been previously formed (Fig. 10D). In addition, another water molecule (the blue-highlighted water molecule) entered the space between the two Aβ peptides and formed hydrogen bonds with the Aβ peptides in Fig. 10E. Even after the red-highlighted water molecule separated from the Aβ peptides, the blue-highlighted water molecule remained there (Fig. 10F). Other water molecules also entered the gap between the Aβ peptides (Fig. 10G). The intermolecular hydrogen bonds between the Aβ peptides could not be reformed before the next laser pulse because the hydrogen bonds between the Aβ peptides were replaced by those between the Aβ peptides and the water molecules. As a result, the intermolecular β-sheet of the Aβ amyloid fibril was disrupted, as in Fig. 10H. This phenomenon occurred throughout the amyloid fibril, and the amyloid fibril was destroyed entirely.

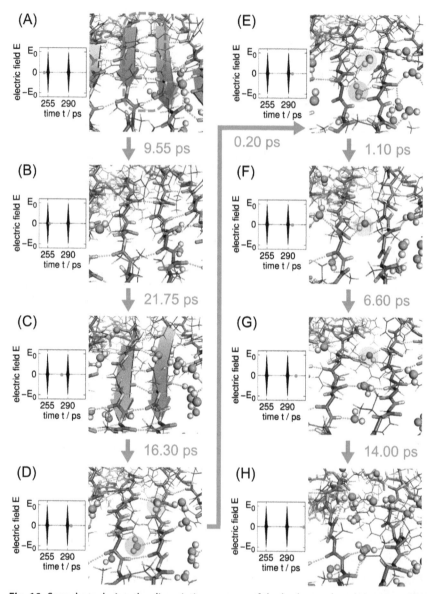

Fig. 10 Snapshots during the dissociation process of the hydrogen bonds between the Aβ peptides and the intensity of the electric field pulse at the corresponding time. Cyan dots mean hydrogen bonds. Two specific water molecules are highlighted with pink and blue circles. *The figure images were created using PyMOL (The PyMOL Molecular Graphics System, V.S., LLC.). Reprinted with permission from Okumura, H., Itoh, S. G., Nakamura, K., & Kawasaki, T. (2021). Role of water molecules and helix structure stabilization in the laser-induced disruption of amyloid fibrils observed by nonequilibrium molecular dynamics simulations.* The Journal of Physical Chemistry B, 125(19), 4964–4976. doi:10.1021/acs.jpcb.0c11491. Copyright 2021 American Chemical Society.

In this way, the advantage of MD simulation is to follow the motion of each molecule in detail in the dissociation of protein aggregates by laser irradiation. Nonequilibrium MD simulation is a method that can be applied to various protein aggregates in general, not only Aβ amyloid fibrils.

4.6 Limitations

The limitation of MD simulation research is that the system under study is only a computer-based virtual system. Therefore, simulation results must always be compared with experimental results to confirm their validity. On top of that, we should elucidate microscopic phenomena at the atomic level that cannot be observed directly by experimental methods and make theoretical predictions before experiments.

For example, in the disruption of the Aβ amyloid fibril by an IR–FEL, the β-sheet structure was destroyed, but the helix structure increased, as shown in Fig. 11. This phenomenon is consistent with experimental results. However, the reason for this phenomenon had not been clarified. Therefore, to elucidate this phenomenon, the equilibrium MD simulations of Aβ peptides in the α-helix and random coil structures were also performed, and the infrared absorption spectra of C=O stretching vibrations were calculated. As a result, as shown in Fig. 12, the resonance wavenumber

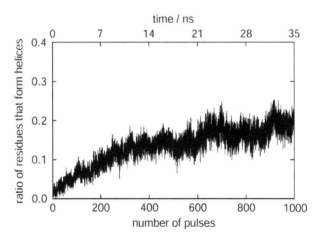

Fig. 11 The ratio of the residues that form helix structures in the nonequilibrium MD simulation of the Aβ amyloid fibril. *Reprinted with permission from Okumura, H., Itoh, S. G., Nakamura, K., & Kawasaki, T. (2021). Role of water molecules and helix structure stabilization in the laser-induced disruption of amyloid fibrils observed by nonequilibrium molecular dynamics simulations.* The Journal of Physical Chemistry B, 125(19), *4964–4976. doi:10.1021/acs.jpcb.0c11491. Copyright 2021 American Chemical Society.*

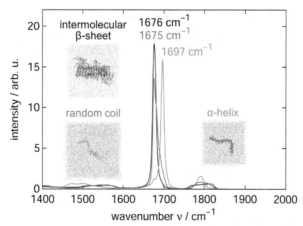

Fig. 12 The Infrared absorption spectra of the backbone C=O stretching vibration in the intermolecular β-sheet (black), α-helix (red), and random-coil (blue) structures of the Aβ peptide. *The figure images were created using PyMOL (The PyMOL Molecular Graphics System, V.S., LLC.). Reprinted with permission from Okumura, H., Itoh, S. G., Nakamura, K., & Kawasaki, T. (2021). Role of water molecules and helix structure stabilization in the laser-induced disruption of amyloid fibrils observed by nonequilibrium molecular dynamics simulations.* The Journal of Physical Chemistry B, 125(19), 4964–4976. doi:10.1021/acs.jpcb.0c11491. *Copyright 2021 American Chemical Society.*

of the random coil structure is $1675\,\mathrm{cm}^{-1}$, which is close to that of the intermolecular β-sheet structure and the infrared laser wavenumber set in this study, while the resonance wavenumber of the α-helix structure is $1697\,\mathrm{cm}^{-1}$, which is far from these wavenumbers. These results suggest that the helix structure can exist stably without breaking the hydrogen bonds between C=O and N—H because the resonance wavenumber is different from that of the laser to destroy the intermolecular β-sheet structure. Rather, the helix structure increases as the intermolecular β-sheet structure and random coil structure decrease due to the laser irradiation.

As we can see in this example, we believe that the significance of MD simulation research is to elucidate phenomena that cannot be observed experimentally and make theoretical predictions while checking whether MD simulations can reproduce phenomena that can be observed experimentally.

5. Summary and conclusions

The simple procedure to prepare the cells having various kinds of aggregates does not require special reagents and transfection but uses only peptides. By changing the experimental conditions, aggregates with different degrees can be produced from a same peptide, which makes it possible to study contribution of aggregation to severity of cellular toxicity caused by aggregates. Using the simple method, therapeutic researches against protein aggregates can be conducted. However, this method is only available to short peptides, and therefore, the obtained results do not reflect those of endogenous aggregates. Combination of gene expression and this procedure complementarily works as tools for research on protein aggregates.

We also showed that IR-FEL is an effective tool for disrupting these protein aggregates. MD simulation can analyze the details of the disruption process at the molecular level. We hope that the combination of these experimental and theoretical methods will lead to the development of new therapies for neurodegenerative diseases in the future.

References

Adegbuyiro, A., Sedighi, F., Pilkington, A. W. T., Groover, S., & Legleiter, J. (2017). Proteins containing expanded polyglutamine tracts and neurodegenerative disease. *Biochemistry, 56*(9), 1199–1217. https://doi.org/10.1021/acs.biochem.6b00936.

Andersson, A., Poline, M., Kodambattil, M., Rebrov, O., Loire, E., Maitre, P., et al. (2020). Structure of proton-bound methionine and tryptophan dimers in the gas phase investigated with IRMPD spectroscopy and quantum chemical calculations. *The Journal of Physical Chemistry A, 124*(12), 2408–2415. https://doi.org/10.1021/acs.jpca.9b11811.

Banez-Coronel, M., Ayhan, F., Tarabochia, A. D., Zu, T., Perez, B. A., Tusi, S. K., et al. (2015). RAN translation in Huntington disease. *Neuron, 88*(4), 667–677. https://doi.org/10.1016/j.neuron.2015.10.038.

Bernacki, J. P., & Murphy, R. M. (2011). Length-dependent aggregation of uninterrupted polyalanine peptides. *Biochemistry, 50*(43), 9200–9211. https://doi.org/10.1021/bi201155g.

Costanzo, M., Abounit, S., Marzo, L., Danckaert, A., Chamoun, Z., Roux, P., et al. (2013). Transfer of polyglutamine aggregates in neuronal cells occurs in tunneling nanotubes. *Journal of Cell Science, 126*(Pt. 16), 3678–3685. https://doi.org/10.1242/jcs.126086.

Crotti, A., Benner, C., Kerman, B. E., Gosselin, D., Lagier-Tourenne, C., Zuccato, C., et al. (2014). Mutant Huntingtin promotes autonomous microglia activation via myeloid lineage-determining factors. *Nature Neuroscience, 17*(4), 513–521. https://doi.org/10.1038/nn.3668.

Davies, J. E., Rose, C., Sarkar, S., & Rubinsztein, D. C. (2010). Cystamine suppresses polyalanine toxicity in a mouse model of oculopharyngeal muscular dystrophy. *Science Translational Medicine, 2*(34), 34ra40. https://doi.org/10.1126/scitranslmed.3000723.

Davies, J. E., Sarkar, S., & Rubinsztein, D. C. (2006). Trehalose reduces aggregate formation and delays pathology in a transgenic mouse model of oculopharyngeal muscular dystrophy. *Human Molecular Genetics, 15*(1), 23–31. https://doi.org/10.1093/hmg/ddi422.

Davies, J. E., Sarkar, S., & Rubinsztein, D. C. (2008). Wild-type PABPN1 is anti-apoptotic and reduces toxicity of the oculopharyngeal muscular dystrophy mutation. *Human Molecular Genetics, 17*(8), 1097–1108. https://doi.org/10.1093/hmg/ddm382.

Davies, J. E., Wang, L., Garcia-Oroz, L., Cook, L. J., Vacher, C., O'Donovan, D. G., et al. (2005). Doxycycline attenuates and delays toxicity of the oculopharyngeal muscular dystrophy mutation in transgenic mice. *Nature Medicine, 11*(6), 672–677. https://doi.org/10.1038/nm1242.

Dieriks, B. V., Park, T. I., Fourie, C., Faull, R. L., Dragunow, M., & Curtis, M. A. (2017). alpha-Synuclein transfer through tunneling nanotubes occurs in SH-SY5Y cells and primary brain pericytes from Parkinson's disease patients. *Scientific Reports, 7*, 42984. https://doi.org/10.1038/srep42984.

Dilna, A., Deepak, K. V., Damodaran, N., Kielkopf, C. S., Kagedal, K., Ollinger, K., et al. (2021). Amyloid-beta induced membrane damage instigates tunneling nanotube-like conduits by p21-activated kinase dependent actin remodulation. *Biochimica et Biophysica Acta. Molecular Basis of Disease, 1867*(12), 166246. https://doi.org/10.1016/j.bbadis.2021.166246.

Dong, G., Ferguson, J. M., Duling, A. J., Nicholas, R. G., Zhang, D., Rezvani, K., et al. (2011). Modeling pathogenesis of Huntington's disease with inducible neuroprogenitor cells. *Cellular and Molecular Neurobiology, 31*(5), 737–747. https://doi.org/10.1007/s10571-011-9679-0.

Dorsman, J. C., Pepers, B., Langenberg, D., Kerkdijk, H., Ijszenga, M., den Dunnen, J. T., et al. (2002). Strong aggregation and increased toxicity of polyleucine over polyglutamine stretches in mammalian cells. *Human Molecular Genetics, 11*(13), 1487–1496. https://doi.org/10.1093/hmg/11.13.1487.

Edwards, G., Logan, R., Copeland, M., Reinisch, L., Davidson, J., Johnson, B., et al. (1994). Tissue ablation by a free-electron laser tuned to the amide II band. *Nature, 371*(6496), 416–419. https://doi.org/10.1038/371416a0.

Elferink, H., Severijnen, M. E., Martens, J., Mensink, R. A., Berden, G., Oomens, J., et al. (2018). Direct experimental characterization of glycosyl cations by infrared ion spectroscopy. *Journal of the American Chemical Society, 140*(19), 6034–6038. https://doi.org/10.1021/jacs.8b01236.

Elonheimo, H. M., Andersen, H. R., Katsonouri, A., & Tolonen, H. (2021). Environmental substances associated with Alzheimer's disease—A Scoping Review. *International Journal of Environmental Research and Public Health, 18*(22). https://doi.org/10.3390/ijerph182211839.

Gaudet, A. D., & Fonken, L. K. (2018). Glial cells shape pathology and repair after spinal cord injury. *Neurotherapeutics, 15*(3), 554–577. https://doi.org/10.1007/s13311-018-0630-7.

Glotin, F., Chaput, R., Jaroszynski, D., Prazeres, R., & Ortega, J. (1993). Infrared sub-picosecond laser pulses with a free-electron laser. *Physical Review Letters, 71*(16), 2587–2590. https://doi.org/10.1103/PhysRevLett.71.2587.

Gonzalez-Alegre, P. (2019). Recent advances in molecular therapies for neurological disease: Triplet repeat disorders. *Human Molecular Genetics, 28*(R1), R80–R87. https://doi.org/10.1093/hmg/ddz138.

Gousset, K., Schiff, E., Langevin, C., Marijanovic, Z., Caputo, A., Browman, D. T., et al. (2009). Prions hijack tunnelling nanotubes for intercellular spread. *Nature Cell Biology, 11*(3), 328–336. https://doi.org/10.1038/ncb1841.

Gupta, S., Jie, S., & Colby, D. W. (2012). Protein misfolding detected early in pathogenesis of transgenic mouse model of Huntington disease using amyloid seeding assay. *The Journal of Biological Chemistry, 287*(13), 9982–9989. https://doi.org/10.1074/jbc.M111.305417.

Harding, A. E. (1983). Classification of the hereditary ataxias and paraplegias. *Lancet, 1*(8334), 1151–1155. https://doi.org/10.1016/s0140-6736(83)92879-9.

Hardy, J., Duff, K., Hardy, K. G., Perez-Tur, J., & Hutton, M. (1998). Genetic dissection of Alzheimer's disease and related dementias: Amyloid and its relationship to tau. *Nature Neuroscience, 1*(5), 355–358. https://doi.org/10.1038/1565.

Hardy, J. A., & Higgins, G. A. (1992). Alzheimer's disease: The amyloid cascade hypothesis. *Science, 256*(5054), 184–185. https://doi.org/10.1126/science.1566067.

Hoang Viet, M., Derreumaux, P., Li, M. S., Roland, C., Sagui, C., & Nguyen, P. H. (2015). Picosecond dissociation of amyloid fibrils with infrared laser: A nonequilibrium simulation study. *The Journal of Chemical Physics, 143*(15), 155101. https://doi.org/10.1063/1. 4933207.

Hughes, J., Piltz, S., Rogers, N., McAninch, D., Rowley, L., & Thomas, P. (2013). Mechanistic insight into the pathology of polyalanine expansion disorders revealed by a mouse model for X linked hypopituitarism. *PLoS Genetics, 9*(3), e1003290. https:// doi.org/10.1371/journal.pgen.1003290.

Hutten, S., & Dormann, D. (2019). RAN translation down. *Nature Neuroscience, 22*(9), 1379–1380. https://doi.org/10.1038/s41593-019-0482-4.

Iizuka, Y., Owada, R., Kawasaki, T., Hayashi, F., Sonoyama, M., & Nakamura, K. (2021). Toxicity of internalized polyalanine to cells depends on aggregation. *Scientific Reports, 11*(1), 23441. https://doi.org/10.1038/s41598-021-02889-6.

Itoh, S. G., Tanimoto, S., & Okumura, H. (2021). Dynamic properties of SARS-CoV and SARS-CoV-2 RNA-dependent RNA polymerases studied by molecular dynamics simulations. *Chemical Physics Letters, 778*, 138819. https://doi.org/10.1016/j.cplett.2021. 138819.

Jafar-Nejad, P., Ward, C. S., Richman, R., Orr, H. T., & Zoghbi, H. Y. (2011). Regional rescue of spinocerebellar ataxia type 1 phenotypes by 14-3-3epsilon haploinsufficiency in mice underscores complex pathogenicity in neurodegeneration. *Proceedings of the National Academy of Sciences of the United States of America, 108*(5), 2142–2147. https://doi.org/10. 1073/pnas.1018748108.

Jayaraman, M., Kodali, R., Sahoo, B., Thakur, A. K., Mayasundari, A., Mishra, R., et al. (2012). Slow amyloid nucleation via alpha-helix-rich oligomeric intermediates in short polyglutamine-containing huntingtin fragments. *Journal of Molecular Biology, 415*(5), 881–899. https://doi.org/10.1016/j.jmb.2011.12.010.

Jayaraman, M., Mishra, R., Kodali, R., Thakur, A. K., Koharudin, L. M., Gronenborn, A. M., et al. (2012). Kinetically competing huntingtin aggregation pathways control amyloid polymorphism and properties. *Biochemistry, 51*(13), 2706–2716. https://doi.org/10. 1021/bi3000929.

Jazurek-Ciesiolka, M., Ciesiolka, A., Komur, A. A., Urbanek-Trzeciak, M. O., Krzyzosiak, W. J., & Fiszer, A. (2020). RAN translation of the expanded CAG repeats in the SCA3 disease context. *Journal of Molecular Biology, 432*(24), 166699. https://doi. org/10.1016/j.jmb.2020.10.033.

Kabsch, W., & Sander, C. (1983). Dictionary of protein secondary structure: Pattern recognition of hydrogen-bonded and geometrical features. *Biopolymers, 22*(12), 2577–2637. https://doi.org/10.1002/bip.360221211.

Kawasaki, T., Fujioka, J., Imai, T., Torigoe, K., & Tsukiyama, K. (2014). Mid-infrared free-electron laser tuned to the amide I band for converting insoluble amyloid-like protein fibrils into the soluble monomeric form. *Lasers in Medical Science, 29*(5), 1701–1707. https://doi.org/10.1007/s10103-014-1577-5.

Kawasaki, T., Fujioka, J., Imai, T., & Tsukiyama, K. (2012). Effect of mid-infrared free-electron laser irradiation on refolding of amyloid-like fibrils of lysozyme into native form. *The Protein Journal, 31*(8), 710–716. https://doi.org/10.1007/s10930-012-9452-3.

Kawasaki, T., Man, V. H., Sugimoto, Y., Sugiyama, N., Yamamoto, H., Tsukiyama, K., et al. (2020). Infrared laser-induced amyloid fibril dissociation: A joint experimental/theoretical study on the GNNQQNY peptide. *The Journal of Physical Chemistry B, 124*(29), 6266–6277. https://doi.org/10.1021/acs.jpcb.0c05385.

Kawasaki, T., Ohori, G., Chiba, T., Tsukiyama, K., & Nakamura, K. (2016). Picosecond pulsed infrared laser tuned to amide I band dissociates polyglutamine fibrils in cells. *Lasers in Medical Science, 31*(7), 1425–1431. https://doi.org/10.1007/s10103-016-2004-x.

Kawasaki, T., Tsukiyama, K., & Irizawa, A. (2019). Dissolution of a fibrous peptide by terahertz free electron laser. *Scientific Reports*, *9*(1), 10636. https://doi.org/10.1038/s41598-019-47011-z.

Kawasaki, T., Yaji, T., Ohta, T., Tsukiyama, K., & Nakamura, K. (2018). Dissociation of beta-sheet stacking of amyloid beta fibrils by irradiation of intense, short-pulsed mid-infrared laser. *Cellular and Molecular Neurobiology*, *38*(5), 1039–1049. https://doi.org/10.1007/s10571-018-0575-8.

Knippels, G. M., van de Pol, M. J., Pellemans, H. P., Planken, P. C., & van der Meer, A. F. (1998). Two-color facility based on a broadly tunable infrared free-electron laser and a subpicosecond-synchronized 10-fs-Ti:sapphire laser. *Optics Letters*, *23*(22), 1754–1756. https://doi.org/10.1364/ol.23.001754.

Lamb, D. C., Tribble, J., Doukas, A. G., Flotte, T. J., Ossoff, R. H., & Reinisch, L. (1999). Custom designed acoustic pulses. *Journal of Biomedical Optics*, *4*(2), 217–223. https://doi.org/10.1117/1.429912.

Landrieu, I., Dupre, E., Sinnaeve, D., El Hajjar, L., & Smet-Nocca, C. (2022). Deciphering the structure and formation of amyloids in neurodegenerative diseases with chemical biology tools. *Frontiers in Chemistry*, *10*, 886382. https://doi.org/10.3389/fchem.2022.886382.

Legleiter, J., Mitchell, E., Lotz, G. P., Sapp, E., Ng, C., DiFiglia, M., et al. (2010). Mutant huntingtin fragments form oligomers in a polyglutamine length-dependent manner in vitro and in vivo. *The Journal of Biological Chemistry*, *285*(19), 14777–14790. https://doi.org/10.1074/jbc.M109.093708.

Lieberman, A. P., Shakkottai, V. G., & Albin, R. L. (2019). Polyglutamine repeats in neurodegenerative diseases. *Annual Review of Pathology*, *14*, 1–27. https://doi.org/10.1146/annurev-pathmechdis-012418-012857.

Luhrs, T., Ritter, C., Adrian, M., Riek-Loher, D., Bohrmann, B., Dobeli, H., et al. (2005). 3D structure of Alzheimer's amyloid-beta(1-42) fibrils. *Proceedings of the National Academy of Sciences of the United States of America*, *102*(48), 17342–17347. https://doi.org/10.1073/pnas.0506723102.

Miller, J., Arrasate, M., Brooks, E., Libeu, C. P., Legleiter, J., Hatters, D., et al. (2011). Identifying polyglutamine protein species in situ that best predict neurodegeneration. *Nature Chemical Biology*, *7*(12), 925–934. https://doi.org/10.1038/nchembio.694.

Mohara, M., Kawasaki, T., Owada, R., Imai, T., Kanetaka, H., Izumi, S., et al. (2018). Restoration from polyglutamine toxicity after free electron laser irradiation of neuron-like cells. *Neuroscience Letters*, *685*, 42–49. https://doi.org/10.1016/j.neulet.2018.07.031.

Nagai, Y., Inui, T., Popiel, H. A., Fujikake, N., Hasegawa, K., Urade, Y., et al. (2007). A toxic monomeric conformer of the polyglutamine protein. *Nature Structural & Molecular Biology*, *14*(4), 332–340. https://doi.org/10.1038/nsmb1215.

Okumura, H. (2012). Temperature and pressure denaturation of chignolin: folding and unfolding simulation by multibaric-multithermal molecular dynamics method. *Proteins*, *80*(10), 2397–2416. https://doi.org/10.1002/prot.24125.

Okumura, H., & Itoh, S. G. (2014). Amyloid fibril disruption by ultrasonic cavitation: Nonequilibrium molecular dynamics simulations. *Journal of the American Chemical Society*, *136*(30), 10549–10552. https://doi.org/10.1021/ja502749f.

Okumura, H., & Itoh, S. G. (2016). Structural and fluctuational difference between two ends of Abeta amyloid fibril: MD simulations predict only one end has open conformations. *Scientific Reports*, *6*, 38422. https://doi.org/10.1038/srep38422.

Okumura, H., Itoh, S. G., Nakamura, K., & Kawasaki, T. (2021). Role of water molecules and helix structure stabilization in the laser-induced disruption of amyloid fibrils observed by nonequilibrium molecular dynamics simulations. *The Journal of Physical Chemistry B*, *125*(19), 4964–4976. https://doi.org/10.1021/acs.jpcb.0c11491.

Okumura, H., Itoh, S. G., & Okamoto, Y. (2007). Explicit symplectic integrators of molecular dynamics algorithms for rigid-body molecules in the canonical, isobaric-isothermal, and related ensembles. *The Journal of Chemical Physics*, *126*(8), 084103. https://doi.org/10.1063/1.2434972.

Olejniczak, M., Urbanek, M. O., & Krzyzosiak, W. J. (2015). The role of the immune system in triplet repeat expansion diseases. *Mediators of Inflammation*, *2015*, 873860. https://doi.org/10.1155/2015/873860.

Oumata, N., Lu, K., Teng, Y., Cave, C., Peng, Y., Galons, H., et al. (2022). Molecular mechanisms in Alzheimer's disease and related potential treatments such as structural target convergence of antibodies and simple organic molecules. *European Journal of Medicinal Chemistry*, *240*, 114578. https://doi.org/10.1016/j.ejmech.2022.114578.

Owada, R., Mitsui, S., & Nakamura, K. (2022). Exogenous polyserine and polyleucine are toxic to recipient cells. *Scientific Reports*, *12*(1), 1685. https://doi.org/10.1038/s41598-022-05720-y.

Pandey, M., & Rajamma, U. (2018). Huntington's disease: The coming of age. *Journal of Genetics*, *97*(3), 649–664. Retrieved from http://www.ncbi.nlm.nih.gov/pubmed/30027901.

Paulson, H. L., Shakkottai, V. G., Clark, H. B., & Orr, H. T. (2017). Polyglutamine spinocerebellar ataxias—From genes to potential treatments. *Nature Reviews. Neuroscience*, *18*(10), 613–626. https://doi.org/10.1038/nrn.2017.92.

Pearce, M. M. P., & Kopito, R. R. (2018). Prion-like characteristics of polyglutamine-containing proteins. *Cold Spring Harbor Perspectives in Medicine*, *8*(2). https://doi.org/10.1101/cshperspect.a024257.

Pearson, C. E. (2011). Repeat associated non-ATG translation initiation: One DNA, two transcripts, seven reading frames, potentially nine toxic entities! *PLoS Genetics*, 7(3), e1002018. https://doi.org/10.1371/journal.pgen.1002018.

Poirier, M. A., Li, H., Macosko, J., Cai, S., Amzel, M., & Ross, C. A. (2002). Huntingtin spheroids and protofibrils as precursors in polyglutamine fibrilization. *The Journal of Biological Chemistry*, *277*(43), 41032–41037. https://doi.org/10.1074/jbc.M205809200.

Ramdzan, Y. M., Trubetskov, M. M., Ormsby, A. R., Newcombe, E. A., Sui, X., Tobin, M. J., et al. (2017). Huntingtin inclusions trigger cellular quiescence, deactivate apoptosis, and lead to delayed necrosis. *Cell Reports*, *19*(5), 919–927. https://doi.org/10.1016/j.celrep.2017.04.029.

Ren, P. H., Lauckner, J. E., Kachirskaia, I., Heuser, J. E., Melki, R., & Kopito, R. R. (2009). Cytoplasmic penetration and persistent infection of mammalian cells by polyglutamine aggregates. *Nature Cell Biology*, *11*(2), 219–225. https://doi.org/10.1038/ncb1830.

Richard, G. F. (2020). Experimenting with trinucleotide repeats: Facts and technical issues. *Methods in Molecular Biology*, *2056*, 1–10. https://doi.org/10.1007/978-1-4939-9784-8_1.

Rodriguez, L., Marano, M. M., & Tandon, A. (2018). Import and export of misfolded alpha-synuclein. *Frontiers in Neuroscience*, *12*, 344. https://doi.org/10.3389/fnins.2018.00344.

Ruiz-Arlandis, G., Pieri, L., Bousset, L., & Melki, R. (2016). Binding, internalization and fate of Huntingtin Exon1 fibrillar assemblies in mitotic and nonmitotic neuroblastoma cells. *Neuropathology and Applied Neurobiology*, *42*(2), 137–152. https://doi.org/10.1111/nan.12258.

Scheltens, P., Blennow, K., Breteler, M. M., de Strooper, B., Frisoni, G. B., Salloway, S., et al. (2016). Alzheimer's disease. *Lancet*, *388*(10043), 505–517. https://doi.org/10.1016/S0140-6736(15)01124-1.

Selkoe, D. J., & Hardy, J. (2016). The amyloid hypothesis of Alzheimer's disease at 25 years. *EMBO Molecular Medicine*, *8*(6), 595–608. https://doi.org/10.15252/emmm.201606210.

Sever, S. (2002). Dynamin and endocytosis. *Current Opinion in Cell Biology*, *14*(4), 463–467. https://doi.org/10.1016/s0955-0674(02)00347-2.

Tan, Z., Dai, W., van Erp, T. G., Overman, J., Demuro, A., Digman, M. A., et al. (2015). Huntington's disease cerebrospinal fluid seeds aggregation of mutant huntingtin. *Molecular Psychiatry*, *20*(11), 1286–1293. https://doi.org/10.1038/mp.2015.81.

Tardivel, M., Begard, S., Bousset, L., Dujardin, S., Coens, A., Melki, R., et al. (2016). Tunneling nanotube (TNT)-mediated neuron-to neuron transfer of pathological Tau protein assemblies. *Acta Neuropathologica Communications*, *4*(1), 117. https://doi.org/10.1186/s40478-016-0386-4.

Trottier, Y., Devys, D., Imbert, G., Saudou, F., An, I., Lutz, Y., et al. (1995). Cellular localization of the Huntington's disease protein and discrimination of the normal and mutated form. *Nature Genetics*, *10*(1), 104–110. https://doi.org/10.1038/ng0595-104.

Velazquez-Perez, L. C., Rodriguez-Labrada, R., & Fernandez-Ruiz, J. (2017). Spinocerebellar ataxia type 2: Clinicogenetic aspects, mechanistic insights, and management approaches. *Frontiers in Neurology*, *8*, 472. https://doi.org/10.3389/fneur.2017.00472.

Verma, A. K., Khan, E., Bhagwat, S. R., & Kumar, A. (2019). Exploring the potential of small molecule-based therapeutic approaches for targeting trinucleotide repeat disorders. *Molecular Neurobiology*. https://doi.org/10.1007/s12035-019-01724-4.

Vest, K. E., Phillips, B. L., Banerjee, A., Apponi, L. H., Dammer, E. B., Xu, W., et al. (2017). Novel mouse models of oculopharyngeal muscular dystrophy (OPMD) reveal early onset mitochondrial defects and suggest loss of PABPN1 may contribute to pathology. *Human Molecular Genetics*, *26*(17), 3235–3252. https://doi.org/10.1093/hmg/ddx206.

Wacker, J. L., Zareie, M. H., Fong, H., Sarikaya, M., & Muchowski, P. J. (2004). Hsp70 and Hsp40 attenuate formation of spherical and annular polyglutamine oligomers by partitioning monomer. *Nature Structural & Molecular Biology*, *11*(12), 1215–1222. https://doi.org/10.1038/nsmb860.

Wu, Y. R., Fung, H. C., Lee-Chen, G. J., Gwinn-Hardy, K., Ro, L. S., Chen, S. T., et al. (2005). Analysis of polyglutamine-coding repeats in the TATA-binding protein in different neurodegenerative diseases. *Journal of Neural Transmission (Vienna)*, *112*(4), 539–546. https://doi.org/10.1007/s00702-004-0197-9.

Yamauchi, M., & Okumura, H. (2017). Development of isothermal-isobaric replica-permutation method for molecular dynamics and Monte Carlo simulations and its application to reveal temperature and pressure dependence of folded, misfolded, and unfolded states of chignolin. *The Journal of Chemical Physics*, *147*(18), 184107. https://doi.org/10.1063/1.4996431.

Yang, W., Dunlap, J. R., Andrews, R. B., & Wetzel, R. (2002). Aggregated polyglutamine peptides delivered to nuclei are toxic to mammalian cells. *Human Molecular Genetics*, *11*(23), 2905–2917. Retrieved from http://www.ncbi.nlm.nih.gov/pubmed/12393802.

Zu, T., Gibbens, B., Doty, N. S., Gomes-Pereira, M., Huguet, A., Stone, M. D., et al. (2011). Non-ATG-initiated translation directed by microsatellite expansions. *Proceedings of the National Academy of Sciences of the United States of America*, *108*(1), 260–265. https://doi.org/10.1073/pnas.1013343108.

CHAPTER FOUR

Optimized protocols for the characterization of Cas12a activities

Lindsie Martin, Saadi Rostami, and Rakhi Rajan*

Department of Chemistry and Biochemistry, Price Family Foundation Institute of Structural Biology, Stephenson Life Sciences Research Center, University of Oklahoma, Norman, OK, United States
*Corresponding author: e-mail address: r-rajan@ou.edu

Contents

Abstract

The CRISPR-associated (Cas) Cas12a is the effector protein for type V-A CRISPR systems. Cas12a is a sequence-specific endonuclease that targets and cleaves DNA containing a cognate short signature motif, called the protospacer adjacent motif (PAM), flanked

Methods in Enzymology, Volume 679
ISSN 0076-6879
https://doi.org/10.1016/bs.mie.2022.08.048

97

by a 20 nucleotide (nt) segment that is complementary to the "guide" region of its CRISPR RNA (crRNA). The guide sequence of the crRNA can be programmed to target any DNA with a cognate PAM and is the basis for Cas12a's current use for gene editing in numerous organisms and for medical diagnostics. While Cas9 (type II effector protein) is widely used for gene editing, Cas12a possesses favorable features such as its smaller size and creation of staggered double-stranded DNA ends after cleavage that enhances cellular recombination events. Collected here are protocols for the recombinant purification of Cas12a and the transcription of its corresponding programmable crRNA that are used in a variety of Cas12a-specific *in vitro* activity assays such as the *cis*, the *trans* and the guide-RNA independent DNA cleavage activities with multiple substrates. Correspondingly, protocols are included for the quantification of the activity assay data using ImageJ and the use of MATLAB for rate constant calculations. These procedures can be used for further structural and mechanistic studies of Cas12a orthologs and other Cas proteins.

1. Introduction

Cas12a is the signature endonuclease protein of the type V-A CRISPR-Cas (clustered regularly interspaced short palindromic repeats-CRISPR associated) adaptive immune system (Zetsche et al., 2015). Cas12a binds to a cognate RNA called the CRISPR RNA (crRNA), specifically by recognizing the 5' end of the crRNA that possesses a pseudoknot structure to form a binary complex (Yamano et al., 2016). The binary complex surveils DNA and binds and cleaves it sequence-specifically if it possesses two specific features: *(i)* a protospacer adjacent motif (PAM), which is a short DNA motif that is specific for each Cas12a ortholog [for example, cognate PAM for Cas12a from *Francisella novicida* Cas12a (FnoCas12a) is 5'-TTN-3']; and *(ii)* complementarity with the 20-nt "guide" region present at the 3' end of the crRNA (Zetsche et al., 2015). After locating the PAM, the guide region of crRNA triggers formation of an R-loop with the DNA, forming the ternary complex. The ternary complex facilitates sequence specific cleavage of DNA [double stranded DNA (dsDNA) or single stranded DNA (ssDNA) that base pairs with the crRNA] using its endonuclease domain, RuvC (Chen et al., 2018; Gao, Yang, Rajashankar, Huang, & Patel, 2016; Zetsche et al., 2015).

Cas12a has a bilobed structure made of the recognition (REC) and nuclease (NUC) lobes (Fig. 1) (Dong et al., 2016; Gao et al., 2016; Yamano et al., 2016). The REC lobe comprises REC1 and REC2 domains, and the NUC lobe contains the RuvC, the PAM-interacting (PI) and the wedge (WED) domains, and the bridge helix (BH) (Yamano et al., 2016). The BH is an arginine-rich helix that bridges the REC and NUC

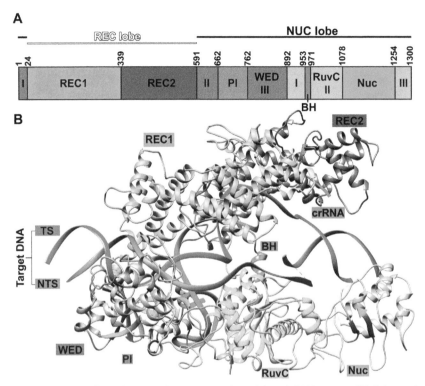

Fig. 1 Structure of FnoCas12a-crRNA in complex with a dsDNA target. (A) Schematic representation of FnoCas12a domain organization. (B) Overall structure of the FnoCas12a-crRNA-target DNA complex. Domains are colored according to the scheme in panel A. (BH: bridge helix; PDB: 6I1K) (Swarts & Jinek, 2019).

lobes and is a conserved feature of type II and type V CRISPR systems (Yamano et al., 2016). When the Cas12a effector transitions from the binary to the ternary complex, the BH undergoes a conformational change that aids in target cleavage (Parameshwaran et al., 2021; Wörle, Jakob, Schmidbauer, Zinner, & Grohmann, 2021). Recent studies have shown the essential role that the BH plays in the Cas12a cleavage mechanism and mismatch tolerance (Parameshwaran et al., 2021; Wörle et al., 2021). The WED and REC1 domains recognize the dsDNA target through non-specific interactions with the sugar-phosphate backbone, while the PI domain sequence-specifically interacts with both strands of the PAM sequence to promote the unwinding of the dsDNA helix (Yamano et al., 2016). While Cas12a orthologs usually have a T-rich cognate PAM sequence (5′-TTN-3′ for FnoCas12a, and 5′-TTTN-3′ for *Lachnospiraceae bacterium* Cas12a (LbCas12a) and

Acidaminococcus sp. Cas12a (AsCas12a) (Fonfara, Richter, Bratovič, Le Rhun, & Charpentier, 2016; Yamano et al., 2016; Zetsche et al., 2015), they can also recognize C-containing PAM sequences through altered interactions with the target DNA (Yamano et al., 2017). As the DNA unwinds upon PAM recognition, the crRNA can hybridize to its complementary sequence, the target DNA strand (TS), leaving a single stranded DNA, the non-target DNA strand (NTS), leading to the formation of the R-loop. The NTS is conducted by the PI domain to the RuvC active site for cleavage (Stella, Alcón, & Montoya, 2017; Swarts, van der Oost, & Jinek, 2017; Yamano et al., 2016). As the R-loop starts to form, Cas12a searches for complementarity with a 3–5-nt seed region near the PAM, which is very crucial in propagating the R-loop formation by further hybridization with the remaining nucleotides of the crRNA guide and the TS. While mismatches between the crRNA guide and the target DNA are tolerated to varying degrees based on the position of the mismatch, mismatches in the seed sequence drastically inhibit R-loop propagation, and by default, the cleavage activity of Cas12a (Stella et al., 2017). While cleaving a dsDNA, Cas12a exhibits a strict order in strand cleavage, NTS followed by TS cleavage (Swarts et al., 2017). After the NTS is cleaved, the REC and NUC lobes undergo conformational changes to position the TS in a single stranded form, with the correct polarity, in the catalytic pocket created by RuvC and a nearby domain called the "Nuc" domain (Cofsky et al., 2020; Swarts et al., 2017; Swarts & Jinek, 2019). The extra distance needed to accommodate the TS in a single stranded form in the catalytic site creates a staggered double stranded cut in the PAM distal end of the R-loop (Stella et al., 2018). This type of DNA cleavage triggered by PAM recognition, base pairing with the crRNA, and sequence-specific dsDNA cleavage is called "*cis*" cleavage in Cas12a terminology (Swarts & Jinek, 2019).

Another interesting feature of Cas12a proteins is its ability to promote collateral ssDNA cleavage following *cis* DNA cleavage (Chen et al., 2018; S.-Y. Li et al., 2018). Based on available mechanisms, it is proposed that the release of the PAM-distal end of the DNA from the complex after *cis* cleavage, leaves the RuvC active site open for indiscriminate ssDNA cleavage (Swarts & Jinek, 2019). This activity is termed *trans* cleavage (Chen et al., 2018; Nguyen, Smith, & Jain, 2020) and has formed the basis of several molecular diagnostic tools where a ssDNA with a measurable probe is cleaved upon recognition of a specific DNA sequence for *cis* cleavage (Nguyen et al., 2020). *Cis* cleavage can be activated by a dsDNA possessing a PAM and a complementary sequence to the guide region of the crRNA or by a ssDNA that is complementary to the guide region of the crRNA (Chen et al., 2018).

Certain Cas proteins (Cas9 and Cas12a) have been shown to possess non-specific DNA cleavage in the absence of cognate crRNAs which is enabled by specific divalent metals, named as RNA-independent or guide-RNA free DNA cleavage (Saha et al., 2020; Sundaresan, Parameshwaran, Yogesha, Keilbarth, & Rajan, 2017). FnoCas12a can nonspecifically nick dsDNA plasmids and degrade ssDNA substrates in the presence of divalent metals such as Mn^{2+} and Co^{2+} (Sundaresan et al., 2017). Interestingly, AsCas12a is very active for RNA-independent DNA cleavage with the activity being triggered by Mg^{2+} as well and dsDNA being degraded following the initial nicking at higher metal concentrations (Li et al., 2020).

Cas12a is also unique since it possesses the ability to process its own crRNA through a unique active site that is different than the RuvC domain (Fonfara et al., 2016; Swarts et al., 2017). The CRISPR array of CRISPR systems (made up of alternating spacer and repeat sequences) is transcribed into one long transcript called the pre-crRNA (Fonfara et al., 2016). The repeat-derived region of the crRNA adopts a pseudoknot conformation that is recognized by the Cas12a protein while binding to the pre-crRNA (Dong et al., 2016; Fonfara et al., 2016). Cas12a then cleaves the repeat sequence upstream of the pseudoknot followed by further trimming to create a mature crRNA (Fonfara et al., 2016). The catalytic site for crRNA processing is located in the WED domain and contains two highly-conserved lysines and a histidine (Swarts et al., 2017). The mature crRNA consists of $\sim$19-nt derived from the repeat sequence at the $5'$ end, and $\sim$24 nt from the spacer sequence at the $3'$ end (Dong et al., 2016; Safari, Zare, Negahdaripour, Barekati-Mowahed, & Ghasemi, 2019; Swarts et al., 2017; Zetsche et al., 2015). The mechanism for trimming of the $3'$ end is unknown currently. This property makes Cas12a unique in being self-sufficient for processing individual crRNAs from a tandem crRNA cassette that can be provided as a continuous stretch of DNA. This has enabled multiplexing, where different sites can be targeted simultaneously, a favorable feature for Cas12a genome applications (Paul & Montoya, 2020).

CRISPR-Cas systems have been engineered to use as genome editing and medical diagnostic tools because of their abilities to specifically target sites in the genome and perform a double-stranded cut which can then be repaired by the host's DNA repair machinery (Bayat, Modarressi, & Rahimpour, 2018). Currently, Cas9 is the main CRISPR system used for gene editing and gene knockouts, but its large size and off-target activity make it a less-desirable tool (Shmakov et al., 2017). Cas12a is increasing in popularity as a genome tool because of its smaller size and multiplexing ability that helps in delivery using viral vectors and the creation of staggered

DNA ends following *cis* cleavage that enhances recombinational events (Paul & Montoya, 2020; Zetsche et al., 2017). In addition, *trans* activity is specifically being used as molecular diagnostic tools, including specific detection of viral samples (Broughton et al., 2020; Mahas et al., 2021). The processive nature of *trans* activity amplifies signal enabling detection as low as 100 aM (Chen et al., 2018; L. Li et al., 2019; S.-Y. Li et al., 2018; Mahas et al., 2021). Catalytically dead Cas12a which is unable to induce dsDNA breaks is being used for gene repression and studies have shown that Cas12a possesses higher gene repression than catalytically dead Cas9 (Kim et al., 2017).

The properties of the Cas12a enzyme make it an excellent candidate for biomedical applications. Because of this, structural and mechanistic studies of Cas12a are very valuable to further enhance the biomedical applicability of Cas12a. The following protocols can be used for the structural and mechanistic studies of FnoCas12a and other Cas12a orthologs (Fig. 2).

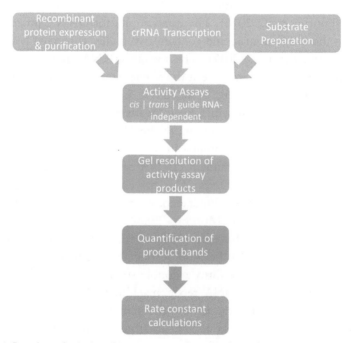

Fig. 2 A flowchart depicting the sequence of events from the combined protocols in this chapter.

2. Recombinant protein purification

Cas12a from several orthologs have been recombinantly expressed in *Escherichia coli* and biochemically and structurally characterized (Dong et al., 2016; Gao et al., 2016; Saha et al., 2020). In this report, protocols for FnoCas12a protein (UniProt ID: A0Q7Q2) will be presented. To recombinantly purify FnoCas12a, it can be expressed with a histidine (His) tag for Nickel-Nitriloacetic Acid (NTA) affinity purification and a maltose-binding protein (MBP) tag to increase protein solubility. The tags can be cleaved off after Ni–NTA column by introducing a protease cut site such as the tobacco etch virus (TEV) protease. The protocol presented below utilizes sonication for lysis, two affinity chromatography columns, nickel-NTA and cation exchange, and a size exclusion chromatography (SEC) (Parameshwaran et al., 2021). All these steps are variable. Alternative protocols use a HiTrap Heparin column in place of a cation exchange column (Mohanraju, Van Der Oost, Jinek, & Swarts, 2018) or use only a nickel-NTA column and an SEC column (Swarts et al., 2017).

2.1 Plasmids and bacterial strains for bacterial expression and purification of FnoCas12a

This protocol uses a pET28-based plasmid holding the Fno *cas12a* gene in the following pattern: His8-3C protease recognition site-MBP-TEV recognition site-FnoCas12a (Sundaresan et al., 2017). His8 tag (eight histidines) increases the affinity of the protein for the immobilized metal affinity chromatography (IMAC) column and the N-terminal Maltose Binding Protein (MBP) tag enhances folding and solubility of the target protein. 3C and TEV are sequence-specific proteases. To remove both tags from the purified protein, TEV protease can be used since it will produce FnoCas12a with minimal extra non-native amino acids. The expression plasmid is transformed into *E. coli* Rosetta strain 2 (DE3) cells for optimizing protein expression following established protocols. This plasmid should be propagated in medium containing both chloramphenicol and kanamycin to maintain the plasmids needed for tRNAs for rare codons and pET28 respectively. Bacterial expression plasmids are also available from the Addgene repository [FnoCas12a, (Addgene 113432), AsCas12a (Addgene 113430), LbCas12a (Addgene 113431)] (Chen et al., 2018), and can be used with the corresponding protocols for each plasmid type, which uses a similar overall procedure for protein purification.

2.2 FnoCas12a recombinant overexpression and purification

1. Grow an overnight primary culture of *E. coli* Rosetta strain 2 (DE3) transformed with the FnoCas12a plasmid in 50 mL of 2xYT broth containing kanamycin (50 μg/mL) and chloramphenicol (34 μg/mL).

2. Prepare secondary cultures in two autoclaved 2-L Erlenmeyer flasks each containing 1 L of autoclaved 2xYT broth containing kanamycin and chloramphenicol with the final concentration of 50 and 34 μg/mL respectively. Use 10 mL of the overnight culture to inoculate 1 L 2xYT broth medium.

3. Grow secondary cultures for ~4 h (until optical density (OD) 600 nm is ~0.6–0.8). Take out 1 mL as an uninduced sample for analysis on a gel. (Pellet the uninduced sample, resuspend in 150 μL water and add 50 μL of 4× gel loading dye (50 mM Tris-HCl pH 6.8, 2% SDS, 10% glycerol, 1% β-mercaptoethanol, 12.5 mM EDTA, 0.02% bromophenol blue). Induce the rest of the culture with 0.2 mM Isopropyl β-D-1-thiogalactopyranoside (IPTG) and let the culture grow overnight (~16 h) at 18 °C. All cell culture is done with constant shaking at 180 RPM (revolutions per minute).

4. Centrifuge the overnight grown culture at 4 °C for 20 min at 5000 relative centrifugal force (RCF). Resuspend the pellet in lysis buffer/Ni-NTA buffer A (20 mM 4-(2-hydroxyethyl)-1-piperazineethanesulfonic acid (HEPES), pH 7.5, 1 M NaCl, 20 mM Imidazole) and store it at −80 °C until further use. A ratio of 1 g of cell pellet to 7 mL of buffer can be used for optimal results.

5. On the day of protein purification, thaw the pellets on ice/water mixture. Pool the thawed pellets into a beaker. Once the pellet is completely thawed, add the protease inhibitors:phenylmethylsulfonyl fluoride (PMSF) and Benzamidine to a final concentration of 1 mM and Pepstatin and Leupeptin to a final concentration of 1 μg/mL.

6. Sonicate the cell suspension for a total of 3 min at 35 Amp with a 3 s ON and 20 s OFF cycle. Take out some sample as the post-sonication lysate.

7. Transfer the lysate into appropriate centrifuge tubes and spin at 30,000 RCF for 45 min. Take a sample of the pellet and the supernatant to run on the gel.

8. Load the supernatant onto a Cytiva HisTrap FF 5 mL column (Ni-NTA column). If using a previously used column, it is ideal to strip the Ni^{2+} off the column and recharge the column with fresh Ni^{2+} as needed. Collect the flow through from the sample application step for analysis on a gel.

9. After loading the supernatant onto the Ni–NTA column, wash with 10 column volumes of Ni–NTA buffer A to remove non-specifically bound proteins. It is a good practice to monitor and ensure that the UV 280 nm is brought to the baseline (i.e., like the column equilibration step) by the end of this step.

10. Elute the bound protein off the Ni–NTA column using a continuous gradient to reach 100% B (Ni–NTA buffer B: 1× concentration is 20 mM HEPES pH 7.5, 1 M NaCl, 500 mM Imidazole) in 15 column volumes. Usually, FnoCas12a elutes as a single major peak at around 175 mM imidazole. Once the run is completed, check every alternate fraction from the major peak, representative fractions from any minor peak, and other samples that were collected during the different steps in the protocol on a 7.5% denaturing sodium dodecyl sulfate–polyacrylamide (SDS) gel (Maizel & Jacob, 2000). The expected size of FnoCas12a with an attached MBP tag is ∼196 kDa.

11. Based on the protein banding pattern on the gel (Fig. 3A), pool the Ni–NTA fractions containing FnoCas12a into a beaker and take a pre-dialysis protein sample for gel analysis. Based on the intensity of the FnoCas12a band, add TEV protease to cleave off the MBP tag (∼1 mg of TEV from a preparation starting with 2 L cell culture; suggested ratio of TEV: protein of interest is 1:100, Raran-Kurussi, Cherry, Zhang, & Waugh, 2017). Transfer this sample to a 30 kDa cut-off dialysis membrane and dialyze it against 1 L of dialysis buffer/ion exchange buffer A (20 mM HEPES pH 7.5, 150 mM KCl, 2 mM ethylenediaminetetraacetic acid (EDTA), 1 mM dithiothreitol (DTT)) overnight with slow stirring at 4 °C. We have experienced that

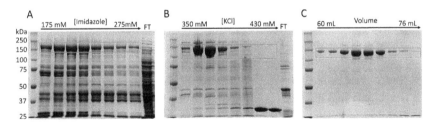

Fig. 3 Purification of FnoCas12a from *E. coli* Rosetta cells by nickel-NTA (A), ion exchange (B), and size exclusion chromatography (C). The band corresponding to Cas12a is shown in red box in each gel. The molecular weight of FnoCas12a is 196 kDa in Ni-NTA elution due to the attached MBP tag. For ion exchange and SEC columns, the expected molecular weight is 152 kDa, after cleavage of the MBP tag. Flow Through (FT).

diluting the pooled fractions with the Ni–NTA buffer B (~50–75 mL for a 2 × 1 L preparation; done before the addition of TEV protease) before dialyzing prevents precipitation of the protein after the overnight dialysis.

12. Spin the overnight dialyzed protein (e.g., at 4 °C for 15 min at 4000 RCF) to remove any trace amounts of precipitates. Load the dialyzed protein onto a Cytiva HiTrapTM SP HP cation exchange column that has been equilibrated with dialysis buffer/ion exchange buffer A. Wash the column with 10 column volumes of ion exchange buffer A and elute with a continuous gradient of 0–100% using ion exchange buffer B (1× concentration: 20 mM HEPES pH 7.5, 1 M KCl, 2 mM EDTA, 1 mM DTT) over 20 column volumes. FnoCas12a elutes from the SP HP column at ~350 mM KCl (Fig. 3B).

13. Check the fractions along with the shoulders of the peaks on a 7.5% SDS gel. The pre-dialysis and post dialysis samples should also be checked on the gel to ensure completeness of the TEV cleavage step. The expected molecular weight of FnoCas12a after TEV cleavage is 152 kDa.

14. Pool the fractions containing pure FnoCas12a and concentrate to ~1.7 mL by centrifugation using a 30 kDa cut off concentrator at 4 °C. Load the sample on to an SEC (HiPrep™ 16/60 Sephacryl® S-300 HR or HiPrep™ 16/60 Sephacryl® S-200 HR) that has been equilibrated with SECbuffer (1× concentration is 20 mM HEPES pH 7.5, 150 mM KCl, 2 mM EDTA and 1 mM TCEP). Elute the protein by isocratic elution using the SEC buffer. FnoCas12a usually elutes at ~60 mL while using an S300 column that has a total column volume of 120 mL (Fig. 3C).

15. Check the SEC fractions on a 7.5% SDS PAGE gel. Also check the Abs (260/280) of each of the fractions to monitor nucleic acid contamination in the protein. Pool the fractions that are devoid of additional bands when visualized on the protein gel and based on the absorbance ratio. (Fractions with an Abs (260/280) ratio between 0.5 and 0.65 will have minimal nucleic acid contamination and are of good quality to be pooled together).

16. Concentrate the fractions using a 30 kDa cut-off spin column at 4 °C. Keep monitoring the concentration of the protein and once it reaches an ideal concentration, aliquot and flash freeze the protein using liquid nitrogen. Store the aliquots at −80 °C. It is a good practice to prepare smaller aliquots such that one frozen sample will be used for

one activity assay. This prevents multiple freeze-thaw cycles of the purified FnoCas12a protein, which will maintain high quality active protein for biochemical assays. The molar extinction coefficient of FnoCas12a that has 1305 amino acids (5 amino acids are left after the TEV cleavage) is $144,330\,M^{-1}\,cm^{-1}$.

3. Preparation of crRNA

3.1 crRNA transcription

Cas proteins are guided by their crRNA to a specific DNA target that possesses a cognate PAM and a complementary sequence to the guide region of the crRNA. Even between Cas orthologs, the crRNA sequence can differ, and the overall secondary structures can be variable (Mir, Edraki, Lee, & Sontheimer, 2018; Yan et al., 2019). The FnocrRNA is usually around ~45 nucleotides (nt) long and contains ~19-nt of a repeat derived sequence and ~24-nt of the spacer sequence in the 5′–3′ direction (Fonfara et al., 2016; Safari et al., 2019; Zetsche et al., 2015). The repeat-derived sequence is specific for each ortholog, while the spacer sequence (i.e., the guide region) can be designed to target any specific DNA of interest possessing a flanking PAM sequence. The easiest option to produce crRNA is by *in vitro* transcription using oligo DNA templates. This requires two oligo DNAs: *(i)* a template strand (non–coding strand) which is used to transcribe crRNA; *(ii)* a T7 promoter strand that is needed for the T7 RNA Polymerase to transcribe the template strand.

i. Template strand: In the 5′–3′ direction, the template strand contains the reverse complement sequences of the following in the specific order as shown: the 24 to 25-nt targeting region of the DNA [i.e., the sequence of the target strand (TS) of the DNA; Fig. 4], the repeat-derived crRNA sequence, and the T7 promoter sequence.

ii. T7 promoter strand: This oligo can be shorter possessing only the T7 polymerase promoter sequence in the 5′–3′ direction (if using T7 RNA Polymerase for *in vitro* transcription). Having only the promoter sequence enables the use of this strand for transcribing different crRNAs with unique sequences.

There are also modifications that can be added to the DNA template to improve RNA transcription. For example, the addition of three guanines to the 5′ terminus of the crRNA sequence can increase the efficiency of *in vitro* transcription by T7 RNA Polymerase. A step-by-step description of crRNA transcription is provided below. The reaction volume can be

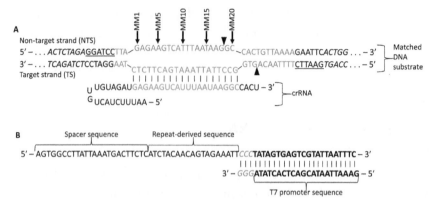

Fig. 4 Sequences related to crRNA-DNA complex. (A) Schematic of the R-loop formed with the matched DNA substrate where the PAM sequence is gold, spacer sequence of the crRNA and the TS of the DNA that are complementary to each other are in red, and restriction enzyme sites are underlined. Arrows above the NTS point to nucleotide positions away from the PAM illustrating the mismatch (MM) positions between the DNA substrate and crRNA. Black triangles indicate Cas12a cleavage sites. (B) Schematic of the T7 promoter strand and template strand oligos for crRNA transcription. Bolded sequence is the T7 promoter sequence and its complement. Blue sequences are extra nucleotides added to increase transcription efficiency.

changed based on the amount of RNA required. A 200 μL transcription reaction following the protocol mentioned below will yield approximately 20–50 μg of RNA.

1. Obtain synthetic oligonucleotides for the template and T7 promoter strands from a vendor (e.g., Integrated DNA Technology, Sigma Aldrich, ThermoFisher Scientific). Prepare working stock solutions of the template strand at 10 μM and the T7 promoter strand at 20 μM concentrations. (Note: based on the scale of transcription required, the concentration of the strands can be varied.)

2. Reconstitute these oligos in molecular-grade water to produce the transcription duplex. For this, the template and T7 promoter strands can be annealed in a 1:1.5 M ratio in an annealing buffer (e.g., 1× annealing buffer: 10 mM Tris-HCl, pH 8, 50 mM NaCl). Heat this mixture to 95 °C for 5 min and allow to cool slowly to room temperature. Measure the concentration of the annealed DNA using extinction coefficient for double stranded DNA. The annealed DNA can be stored at −20 °C and used as needed to perform transcription reactions.

3. Prepare 200 μL *in vitro* transcription reaction by mixing 1× transcription buffer (40 mM Tris-HCl, pH 8, 1 mM spermidine, 50 μg bovine serum

albumin, 20 mM $MgCl_2$, 5 mM DTT), 400 ng annealed transcription template, 8 mM ATP, 8 mM CTP, 8 mM UTP, 9 mM GTP, 50 µg RNasin (New England BioLabs), 1 µg inorganic phosphatase (New England BioLabs), and 40 µg T7 RNA polymerase. Incubate the reaction at 37 °C for 4 h. T7 RNA polymerase can be purified in house (Rio, 2013) or can be purchased from a vendor (use amounts specified by the vendor in this case). Adding T7 RNA polymerase in two steps can increase the efficiency of transcription (for example, 30 µg at the start of the reaction and an additional 10 µg after 3 h).

4. Since the DNA template for transcription and the expected RNA product are of similar size, removing the DNA template will improve the quality of the RNA product. This can be achieved by DNase treatment at 37 °C for 30 min [add 0.01 mg/mL DNase I (New England Biolabs) and 1× DNase buffer (10 mM Tris-HCl, pH 7.5, 2.5 mM $MgCl_2$, 0.5 mM $CaCl_2$) to the 200 µL transcription reaction].

5. The RNA in the transcription reaction can be precipitated for further purification. For this, add 3-volumes of cold 100% ethanol to the above reaction. Mix the tube by gentle inversions and store the mixture overnight (16–18 h) at −20 °C to precipitate the RNA out of solution. In 100% ethanol, the RNA can be stored at −20 °C for longer periods of time. As needed, a small reaction volume can be tested on a 12% denaturing urea-formamide acrylamide gel along with an appropriate ladder to assess the success of transcription and the size of the RNA band (Beckert & Masquida, 2011).

3.2 crRNA purification

To ensure a good quality RNA product (i.e., a single band of the right size), it is a good practice to extract the *in vitro* transcribed RNA out of a denaturing urea acrylamide gel. A suggested procedure is below. (Note: while for quality analysis of RNA a urea-formamide acrylamide gel can be used, for RNA extraction and purification, urea acrylamide gels should be used. The pelleted RNA should be dissolved in an 8 M urea-containing buffer to ensure proper unfolding of RNA that is devoid of secondary structures.)

1. Centrifuge the RNA sample to pellet the RNA precipitate (for example, 21,130 RCF at 4 °C for 30 min or as appropriate). The precipitated RNA will form a small white pellet. Wash the pellet with two successive cold 80% ethanol washes. Flick the tube to dislodge the pellet while performing 80% ethanol washes. After the second wash, remove as much of

the supernatant as possible without disturbing the RNA pellet. Completely evaporate all liquid from the pellet. This can be achieved by air drying for an appropriate time or by assisted drying using a vacufuge [e.g., Eppendorf Vacufuge plus 5305 V-AL (vacuum alcoholic solutions) set at 30 °C]. During the washes and vacufuging, the RNA pellet will change its appearance from bright white to a translucent white indicating that salts are being washed out of the RNA pellet. Resuspend the RNA pellet by adding 200 μL of 1× gel loading solution [8 M urea, 1× TBE (89 mM Tris-HCl, 89 mM boric acid, 2 mM EDTA), 0.02% bromophenol blue, 0.02% xylene cyanol] to the dried RNA pellet. Resuspension can be enhanced by frequent mixing of RNA in the gel loading solution and by slightly warming the solution. Ensure that the RNA pellet is completely dissolved before loading on the gel. (Note: to visualize RNA dissolution, it is a common practice to resuspend the RNA pellet in the gel loading dye devoid of the dyes, followed by adding the required amounts of dyes before loading the sample on the gel).

2. Prepare a 12% polyacrylamide gel containing 8 M urea. The length of the gel depends on the size of the RNA product and the number of products after transcription. For crRNA of 45-nt size, gels with the dimensions 320 mm × 165 mm × 1.5 mm (h × w × d) are effective. Preheat the gel to 50 °C and clean settled urea out of the wells before loading sample. Load the sample and run at 35 V for ~50–65 min until the RNA bands resolve. The bromophenol blue dye front can be used to track the gel running time (for 45-nt RNA product, it is advisable to stop the gel once the bromophenol blue band reaches the bottom of the gel. In a 12% polyacrylamide gel, the bromophenol blue dye migrates comparative to 12-nt, and the xylene cyanol dye migrates comparative to 55-nt (Ausubel et al., 1987). Running the bromophenol blue dye to the bottom of the gel ensures maximum separation between the 45-nt RNA band and the xylene cyanol band without the RNA band running out of the gel.

3. Wrap the gel in plastic wrap and visualize the RNA band by UV shadowing (for example, using a hand-held UV imager like the VWR UV-AC Dual Hand Lamp at wavelength 254 nm). It is very important to use a long wavelength UV and perform the visualization quickly to prevent RNA damage from UV. Use a marker to outline the RNA band (on the plastic wrap) and cut it around the marker. Transfer the gel piece into a 50 mL conical tube, crush the gel piece, and add enough volume of

RNA elution buffer (50 mM potassium acetate, 20 mM KCl, adjust to pH 7) to cover the gel piece. Rotate the sample at 4°C overnight (16–18 h) to elute RNA into the solution. (Note: electroelution (for example using the CBS Scientific Electro-Eluter Concentrator ECU-040) from the RNA band is another option for this process).

4. Briefly centrifuge the 50 mL conical tube at a low centrifugal force to collect all the liquid off the sides. Filter the liquid with a 0.22 μm filter to separate the gel pieces from the solution (e.g., using the 50 mL disposable Steriflip Vacuum-driven Filtration System with a 0.22 μm membrane). Add 2.5–3 volumes of cold 100% ethanol to precipitate RNA and store at −20°C, which will provide elution-1 of the RNA sample. Note: the gel pieces left behind after elution-1 can be returned to the 50 mL conical tube for another one or two elutions based on the amount of RNA present in the transcription reaction (estimate this based on the band intensity during ultraviolet (UV) shadowing). The elutions are typically stored in 100% ethanol at −20°C until further use to prevent RNA degradation.

5. To extract RNA from 100% ethanol, centrifuge the sample [for example, 3220 RCF at 4°C for 30 min (variable based on the centrifuge being used)]. Remove the supernatant and wash the pellet successively with two, 80% cold ethanol washes. After the second 80% ethanol wash, dry the RNA pellet and resuspend in RNase-free water or an appropriate buffer (1× annealing buffer, 10 mM Tris-HCl, pH 8, 50 mM NaCl). Anneal the RNA by incubating it at 95°C for 2 min and slowly cooling to room temperature to allow the crRNA to attain its secondary fold. Quantify the folded RNA by measuring the absorbance at 260 nm. The folded FnoCas12a crRNA can be stored at 4°C for up to a week without significant activity reduction. The stability of crRNA should be closely monitored for specific Cas12a orthologs. Addition of divalent metal ions, (usually Mg^{2+} at 1 mM concentration) in the annealing reaction buffer is also an option for aiding proper crRNA folding (Babu et al., 2019). Inclusion of Mg^{2+} in the annealing reaction will require slight modifications in the annealing process, for example heating to a lower temperature (<60°C) or adding Mg^{2+} only during the slow cooling process when the temperature reaches around 60°C (Babu et al., 2019).

In addition to *in vitro* transcription using oligo DNA templates as mentioned in this protocol, other methods such as using a vendor to synthesize RNA (e.g., Integrated DNA Technologies, ThermoFisher Scientific, Sigma Aldrich) (Swarts & Jinek, 2019), *in vitro* transcription using a linearized plasmid holding the crRNA gene as the template using run–off transcription (Zoephel, Dwarakanath, Richter, Plagens, & Randau, 2012), and modification of oligo DNA template lengths for *in vitro* transcription can be adopted as needed (Karvelis et al., 2013).

4. Activity assays to characterize Cas12a properties

Previous studies have shown that FnoCas12a is capable of three different forms of DNA cleavage: *(i) cis* cleavage, *(ii) trans* cleavage, and *(iii)* RNA-independent or guide-free DNA cleavage (Fig. 5) (Chen et al., 2018; Sundaresan et al., 2017; Swarts & Jinek, 2019). In addition, Cas12a is able to process its crRNA to mature forms needed for DNA cleavage (Swarts et al., 2017).

Cis cleavage is the guide RNA-dependent, sequence specific double stranded cleavage of a DNA target that possesses a complementary sequence as the guide region of the crRNA and a cognate PAM (Fig. 4). NTS is cleaved by the RuvC domain, while TS is cleaved by RuvC along with the assistance of Nuc domain (Yamano et al., 2016). There is a sequential

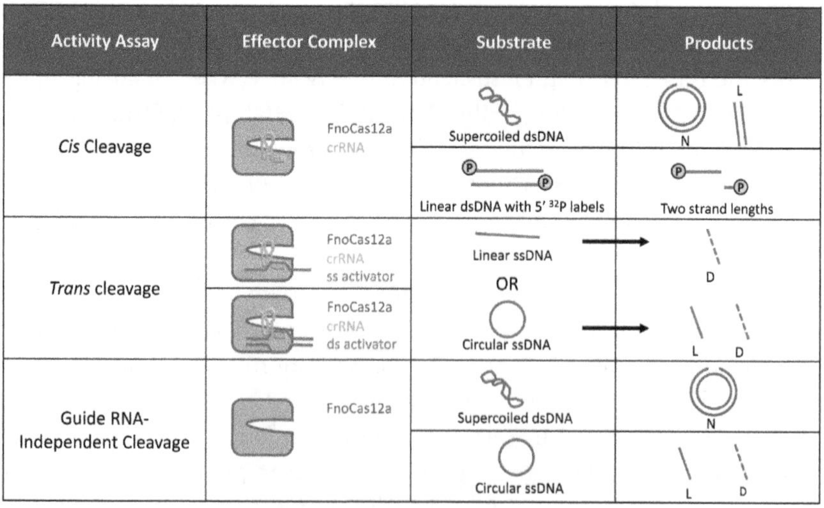

Activity Assay	Effector Complex	Substrate	Products
Cis Cleavage	FnoCas12a crRNA	Supercoiled dsDNA	N, L
		Linear dsDNA with 5′ ³²P labels	Two strand lengths
Trans cleavage	FnoCas12a crRNA ss activator	Linear ssDNA OR	D
	FnoCas12a crRNA ds activator	Circular ssDNA	L, D
Guide RNA-Independent Cleavage	FnoCas12a	Supercoiled dsDNA	N
		Circular ssDNA	L, D

Fig. 5 Illustration of the components, substrates and products produced from the *cis*, *trans*, and guide-RNA independent cleavage assays. (Note: the products for the guide RNA-independent activity assay depend on the Cas12a ortholog and the divalent metal ion present (Li et al., 2020; Parameshwaran et al., 2021)) For assays with labels, only visible products are shown.

order to cleave DNA by Cas12a, with NTS preceding TS cleavage, even though studies have shown that nicks or other destabilization of the NTS strand will allow TS cleavage without the prerequisite of NTS cleavage (Swarts & Jinek, 2019). *Cis*-DNA cleavage is the basis for gene editing applications using Cas12a.

Trans cleavage is the non-specific cleavage of ssDNA by the RuvC domain, which occurs as an after-effect of *cis* DNA cleavage. This cleavage needs crRNA and can be activated under the following conditions: *(i)* after cleavage of a dsDNA target by Cas12a followed by release of the PAM-distal DNA product that exposes RuvC active site for ssDNA binding and cleavage; *(ii)* Cas12a-crRNA bound to a TS strand representing a post-cleavage product (*i.e.,* a 23-mer ssDNA bearing at least 15-nt complementarity with the guide-region of the crRNA and a cognate PAM) (Chen et al., 2018).

RNA-independent cleavage is the non-specific degradation of ssDNA or nicking of dsDNA in the presence of certain divalent metal ions such as Mn^{2+} and Co^{2+}. This activity is crRNA-independent (Sundaresan et al., 2017). Recent literature shows that RNA-independent DNA cleavage activity can be promoted by Mg^{2+} as well in other Cas12a orthologs such as AsCas12a (Li et al., 2020). Guide RNA free DNA cleavage by Cas9 was shown to induce DNA damage in human cells transfected with Cas9 devoid of guide RNA (Saha et al., 2020), and hence must be assessed for Cas protein orthologs while developing genome tools.

All the three *in vitro* DNA cleavage assays require purified Cas protein, but both *cis* and *trans* activity assays also require purified crRNA. The different cleavage mechanisms require the preparation of different DNA substrates. Protocols for each specific activity are mentioned in the following sections.

4.1 *Cis* DNA cleavage assay

Cis DNA cleavage is usually performed using a plasmid or a linear dsDNA substrate possessing a complementary sequence as the guide-region of the crRNA and flanked on the upstream by a cognate PAM (Fig. 4). In addition to analyzing cleavage of a target DNA possessing a completely complementary region as the crRNA guide (called matched DNA), sensitivity to cleave DNA possessing mismatches with the guide-region is also routinely performed to characterize new variants and orthologs of Cas12a. We describe the protocol for plasmid based *cis* cleavage assays below.

4.1.1 Preparation of matched and mismatched plasmid DNA substrates

1. Order NTS and TS oligos from IDT that consists of a sequence comprising a 24-nt region complementary to the crRNA-guide region and a 3-nt cognate PAM sequence. Since the protospacer sequence needs to be inserted into a plasmid, such as pUC19, the oligo ends should hold restriction site sequences that will be used to ligate the oligo into the plasmid in the correct direction (e.g., BamHI and EcoRI sites) (Fig. 4). (Note: It is advisable to order this oligo to resemble the post cleavage sequence left after restriction digestion, along with a 5′-phosphate on both ends to enable easy ligation into the plasmid.)

2. Anneal the TS and NTS oligos by mixing equal molars (for example: 10 μM each) of each of the strands in 1× annealing buffer (10 mM Tris-HCl, pH 8, 50 mM NaCl). Heat the reaction to 95 °C for 2 min and then allow it to slowly cool to room temperature.

3. Phosphorylate the newly annealed DNA target insert following manufacture recommendation for the kinase enzyme. An example reaction includes mixing up to 300 pmol of 5′ ends with 1× T4 polynucleotide kinase (PNK) buffer, 1 mM ATP and 10 units of T4 PNK enzyme (New England BioLabs) to a final volume of 50 μL. Incubate the mixture at 37 °C for 30 min. Inactivate the enzyme by incubating at 65 °C for 20 min. (Note: If the oligos were ordered from IDT with phosphorylated 5′-ends, this step can be skipped.)

4. Digest 1 μg of pUC19 plasmid containing the ampicillin selection marker by adding 20 units of EcoRI (New England BioLabs), 20 units of BamHI (New England BioLabs) and NEBuffer r3.1 in a total reaction volume of 50 μL. Incubate at 37 °C for 1 h.

5. To prevent self-ligation of the linearized pUC19, dephosphorylate the product after restriction digestion by adding 5 μL of Antarctic Phosphatase Reaction Buffer 10× (New England Biolabs) and 5 units of antarctic phosphatase (New England BioLabs) directly to the 1 μg of digested plasmid. Incubate the mixture at 37 °C for 30 min. Deactivate the enzyme by incubating at 80 °C for 2 min.

6. To remove the presence of trace amounts of uncut plasmid that can produce negative colonies after ligation, extract the linearized pUC19 by resolving the digested products on a 0.8% agarose gel pre-stained with ethidium bromide. Cut out the digested band of the right size and use a gel extraction kit (Omega Biotek E.Z.N.A. Gel Extraction Kit, QIAGEN QIAquick Gel Extraction Kit) to extract the linearized plasmid from the gel.

7. Ligate the phosphorylated insert and the dephosphorylated plasmid vector following the recommendations for the specific ligase being used. For example, T4 ligase (New England BioLabs) recommends mixing 50 ng of vector with 37.5 ng of insert, 1× T4 Ligase Buffer (New England BioLabs) and 1 μL T4 DNA ligase (New England BioLabs) in a total reaction volume of 20 μL.

8. Transform 5–7 μL of the ligation reaction into 50 μL of *E. coli* DH5α cells following manufacture recommendations and plate on a Luria-Bertani-agar (LB-agar) culture plate containing 0.1 mg/mL ampicillin and grow overnight at 37 °C. Choose a few single colonies from this plate and confirm insertion of the protospacer by Sanger sequencing. Maintain glycerol and plasmid stocks of the positive clone.

9. The positive clone for the "matched/wild-type" plasmid substrate can be used as a template for site-directed mutagenesis (SDM) (Bachman, 2013) to develop the mismatched DNA substrates using a primer that introduces a single/multiple nucleotide "mismatch(es)" in the protospacer region (i.e., changing the DNA sequence to introduce non-complementarity with the guide-region of the crRNA). The mismatch positions are numbered according to the nucleotide's position downstream of the PAM on the non-target strand (NTS). Alternatively, mismatches can be introduced by changing the nt of the guide region of the crRNA and keeping the protospacer sequence constant (Zetsche et al., 2015). In this case, *in vitro* transcriptions with different crRNA template strands need to be performed to create mismatch containing crRNA.

4.1.2 Cis DNA cleavage reaction

Cas12a is a single turn over enzyme under *in vitro* conditions for *cis* DNA cleavage. Most of the protocols assessing DNA cleavage properties of Cas12a orthologs and variants uses an excess of protein compared to the DNA substrate. Recombinantly purified Cas12a protein is bound with purified and annealed crRNA and incubated with DNA substrate in an appropriate buffer containing a divalent metal ion such as Mg^{2+} for a required amount of time. A protocol for *cis* DNA cleavage is as follows.

1. A reaction volume of 10 μL will contain the following: 1× cleavage buffer (20 mM HEPES, pH 7.5, 150 mM KCl, 5% glycerol, 0.5 mM DTT), 5 mM $MgCl_2$, 25 nM Cas12a, 30 nM crRNA, (1:1.2 M ratio of Cas12a and crRNA), and 100 ng of a plasmid DNA substrate (~5 nM

for pUC19-based substrate) containing the protospacer and PAM as mentioned in Section 4.1.1. Omit Cas protein in the control sample.

2. Incubate all the reaction components except the DNA substrate at 37 °C for 10 min to form the ribonucleoprotein (RNP) complex.

3. Following this, add the DNA substrate and incubate at 37 °C for 30 min or another desired time.

4. Stop the reaction by adding the stop dye (1× composition: 50 mM EDTA, 1% SDS, 10% glycerol, 0.08% orange G).

5. Resolve the reaction products on a 1% agarose gel and visualize the DNA bands by post-staining with 100 mL of 1 μg/mL ethidium bromide solution (Fig. 6). Post-staining is recommended to resolve linear, nicked, and supercoiled DNA bands, which need to be quantified separately to characterize the DNA cleavage properties of Cas proteins. Destaining with 100 mL water following ethidium bromide staining is advisable to reduce background staining and saturating DNA band signal. The band intensities can be quantified as mentioned in Section 5 to assess the DNA cleavage properties of Cas12a protein. (Note: a user can optimize the staining and destaining protocol for ideal DNA band qualities as needed.)

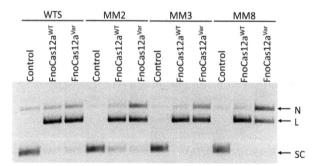

Fig. 6 Agarose gel showing *cis* DNA cleavage using a ds plasmid DNA substrate. *Cis* activity assay using a wild-type FnoCas12a (FnoCas12a^{WT}) and a variant FnoCas12a (FnoCas12a^{Var}; FnoCas12a-K969P/D970P) on a substrate with 100% complementarity with the guide sequence of the crRNA (WTS) or a substrate with a mismatch with the guide sequence of the crRNA and DNA target (MM2, MM3, MM8, numbers represent mismatch position downstream of PAM) are shown. See Fig. 4 for more details about mismatch positions. A 10 μL reaction with 100 ng of substrate, as mentioned in Section 4.1.2, was resolved on a 1% agarose gel. The control reaction was done in the absence of FnoCas12a protein. Arrows on the right show to where the nicked (N), linear (L) and supercoiled (SC) products migrated. FnoCas12a^{Var} shows lower DNA linearization compared to FnoCas12a^{WT}, which is more pronounced in mismatch containing DNA indicating the sensitivity of the variant in cleaving mismatch containing DNA.

Alternative methods to characterize Cas12a's DNA cleavage activity includes use of linearized plasmid or use of annealed target strand (TS) and non-target strand (NTS) which can be labeled with ^{32}P or fluorescein amidite (FAM) at both or one of the 5′ ends (Parameshwaran et al., 2021). Linearized substrate is created by restriction digestion of the plasmid substrate containing the protospacer matching the guide region of the crRNA flanked by a cognate PAM. Ideally, the restriction enzyme should be selected such that double stranded cleavage by Cas12a will create two linear bands with distinct sizes that can be resolved by an agarose gel. Similar rationale should be used while designing TS and NTS strands for oligo DNA cleavage assay. For oligo cleavage assays, the products should be resolved on a urea-formamide gel of appropriate percentage needed to resolve the different DNA bands (Parameshwaran et al., 2021). Another modification for oligo cleavage is labeling only one of the strands or using two distinct labels for each of the two DNA oligo strands to enable monitoring of NTS and TS cleavages separately.

4.2 *Trans* DNA cleavage assay

The *trans* activity assay requires ssDNA substrates. Both single stranded M13mp18 plasmid (New England BioLabs) and a ssDNA oligo (length can vary) can be used as substrates for *trans* cleavage. In addition to the ss DNA substrate, the *trans* activity assay also requires a ssDNA or dsDNA activator. The activator has a complementary sequence to the guide-region of the crRNA, and when they pair, a conformational change is induced in the protein which opens the RuvC catalytic pocket to allow ssDNA to enter and to be processively cleaved (Swarts & Jinek, 2019). (Note: Though *trans* activity is robust with ssDNA substrates, it has been reported to occur on dsDNA and RNA substrates with different Cas12a orthologs with appropriate reaction conditions (Fuchs, Curcuru, Mabuchi, Yourik, & Robb, 2019).)

4.2.1 *Substrates*

1. Procure single stranded M13mp18 plasmid (New England BioLabs) to use as a single-stranded, circular substrate.
2. Order a 54-nt oligo from IDT to be used as a single-stranded, linear substrate. This DNA can be labeled with ^{32}P or FAM to specifically monitor the cleavage of this DNA since the reaction mix contains several other DNA strands (e.g., activator strands). (Note: the length of the DNA substrate can vary.)

4.2.2 Activators

1. For the single stranded activator, order an oligo (IDT) that contains a PAM sequence and at least 15-nt of complementarity with the guide region of the crRNA (Chen et al., 2018).
2. For the double stranded activator, order forward and reverse activator strands that contain Cas12a's cognate PAM sequence and at least 15-nt of complementarity with the crRNA (Chen et al., 2018). Anneal the strands by mixing them in a 1:1 M ratio in 1× annealing buffer (10 mM Tris-HCl, pH 8, 50 mM NaCl). Incubate at 95 °C for 2 min and allow to slowly cool to room temperature.

4.2.3 Reaction

1. A reaction volume of 10 μL will contain the following: 1× cleavage buffer (20 mM HEPES, pH 7.5, 150 mM KCl, 5% glycerol, 0.5 mM DTT), 5 mM $MgCl_2$, 25 nM Cas12a, 30 nM crRNA, (1:1.2 M ratio of Cas12a and crRNA), 30 nM of activator, and 100 ng of DNA substrate. Omit Cas protein in the control sample.
2. Incubate all the reaction components except the activator and the DNA substrate at 37 °C for 10 min to form the ribonucleoprotein (RNP) complex.
3. Following this, add the activator and DNA substrate, and incubate at 37 °C for 60 min.
4. Stop reaction by adding the stop dye (1× composition: 50 mM EDTA, 1% SDS, 10% glycerol, 0.08% orange G).
5. Resolve the reaction products on a 1% agarose gel (if using M13mp18 substrate) and visualize the DNA bands by post-staining with ethidium bromide as mentioned in Section 4.1.2 (Fig. 7). If using oligo DNA, resolve the products on an acrylamide gel appropriate for the size of the ssDNA substrate and visualize product formation by monitoring ^{32}P or FAM or another appropriate label's signal intensity. The band intensities can be quantified as mentioned in Section 5 to characterize the *trans* DNA cleavage of Cas12a.

4.3 RNA-independent DNA cleavage assay

Several Cas proteins can perform non-specific DNA cleavage through the RNA-independent DNA cleavage in the absence of a cognate crRNA/tracrRNA (*trans* activating crRNA). The types of DNA substrates that can be cleaved (e.g., ssDNA, dsDNA, plasmid, linear DNA) varies between different Cas proteins and the divalent metal ions that can promote the

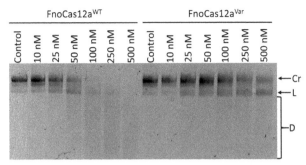

Fig. 7 A *trans* activity assay performed with increasing concentrations of wild-type FnoCas12a (FnoCas12a^WT) or a variant FnoCas12a (FnoCas12a^Var; FnoCas12a-K969P/D970P) complexed with crRNA (RNP). Activator was 24 bp of dsDNA. A 10 μL reaction with 100 ng of single-stranded substrate, as mentioned in Section 4.2.3, was resolved on a 1% agarose gel. Arrows on the right show to where the circular (Cr), linear (L), and degraded (D) products migrated. In the example shown, FnoCas12a^WT has robust *trans* cleavage activity, while it is reduced in the variant (FnoCas12a^Var).

RNA-independent DNA cleavage also varies between different Cas proteins and orthologs (Li et al., 2020; Sundaresan et al., 2017). So, it is advisable to test different DNA substrates and different divalent ions while testing this activity in a Cas protein.

4.3.1 Substrates

Procure M13mp18 single-stranded circular plasmid (New England BioLabs), double-stranded plasmid (e.g., pUC19), linearized ss and ds plasmids using appropriate restriction enzymes, PCR amplified dsDNA of appropriate lengths that need to be tested, or oligos of appropriate length from a vendor that can be used for ssDNA or annealed to create short dsDNA substrates.

4.3.2 Reaction

1. A reaction volume of 10 μL will contain the following: 1× buffer (20 mM HEPES, pH 7.5, 150 mM KCl, 2 mM TCEP), 100 nM purified Cas12a, 100 ng DNA substrate and 10 mM divalent metal.
2. Omit Cas12a in control sample to obtain the intensity of the uncleaved substrate. While using ds plasmid substrate (e.g., pUC19), create nicked and linear markers by digesting pUC19 with Nt.*Bsp*QI (one nicking site in pUC19) and *Eco*RI (one linearization site in pUC19) restriction enzymes respectively.
3. Incubate the reaction mix at 37 °C for 30 min.

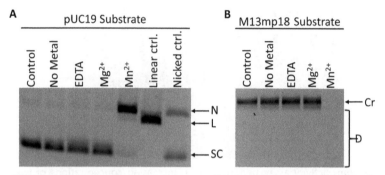

Fig. 8 RNA-independent activity assay performed with wild-type FnoCas12a (FnoCas12a^{WT}) with no added divalent metal or in the presence of EDTA, Mg^{2+}, or Mn^{2+} using a dsDNA plasmid [pUC19, (A)] or a ss DNA plasmid [M13mp18, (B)]. A 10 μL reaction with 100 ng of substrate, as mentioned in Section 4.3.2, was resolved on a 1.2% agarose gel. The control reaction was done in the absence of FnoCas12a protein. The nicked and linear controls were made by digesting pUC19 plasmid with Nt. BspQI and EcoRI respectively. Arrows on the sides show to where the nicked (N), linear (L), supercoiled (SC), circular (Cr), and degraded (D) products migrated. The gels show that in FnoCas12a, Mn^{2+} induces nicking of dsDNA plasmid and degradation of ssDNA when guide-RNA is absent.

4. Stop reaction with 1× stop dye (50 mM EDTA, 1% SDS, 10% glycerol, 0.08% *w/v* Orange G).
5. Resolve the reaction products on a 1% agarose gel and post stain with ethidium bromide (Fig. 8).

4.4 crRNA processing assay

Cas12a proteins have the intrinsic ability to process their crRNAs with an active site in the WED domain that is independent of the RuvC active site (Fonfara et al., 2016; Zetsche et al., 2017). Since crRNA processing does not involve the RuvC active site, this reaction can be performed with a "dead" Cas12a (dCas12a) that has an inactive RuvC site or a Cas12a protein with a functional RuvC active site. The following protocol is from the publication by Swarts et al. (2017) in *Molecular Cell*.

4.4.1 RNA substrates

Cas12a processes pre-crRNA into mature crRNAs and there is direct experimental evidence of Cas12a trimming the 5′ end of pre-crRNA (Safari et al., 2019). While the catalytic mechanism by which Cas12a processes pre-crRNA is metal-independent, it has been shown that divalent metal ions increase Cas12a's affinity for crRNA by associating with the crRNA pseudoknot (Dong et al., 2016; Fonfara et al., 2016; Swarts et al., 2017). Possible RNA substrates include a wild-type pre-crRNA of appropriate length and sequence for the specific Cas12a ortholog, and a mature crRNA as a control for the size of processed crRNA. Other forms such as a pre-crRNA where the $2'$-hydroxyl group upstream of the trimming site is disrupted to prevent its nucleophilic attack on the scissile phosphate can be used for mechanistic characterization of RNA cleavage. These RNA substrates can be produced using the protocol from Section 2 or through custom synthesis by vendors for modified RNAs, for example without the $2'$-hydroxyl group.

4.4.2 Reaction

1. A reaction volume of 20 μL will contain the following: 2.5 μM purified Cas12a, 0.5× SEC buffer (1× SEC buffer, 20 mM HEPES-KOH, pH 7.5, 500 mM KCl, 1 mM DTT), 1 μM crRNA substrate, and 5 mM of a chelating agent [e.g., EDTA, EGTA (ethylene glycol-bis (β-aminoethyl ether)-N,N,N′,N′-tetraacetic acid)] or divalent cation (e.g., $CaCl_2$, $MgCl_2$, $MnCl_2$, $NiCl_2$). Incubate at 37 °C for 10 min.
2. Stop the reaction by adding 80 mM EDTA and 0.8 mg/mL proteinase K, and incubate at 37 °C for 30 min. (Note: EGTA is better at chelating Ca^{2+} compared to EDTA.)
3. Add 20 μL of a 2× RNA loading dye, incubate at 95 °C for 10 min and then resolve on a 7 M urea, 20% polyacrylamide gel.

Alternatively, fluorophores can be attached to the $5'$ or $3'$ end of the pre-crRNA substrate, and the reaction products can be resolved on a denaturing gel and fluorescence detected with a gel imager (Nguyen et al., 2020).

4.5 Time course activity assay

An activity assay where the reaction is stopped at progressive time points results in data that can be used to calculate rate constants of the reactions. The setup is the same as in the cleavage assays mentioned above except for stopping the reaction at the required time points.

1. Prepare a 10 μL reaction for each time point of interest (an example for time points is 0 s to 1 h, with a few time points in seconds (15 and 30 s) and the rest in minutes that are appropriate to capture the reaction kinetics) (Parameshwaran et al., 2021).
2. Stop the reaction at the desired time point with 1× stop dye.
3. Resolve the reaction products on an appropriate gel, and image the gels with an appropriate method (Fig. 9).

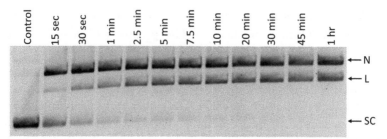

Fig. 9 Time course *cis* DNA cleavage assay performed with FnoCas12a^{WT} and a double-stranded substrate with a mismatch in position 8 (MM8). A 10 µL reaction with 100 ng of substrate, as mentioned in Section 4.5, was resolved on a 1% agarose gel. Arrows on the right show to where the nicked (N), linear (L) and supercoiled (SC) products migrated. Nicked DNA is the intermediate in the reaction and linear DNA is the final product. Note that the amount of nicked DNA raises first and then decreases when it is converted to linear product at longer time points.

4.6 Concentration titration assay

While characterizing a new Cas protein ortholog or a variant, it is advisable to test different concentrations of the protein-RNA complex to analyze differences in reaction saturation conditions when compared to a well characterized ortholog or the wild-type protein. The setup is the same as in the cleavage assays mentioned above except for using different concentrations of RNP or other reaction components (e.g., concentration of divalent metal, activator for *trans* cleavage).

1. Prepare a 10 µL reaction for each concentration point of interest (examples include protein titration from 0 to 1 µM; divalent metal analysis at 1, 5, and 10 mM) (Parameshwaran et al., 2021).
2. Stop the reaction at the desired time point with 1× stop dye.
3. Resolve the reaction products on a gel and image appropriately (Fig. 7).

5. Quantification of Cas12a activity

Visual identification of the activity assay gels is not accurate enough to form conclusions. Quantification of DNA bands corresponding to substrates and products can provide accurate measurements of Cas12a activities. The DNA bands can be quantified using the ImageJ program (Rasband, 1997) and then used in calculations with normalization against the no protein control lane to obtain quantitative measurement of activity. To avoid experimental errors, it is important to include control lanes for each reaction to account for any variation in reaction components. Similarly, to account for differences in active protein fractions, it is recommended that different replications be performed for each experiment using at least two independent protein preparations, for example, three reaction replicates using two independent protein preparations. The following protocol and calculations refer to the possible quantification products produced by the activity assays.

For *cis* cleavage using a plasmid substrate, there are three different measurements that need to be taken for quantification: nicked (N) product [refers to a double-stranded circular DNA substrate that has undergone cleavage of only one strand, which is an intermediate during the linearization of supercoiled DNA], linear (L) product [refers to a double-stranded DNA substrate that has undergone cleavage of both strands], leftover supercoiled (SC) product (refers to uncut dsDNA substrate that was unused in the reaction).

For activity assays measuring *trans* cleavage using single-stranded circular DNA substrates, the gel will have the following forms of DNA: uncut circular (Cr, leftover unused substrate), linear (L, intermediate where DNA is linearized, but not progressively degraded) or degraded product (D, DNA substrate that has undergone progressive degradation). For *trans* cleavage quantification, following the intensity reduction of circular band is ideal for quantification.

5.1 *Cis* cleavage assay with dsDNA plasmid

1. For a double-stranded plasmid substrate with multiple cleavage products, measure intensities (I) of nicked (I_N), linear (I_L) and supercoiled (I_{SC}) bands using ImageJ.
2. Perform a background correction as recommended by the ImageJ software.
3. Subtract the intensities of nicked and linear DNA bands from the control lane (i.e., no protein lane) to account for the presence of nicked and linear bands in the supercoiled substrate by using the following equations:

$$Nicked = \left[\frac{I_N}{I_N + I_L + I_{SC}} - \left(\frac{I_N}{I_N + I_L + I_{SC}} \right)_C \right] \tag{1}$$

$$Linear = \left[\frac{I_L}{I_N + I_L + I_{SC}} - \left(\frac{I_L}{I_N + I_L + I_{SC}} \right)_C \right] \tag{2}$$

where the subscript "C" represents the intensities of the control lane containing no Cas protein.

4. Calculate total cleavage as:

$$Total\ cleavage = Nicked + Linear \tag{3}$$

5. Calculate the fraction of remaining uncleaved (supercoiled) substrate (Frac[SC]) as:

$$Frac[SC] = \left[\left(\frac{I_{SC}}{I_N + I_L + I_{SC}} \right) \bigg/ \left(\frac{I_{SC}}{I_N + I_L + I_{SC}} \right)_C \right] \tag{4}$$

where the subscript "C" represents the intensities of the control lane containing no Cas protein. An alternate method to calculate remaining substrate is [1-(nicked + linear)] for each lane (see reference Parameshwaran et al., 2021 for details).

6. Calculate the standard deviation (SD) as:

$$SD = \sqrt{\Sigma\big((R - R_{AV})^2 \div (n-1)\big)} \tag{5}$$

where R is the value from each replication, R_{AV} is the average of data values from all the replications, and n is the number of replications. Note that R depends on what is being plotted (e.g., total cleavage (Eq. 3), linear product formation (Eq. 2), or monitoring the usage of supercoiled substrate (Eq. 4)).

7. Calculate the standard error of mean (SEM) as:

$$SEM = SD \div \sqrt{n} \tag{6}$$

8. Plot appropriate graphs (bar or line) for visual representation of the data. Bar graph is used for single time or concentration points, while line graphs are used for time or concentration gradients.

Nicked, linear, or total cleavage can be used for plotting as required to analyze the results from activity assays. This can be useful while characterizing Cas12a variants or orthologs that shows differences in linearization and nicking compared to already characterized Cas12a proteins. For off-target analysis, reduction in linearization, while maintaining efficient nicking, is considered to have better mismatch sensitivity since nicked products are repaired during genome editing applications.

For modified *cis* cleavage assays that use labeled oligo DNA or linearized plasmid DNA as substrates, there are slight variations in the products being quantified. Details of these calculations can be referred from our Cas12a publication that tested different types of DNA substrates for *cis* cleavage (Parameshwaran et al., 2021).

5.2 *Trans* cleavage assay with ssDNA

For single stranded DNA where the DNA is degraded, calculate the fraction of remaining uncleaved (circular) substrate (Frac[CIR]) as:

$$Frac[CIR] = \frac{I_{CIR}}{(I_{CIR})_C} \tag{7}$$

where I_{CIR} is the intensity of the uncleaved (circular) band left in the reaction with protein, and the subscript "C" represents the intensity of the uncleaved band in the control lane containing no Cas protein. SD and SEM can be calculated as mentioned in Eqs. (5) and (6).

6. Kinetic analysis

The data from time course assays can be fit with appropriate equations to obtain reaction rate constants. The choice of single exponential or double exponential fit depends on the properties of each protein being tested and can also vary based on experimental parameters. Details of different equations that can be used for kinetic analysis of Cas protein assays have been recently reported (Babu et al., 2019; Parameshwaran et al., 2021). Other kinetic analysis methods have been reported using oligo assays to monitor conformational changes along the reaction pathway by careful manipulation of the reaction conditions (Liu et al., 2020; Raper, Stephenson, & Suo, 2018; Strohkendl, Saifuddin, Rybarski, Finkelstein, & Russell, 2018). Equations for single and double exponential fit of the time course data is as follows:

Single-exponential decay equation:

$$y = 1 - a \cdot [1 - \exp{(-k_{obs} \cdot t)}] \tag{8}$$

y is the measured parameter (for example: disappearance of precursor), k_{obs} is the reaction rate constant, *t* is time and *a* is the total active fraction (*i.e.,* the amount of substrate being used up in the reaction).

Double-exponential decay equation:

$$Frac[SC] = 1 - a1 \cdot [1 - \exp{(-k_1 \cdot t)}] - a2 \cdot [1 - \exp{(-k_2 \cdot t)}] \tag{9}$$

Frac[SC] is specified by Eq. (4), k_1 and k_2 are the reaction rate constants, *t* is time and *a*1 and *a*2 are the fraction of supercoiled substrate that reacted respectively with the k_1 and k_2 rate constants. The total active fraction $a = a1 + a2$.

Time-course measurements can be fit with OriginLab (Origin(Pro) (Version 2022b), 2021) or MATLAB (MATLAB and Statistics Toolbox (Version 2021b), 2021) graphing and analysis software.

7. Conclusions

Cas12a is a versatile protein that possesses diverse activities: both DNA and RNA cleavages. DNA cleavage itself can occur by diverse reaction mechanisms that have led to unique biotechnology tool developments using Cas12a. Thus Cas12a provides a unique system for not only improved biotechnology and biomedical applications, but also characterization of unique reaction mechanisms from a fundamental mechanistic perspective. The protocols mentioned in this report can be easily adapted to different Cas12a orthologs and variants for their biochemical characterization.

Acknowledgments

The kinetic analysis methods were developed through our collaborative work with Dr. Peter Z. Qin at the University of Southern California. We thank Dr. Qin and his laboratory for kinetic analysis methods that are being reported in this methods manuscript. We thank the use of OU Protein Production and Characterization Core (PPC Core) for protein purification and instrument support. This core is supported by an IDeA grant from the NIGMS (grant number P20GM103640).

Conflicts of interest

Two US patents have been filed for a bridge helix variant of FnoCas12a protein with the patent numbers, US20200332275A1 and US20220213459A1. The authors declare no other competing financial interest.

Funding

The work reported here was partially supported by the Oklahoma Center for the Advancement of Science and Technology (OCAST) award [grant number HR20–103, awarded to RR] and partially by a grant from the Research Council and the Office of the Vice President for Research and Partnerships of the University of Oklahoma Norman Campus to RR.

References

Ausubel, F. M., Brent, R., Kingston, R. E., Struhl, K., Smith, J. A., Moore, D. D., et al. (1987). *Current protocols in molecular biology*. Greene Pub. Associates, J. Wiley. order fulfillment.

Babu, K., Amrani, N., Jiang, W., Yogesha, S. D., Nguyen, R., Qin, P. Z., et al. (2019). Bridge helix of Cas9 modulates target DNA cleavage and mismatch tolerance. *Biochemistry*, *58*(14), 1905–1917. https://doi.org/10.1021/acs.biochem.8b01241.

Bachman, J. (2013). Site-directed mutagenesis. *Methods in Enzymology*, *529*, 241–248. https://doi.org/10.1016/B978-0-12-418687-3.00019-7.

Bayat, H., Modarressi, M. H., & Rahimpour, A. (2018). The conspicuity of CRISPR-Cpf1 system as a significant breakthrough in genome editing. *Current Microbiology*, *75*(1), 107–115. https://doi.org/10.1007/s00284-017-1406-8.

Beckert, B., & Masquida, B. (2011). Synthesis of RNA by in vitro transcription. *Methods in Molecular Biology (Clifton, N.J.)*, *703*, 29–41. https://doi.org/10.1007/978-1-59745-248-9_3.

Broughton, J. P., Deng, X., Yu, G., Fasching, C. L., Servellita, V., Singh, J., et al. (2020). CRISPR–Cas12-based detection of SARS-CoV-2. *Nature Biotechnology*, *38*(7), 870–874. https://doi.org/10.1038/s41587-020-0513-4.

Chen, J. S., Ma, E., Harrington, L. B., Da Costa, M., Tian, X., Palefsky, J. M., et al. (2018). CRISPR-Cas12a target binding unleashes indiscriminate single-stranded DNase activity. *Science*, *360*(6387), 436. https://doi.org/10.1126/science.aar6245.

Cofsky, J. C., Karandur, D., Huang, C. J., Witte, I. P., Kuriyan, J., & Doudna, J. A. (2020). CRISPR-Cas12a exploits R-loop asymmetry to form double-strand breaks. *eLife*, *9*. https://doi.org/10.7554/eLife.55143.

Dong, D., Ren, K., Qiu, X., Zheng, J., Guo, M., Guan, X., et al. (2016). The crystal structure of Cpf1 in complex with CRISPR RNA. *Nature*, *532*(7600), 522–526. https://doi.org/10.1038/nature17944.

Fonfara, I., Richter, H., Bratovič, M., Le Rhun, A., & Charpentier, E. (2016). The CRISPR-associated DNA-cleaving enzyme Cpf1 also processes precursor CRISPR RNA. *Nature*, *532*(7600), 517–521. https://doi.org/10.1038/nature17945.

Fuchs, R. T., Curcuru, J., Mabuchi, M., Yourik, P., & Robb, G. B. (2019). Cas12a trans-cleavage can be modulated in vitro and is active on ssDNA, dsDNA, and RNA. *bioRxiv*, 600890. https://doi.org/10.1101/600890.

Gao, P., Yang, H., Rajashankar, K. R., Huang, Z., & Patel, D. J. (2016). Type V CRISPR-Cas Cpf1 endonuclease employs a unique mechanism for crRNA-mediated target DNA recognition. *Cell Research*, *26*(8), 901–913. https://doi.org/10.1038/cr.2016.88.

Karvelis, T., Gasiunas, G., Miksys, A., Barrangou, R., Horvath, P., & Siksnys, V. (2013). CrRNA and tracrRNA guide Cas9-mediated DNA interference in Streptococcus thermophilus. *RNA Biology*, *10*(5), 841–851. https://doi.org/10.4161/rna.24203.

Kim, S. K., Kim, H., Ahn, W.-C., Park, K.-H., Woo, E.-J., Lee, D.-H., et al. (2017). Efficient transcriptional gene repression by type V-A CRISPR-Cpf1 from Eubacterium eligens. *ACS Synthetic Biology*, *6*(7), 1273–1282. https://doi.org/10.1021/acssynbio.6b00368.

Li, S.-Y., Cheng, Q.-X., Liu, J.-K., Nie, X.-Q., Zhao, G.-P., & Wang, J. (2018). CRISPR-Cas12a has both cis- and trans-cleavage activities on single-stranded DNA. *Cell Research*, *28*(4), 491–493. https://doi.org/10.1038/s41422-018-0022-x.

Li, S.-Y., Cheng, Q.-X., Wang, J.-M., Li, X.-Y., Zhang, Z.-L., Gao, S., et al. (2018). CRISPR-Cas12a-assisted nucleic acid detection. *Cell Discovery*, *4*(1), 1–4. https://doi.org/10.1038/s41421-018-0028-z.

Li, L., Li, S., Wu, N., Wu, J., Wang, G., Zhao, G., et al. (2019). HOLMESv2: A CRISPR-Cas12b-assisted platform for nucleic acid detection and DNA methylation quantitation. *ACS Synthetic Biology*, *8*(10), 2228–2237. https://doi.org/10.1021/acssynbio.9b00209.

Li, B., Yan, J., Zhang, Y., Li, W., Zeng, C., Zhao, W., et al. (2020). CRISPR-Cas12a possesses unconventional DNase activity that can be inactivated by synthetic oligonucleotides. *Molecular Therapy- -Nucleic Acids*, *19*, 1043–1052. https://doi.org/10.1016/j.omtn.2019.12.038.

Liu, M.-S., Gong, S., Yu, H.-H., Jung, K., Johnson, K. A., & Taylor, D. W. (2020). Engineered CRISPR/Cas9 enzymes improve discrimination by slowing DNA cleavage to allow release of off-target DNA. *Nature Communications, 11*(1), 3576. https://doi.org/10.1038/s41467-020-17411-1.

Mahas, A., Hassan, N., Aman, R., Marsic, T., Wang, Q., Ali, Z., et al. (2021). LAMP-coupled CRISPR–Cas12a module for rapid and sensitive detection of plant DNA viruses. *Viruses, 13*(3), 466. https://doi.org/10.3390/v13030466.

Maizel, J., & Jacob, V. (2000). SDS polyacrylamide gel electrophoresis. *Trends in Biochemical Sciences, 25*(12), 590–592. https://doi.org/10.1016/S0968-0004(00)01693-5.

MATLAB and Statistics Toolbox (Version 2021b). (2021). The MathWorks, Inc.

Mir, A., Edraki, A., Lee, J., & Sontheimer, E. J. (2018). Type II-C CRISPR-Cas9 biology, mechanism and application. *ACS Chemical Biology, 13*(2), 357–365. https://doi.org/10.1021/acschembio.7b00855.

Mohanraju, P., Van Der Oost, J., Jinek, M., & Swarts, D. C. (2018). Heterologous expression and purification of the CRISPR-Cas12a/Cpf1 protein. *Bio-Protocol, 8*(9), e2842–e2842.

Nguyen, L. T., Smith, B. M., & Jain, P. K. (2020). Enhancement of trans-cleavage activity of Cas12a with engineered crRNA enables amplified nucleic acid detection. *Nature Communications, 11*(1), 4906. https://doi.org/10.1038/s41467-020-18615-1.

Origin(Pro) (Version 2022b). (2021). OriginLab Corporation.

Parameshwaran, H. P., Babu, K., Tran, C., Guan, K., Allen, A., Kathiresan, V., et al. (2021). The bridge helix of Cas12a imparts selectivity in cis-DNA cleavage and regulates trans-DNA cleavage. *FEBS Letters, 595*(7), 892–912. https://doi.org/10.1002/1873-3468.14051.

Paul, B., & Montoya, G. (2020). CRISPR-Cas12a: Functional overview and applications. *Biomedical Journal, 43*(1), 8–17. https://doi.org/10.1016/j.bj.2019.10.005.

Raper, A. T., Stephenson, A. A., & Suo, Z. (2018). Functional insights revealed by the kinetic mechanism of CRISPR/Cas9. *Journal of the American Chemical Society, 140*(8), 2971–2984. https://doi.org/10.1021/jacs.7b13047.

Raran-Kurussi, S., Cherry, S., Zhang, D., & Waugh, D. S. (2017). Removal of affinity tags with TEV protease. *Methods in Molecular Biology, 1586*, 221–230. https://doi.org/10.1007/978-1-4939-6887-9_14.

Rasband, W. S. (1997). *ImageJ.* U.S. National Institutes of Health. https://imagej.nih.gov/ij/.

Rio, D. C. (2013). Expression and purification of active recombinant T7 RNA polymerase from *E. coli. Cold Spring Harbor Protocols, 2013*(11), pdb.prot078527. https://doi.org/10.1101/pdb.prot078527.

Safari, F., Zare, K., Negahdaripour, M., Barekati-Mowahed, M., & Ghasemi, Y. (2019). CRISPR Cpf1 proteins: Structure, function and implications for genome editing. *Cell & Bioscience, 9*(1), 36. https://doi.org/10.1186/s13578-019-0298-7.

Saha, C., Mohanraju, P., Stubbs, A., Dugar, G., Hoogstrate, Y., Kremers, G.-J., et al. (2020). Guide-free Cas9 from pathogenic Campylobacter jejuni bacteria causes severe damage to DNA. *Science Advances, 6*(25), eaaz4849. https://doi.org/10.1126/sciadv.aaz4849.

Shmakov, S., Smargon, A., Scott, D., Cox, D., Pyzocha, N., Yan, W., et al. (2017). Diversity and evolution of class 2 CRISPR-Cas systems. *Nature Reviews. Microbiology, 15*(3), 169–182. https://doi.org/10.1038/nrmicro.2016.184.

Stella, S., Alcón, P., & Montoya, G. (2017). Structure of the Cpf1 endonuclease R-loop complex after target DNA cleavage. *Nature, 546*(7659), 559–563. https://doi.org/10.1038/nature22398.

Stella, S., Mesa, P., Thomsen, J., Paul, B., Alcón, P., Jensen, S. B., et al. (2018). Conformational activation promotes CRISPR-Cas12a catalysis and resetting of the endonuclease activity. *Cell, 175*(7), 1856–1871. e21 https://doi.org/10.1016/j.cell.2018.10.045.

Strohkendl, I., Saifuddin, F. A., Rybarski, J. R., Finkelstein, I. J., & Russell, R. (2018). Kinetic basis for DNA target specificity of CRISPR-Cas12a. *Molecular Cell, 71*(5), 816–824.

Sundaresan, R., Parameshwaran, H. P., Yogesha, S. D., Keilbarth, M. W., & Rajan, R. (2017). RNA-independent DNA cleavage activities of Cas9 and Cas12a. *Cell Reports*, *21*(13), 3728–3739. https://doi.org/10.1016/j.celrep.2017.11.100.

Swarts, D. C., & Jinek, M. (2019). Mechanistic insights into the cis- and trans-acting DNase activities of Cas12a. *Molecular Cell*, *73*(3), 589–600. e4. https://doi.org/10.1016/j.molcel.2018.11.021.

Swarts, D. C., van der Oost, J., & Jinek, M. (2017). Structural basis for guide RNA processing and seed-dependent DNA targeting by CRISPR-Cas12a. *Molecular Cell*, *66*(2), 221–233. e4. https://doi.org/10.1016/j.molcel.2017.03.016.

Wörle, E., Jakob, L., Schmidbauer, A., Zinner, G., & Grohmann, D. (2021). Decoupling the bridge helix of Cas12a results in a reduced trimming activity, increased mismatch sensitivity and impaired conformational transitions. *Nucleic Acids Research*, *49*(9), 5278–5293. https://doi.org/10.1093/nar/gkab286.

Yamano, T., Nishimasu, H., Zetsche, B., Hirano, H., Slaymaker, I. M., Li, Y., et al. (2016). Crystal structure of Cpf1 in complex with guide RNA and target DNA. *Cell*, *165*(4), 949–962. https://doi.org/10.1016/j.cell.2016.04.003.

Yamano, T., Zetsche, B., Ishitani, R., Zhang, F., Nishimasu, H., & Nureki, O. (2017). Structural basis for the canonical and non-canonical PAM recognition by CRISPR-Cpf1. *Molecular Cell*, *67*(4), 633–645. e3. https://doi.org/10.1016/j.molcel.2017.06.035.

Yan, W. X., Hunnewell, P., Alfonse, L. E., Carte, J. M., Keston-Smith, E., Sothiselvam, S., et al. (2019). Functionally diverse type V CRISPR-Cas systems. *Science*, *363*(6422), 88–91. https://doi.org/10.1126/science.aav7271.

Zetsche, B., Gootenberg, J. S., Abudayyeh, O. O., Slaymaker, I. M., Makarova, K. S., Essletzbichler, P., et al. (2015). Cpf1 is a single RNA-guided endonuclease of a Class 2 CRISPR-Cas system. *Cell*, *163*(3), 759–771. https://doi.org/10.1016/j.cell.2015.09.038.

Zetsche, B., Heidenreich, M., Mohanraju, P., Fedorova, I., Kneppers, J., DeGennaro, E. M., et al. (2017). Multiplex gene editing by CRISPR–Cpf1 using a single crRNA array. *Nature Biotechnology*, *35*(1), 31–34. https://doi.org/10.1038/nbt.3737.

Zoephel, J., Dwarakanath, S., Richter, H., Plagens, A., & Randau, L. (2012). Substrate generation for endonucleases of CRISPR/Cas systems. *Journal of Visualized Experiments: JoVE*, *67*, 4277. https://doi.org/10.3791/4277.

CHAPTER FIVE

Proximity Labeling and Proteomics: Get to Know Neighbors

Norihiro Kotani[a,b,*], Tomoyuki Araki[b], Arisa Miyagawa-Yamaguchi[c], Tomoko Amimoto[d], Miyako Nakano[e], and Koichi Honke[c]

[a]Medical Research Center, Saitama Medical University, Saitama, Japan
[b]Department of Biochemistry, Saitama Medical University, Saitama, Japan
[c]Department of Biochemistry, Kochi University Medical School, Nankoku, Japan
[d]Natural Science Center for Basic Research and Development, Hiroshima University, Higashi-Hiroshima, Japan
[e]Graduate School of Integrated Sciences for Life, Hiroshima University, Higashi-Hiroshima, Hiroshima, Japan
*Corresponding author: e-mail address: kotani@saitama-med.ac.jp

Contents

Abstract

Protein–protein interactions are essential in biological reactions and fundamental to cell–cell communication (e.g., the binding of secreted proteins, such as hormones, to cell membrane receptors) and the subsequent intracellular signal transduction cascade. Several studies have been extensively carried out on protein–protein interactions because they have the potential to resolve various problems in molecular biology. Biochemical methods, such as chemical cross-linking and immunoprecipitation, have long been used to analyze which proteins interact with each other. However, there are some problems, such as unphysiological states and non-specific binding, that require the development

Methods in Enzymology, Volume 679
ISSN 0076-6879
https://doi.org/10.1016/bs.mie.2022.07.031
131

of more useful experimental methods. This chapter discusses the "proximity labeling (Proteomics)" analysis technique, which has been attracting attention in protein–protein interaction analysis in recent years and is used in many biological studies. "Membrane proximity labeling (proteomics)," which analyzes the interaction of cell membrane proteins, and "intracellular proximity labeling (proteomics)" will be explained in-depth.

1. Introduction

Proteins are one of the basic components of tissues and cells. Sometimes they do not function alone and instead contribute to various biological functions by interacting with each other (protein–protein interactions) or other components, such as lipids and nucleic acids. Among them, protein–protein interactions are an indispensable event in biological reactions (Frieden, 1971; Stelzl & Wanker, 2006). Studies have found that several molecules involved in many signaling pathways specifically interact in various organelles (Virkamäki, Ueki, & Kahn, 1999; Wodak, Vlasblom, Turinsky, & Pu, 2013).

In vivo protein–protein interactions occur in two major environments: the cell membrane surface or intracellular (extracellular free) space. The former is a state in which at least one protein is immobilized in the cell membrane, and the latter is a state in which two or more protein groups specifically interact with each other by free movement in the fluid. The physical condition is critically different between these two environments. Therefore, they should be analyzed using separate strategies. Until now, the chemical cross-linking (Lutter & Kurland, 1975) and immunoprecipitation method has been used as the main protein–protein interaction analysis form. However, it has been used in both environments. The immunoprecipitates contain non-interacting molecules recovered by non-specific binding in both environments. More non-specific molecules are detected in interaction analyses on the cell membrane surface. In addition, some researchers have doubts whether physiological protein–protein interactions are observed because the immunoprecipitation method occurs under a state where target biological materials, such as tissues and cells, are mostly homogenized. Hence, we endeavored to develop a novel strategy with as few non-specific absorptions as possible, resulting in greater detection of more physiological protein–protein interactions.

In addition to the immunoprecipitation method described above, "detergent-insoluble membrane analysis (DIM)" has been frequently used as the interaction assay on the cell membrane surface. In the case of interactions on the cell membrane, a molecular complex, referred to as a "lipid raft (microdomain)," may form (Lingwood & Simons, 2010; Vereb et al., 2003). It functions as a platform for important biological phenomena, such as signal transduction and intracellular protein transport (Brown & London, 1998; Simons & Toomre, 2000). A low-density detergent-insoluble fraction that floats when the cell is homogenized with a neutral detergent followed by sucrose density gradient ultracentrifugation is considered to contain a lipid raft fraction and can be collected and analyzed (Brown & Rose, 1992). This method is simple and has been widely used, but it still requires a new strategy for the interaction assay on the cell membrane surface. This is because:

(i) Irrelevant membrane proteins are often recovered together with the cell membrane and skeletal structure in this fraction

(ii) The protein–protein interaction on the cell membrane depends on the fluidity of the lipid bilayer. Therefore, the analysis should be performed under more physiological conditions

(iii) Each lipid raft structure is composed of interactions between different molecules. Hence, accurate information about the protein–protein interaction cannot be obtained if they are recovered in a lump sum by DIM (Fig. 1)

In 2008, we found that the aryl azide compound, typically used as a cross-linking reagent, was converted to aryl nitrene radicals (active form)

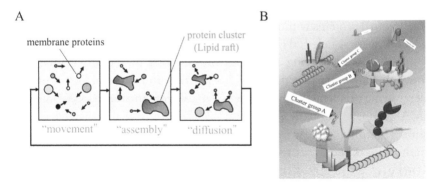

Fig. 1 Summary of membrane protein clusters (lipid raft) on the cell surface. (A) The model displays how protein clusters are formed in the membrane. Membrane proteins are constantly moving around on the cell membrane (movement), and some molecules form protein complexes, resulting in biological functions (assembly). Complex formation is reversible, and each membrane protein diffuses and disperses again. (B) Each cluster contains different membrane proteins. Therefore, if each cluster is collected and analyzed together under DIM, important information regarding differences in clustered molecules cannot be observed.

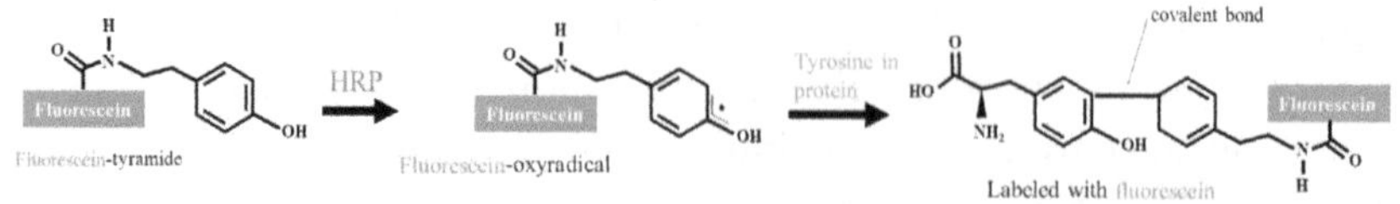

Fig. 2 EMARS radical reaction scheme. An example of an EMARS reaction with fluorescein tyramide. Part of the tyramide structure undergoes oxyradical catalyzation by HRP, and radicals covalently bind to tyrosine residues in proteins. This radical reaction proceeds under physiological conditions.

by horseradish peroxidase (HRP). This is now referred to as an "enzyme-mediated activation of radical sources" (EMARS) reaction using aryl azide and tyramide reagents (Fig. 2) (Kotani et al., 2008).

Radicals typically react with water molecules and disappear. They can only exist in very close proximity to HRP, resulting in a short lifespan. Based on the radical properties, we established a protocol to label proximity molecules around given molecules on the surface of living cells. The HRP-conjugated recognizing molecule (antibody, lectin, etc.), which binds to given molecules located on the cell surface, is needed to initiate EMARS reactions. After the HRP-conjugated recognizing molecules were treated, the cells were subsequently treated with the aryl azide or tyramide compound to generate radicals, resulting in the EMARS reaction. They reacted only with protein molecules in the proximity of HRP with covalently bonded tags (Fig. 3).

Proximal proteins within 300 nm (aryl azide reagent) or 20 nm (tyramide reagent) were labeled (Honke & Kotani, 2011, 2012). This area size is suitable for identifying molecules that interact in respective lipid raft domains, suggesting that the EMARS reaction is an important technique for resolving problems (i) to (iii) above. Typical proteomic analysis can identify EMARS-labeled proteins (Jiang et al., 2012). We have investigated several issues regarding cell surface molecular interactions using the EMARS method (Esaki et al., 2018; Hashimoto et al., 2012; Ishiura et al., 2010; Kaneko et al., 2016; Kotani, Ishiura, Yamashita, Ohnishi, & Honke, 2012; Miyagawa-Yamaguchi, Kotani, & Honke, 2014, 2015; Yamashita, Kotani, Ishiura, Higashiyama, & Honke, 2011). Interestingly, it was revealed that protein–protein interactions on the extracellular vesicle membrane could be analyzed (Kaneda et al., 2021). In addition, novel technologies using fluorophore and photochemistry instead of EMARS have recently been developed (Sjöstedt et al., 2020).

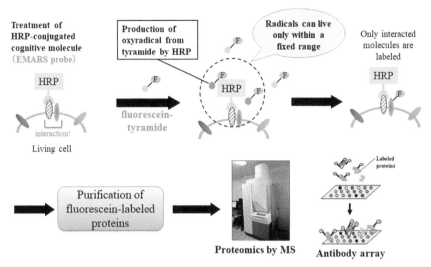

Fig. 3 Procedure for membrane proximity labeling using EMARS. The protocols for membrane proximity labeling and proteomics with fluorescein-tyramide. The HRP-conjugated antibody recognizing a given molecule you want to analyze is treated, followed by a tyramide-fluorescein treatment. The reaction proceeds under physiological conditions, and the fluorescein group attaches to proximity proteins within approximately 10 min. They are purified using anti-fluorescein antibody-Sepharose and identified via an antibody array or MS-based proteomics.

A few years after developing and publishing the EMARS method, some techniques for analyzing protein–protein interactions in the intracellular (extracellular free) space have been developed (Qin, Cho, Cavanagh, & Ting, 2021). Similar to the EMARS concept, the methods conjugate the special enzyme to a target protein molecule and then label nearby proteins via the specific enzymatic reaction. Some researchers, including our group, have suggested that the EMARS method does not work in the intracellular environment (the reason is unknown). Hence, we need methods other than EMARS for intracellular labeling when observing intracellular protein interactions in vivo.

Martell et al. (Martell et al., 2012) and Lam et al. (Lam et al., 2015) reported on an engineered monomeric peroxidase, "APEX," that, like HRP, produces phenoxy radicals to label interacting molecules in protein complexes in vivo. Roux et al. reported on the in vivo intracellular labeling of proximity molecules around biotin ligase, which is in charge of the biotinylation of *Escherichia coli* proteins (Roux, Kim, Raida, & Burke, 2012). This method using biotin ligase has been called BioID, and subsequent technologies, such as TurboID (Branon et al., 2018) and AirID (Kido et al., 2020), devised to improve labeling efficiency, are widely used for protein–protein interaction analysis.

In this chapter, we introduce experimental procedures for in vivo proximity labeling in the cell surface (cell membrane) and intracellular space. The first section introduces membrane proximity labeling, a protein–protein interaction assay on the cell surface using EMARS, and the second introduces intracellular proximity labeling, which labels proximity proteins of intracellular Rho A by AirID. Lastly, we discuss mass spectrometry-based identification of labeled proteins by EMARS for proteomics analysis of these labeling proteins.

2. Membrane proximity labeling

Here, we introduce a protocol for EMARS reactions on the cell surface, developed in 2008. As described above, it can be applied to the cell membrane and the surface of the extracellular vesicle membrane. In the latter case, it is necessary to prepare extracellular vesicles purified from the biological sample in advance. Since analyzing extracellular vesicles is practically the same as analyzing the cell surface, we will only describe the protocols applied to the cell surface. In addition to this protocol, we would appreciate you selecting a protocol suitable for your analysis by referring to papers by other research groups using EMARS (Chang et al., 2017; Imamaki et al., 2018; Ito, Honda, & Igarashi, 2018; Iwamaru et al., 2016; Killinger et al., 2022; Ohkawa et al., 2015).

2.1 Preparing EMARS probes

EMARS reactions require a tool for conjugating HRP to the target membrane protein because the basic radical reaction needs HRP. This tool has been called the EMARS probe and is usually an HRP-labeled antibody or analog thereof (e.g., lectin) recognizing the target molecule to be analyzed. Since many molecular biology experiments use HRP, it can be purchased and used if it is commercially available. In this case, it is unnecessary to prepare a special EMARS probe in advance. Here, as an example of preparing self-made EMARS probes, we introduce how to prepare the EMARS probe that recognizes close homolog of L1 (CHL1) molecules on the cell surface (Kotani et al., 2019).

2. Since making the monovalent probe requires partially treating the antibody with a reducing reagent, the antibody may sometimes lose its binding ability and may not work. Take care when selecting antibodies

2.2 EMARS reaction

Once the appropriate EMARS probe is ready, you can perform the EMARS reaction. As mentioned above, the main purpose is proximity labeling cell-surface proteins, but in some cases, it can also be applied to living organisms composed of lipid bilayers, such as extracellular vesicles. This section introduces the proximity labeling protocol of the GM1 lipid raft-resident proteins (proximal proteins around GM1) by EMARS using cholera toxin unit B (CTxB), which recognizes the glycolipid GM1 expressed on the cell membrane.

2.2.1 Key resources table

Reagent or Resource	Source	Identifier
Antibodies		
Anti-fluorescein antibody	Rockland	Cat#600-101-096
HRP-conjugated anti-goat immunoglobulin G (IgG)	Santa Cruz	Cat#sc-2020
Chemicals, Peptides, and Recombinant Proteins		
HRP-conjugated Cholera Toxin B Subunit	LIST Biological Lab.	Cat#105
Tyramine	Sigma-Aldrich	Cat#T90344
NHS-fluorescein	Thermo Fisher Scientific	Cat#46410
Dimethylacetamide	Tokyo Chemical Industry	Cat#D0641
Tris(hydroxymethyl)aminomethane	Nacalai tesque	Cat#35406-75
Nonidet P-40	Nacalai tesque	Cat#25223-04
Protease Inhibitor Cocktail	Nacalai tesque	Cat#25955-24
H_2O_2	Nacalai tesque	Cat#20779-65
TLC Silica gel 60 F254	Merck	Cat#1.05554.0001
Tyramide-fluorescein	R&D	Cat#6456/1
Biotinyl tyramide	Merck	Cat#SML2135

Continued

—cont'd

Reagent or Resource	Source	Identifier
Critical Commercial Assays		
Pierce™ BCA Protein Assay Kit	Thermo Fisher Scientific	Cat#23225
AG® 50 W-X8 Cation Exchange Resin	Bio-Rad	Cat#1421421
Experimental Models: Cell Lines		
Human HeLaS3	JCRB cell bank	JCRB9010
Other		
ChemiDoc MP image analyzer	BIO-RAD	Cat#170-8280J1

2.2.2 Materials and equipment

- Tyramide-fluorescein (Dimethylacetamide: DMAA solution)
- HRP-conjugated CTxB (CTxB-HRP) (1 µg/µL)
- PBS (pH 7.2–7.4)
- 100 mM Tris–HCl (pH 7.4)
- Plastic syringe with a 21–26 G needle
- NP-40 lysis buffer (20 mM Tris–HCl (pH 7.4), 150 mM NaCl, 5 mM ethylenediaminetetraacetic acid (EDTA), 1% NP-40, 10% glycerol, and a protease inhibitor cocktail)
- Bicinchoninic acid (BCA) protein assay kit (Pierce)
- Anti-fluorescein antibody
- HRP-conjugated anti-goat immunoglobulin G (IgG)

2.2.3 Step-by-step method details

1. Prepare (primary) cultured cells (approximately 80% of the confluent culture is recommended in a 10 cm plastic dish). In this section, we use cervical epithelioid carcinoma (HeLaS3) cells (see Section 2.2.4)
2. Wash the cells once with PBS
3. Add CTxB-HRP dissolved in an appropriate buffer (usually in PBS, 1–2 µg/mL; see Section 2.2.4)
4. Incubate at a suitable temperature (4 °C, 25 °C, or 37 °C) for 20 min (see Section 2.2.4)
5. Wash the cells three times with PBS
6. Add 0.1 mM of tyramide-fluorescein with H_2O_2 (final concentration: 0.00004% to 0.00008%) in PBS (see Section 2.2.4)

7. Incubate at a suitable temperature (4 °C, 25 °C, or 37 °C) for 10 min
8. Wash the cells three times with PBS
9. Add 300 μL of 100 mM Tris–HCl (pH 7.5) with a protease inhibitor cocktail (optional)
10. Harvest the cells using a scraper
11. Transfer the harvested cells to a microtube
12. Homogenize the cells using a disposable plastic syringe equipped with a 21–26 G needle (10–15 strokes) or Dounce homogenizer
13. Transfer the lysate to a new microtube. If you need to remove the nuclei, centrifuge at $800 \times g$ for 5 min at 4 °C, then transfer the supernatant to new tubes
14. Centrifuge at $20,000 \times g$ for 15 min at 4 °C to collect the microsome fraction containing the membrane
15. Discard the supernatant and wash the pellet (microsome fraction) once with 500 μL of 100 mM Tris–HCl (pH 7.5)
16. Solubilize the pellet with the NP-40 lysis buffer or add a chloroform/methanol solution for mass spectrometry analysis (see Section 4)
17. Measure the total protein concentration in each sample using a BCA protein assay kit (optional)
18. Apply the samples to several subsequent experiments (sodium dodecyl-sulfate polyacrylamide gel electrophoresis (SDS-PAGE), western blot, immunoprecipitation, or antibody array; see Section 2.2.4) (Fig. 4)

2.2.4 Notes

1. The attached cells in a culture dish and free cells in a microtube are both EMARS applicable because EMARS can be performed in both a cultured dish and a microtube
2. The sample cell number is arbitrary. Approximately 1×10^5–1×10^7 cells per microtube are suitable for the free cell EMARS method
3. EMARS reactions can be performed under several temperatures. We typically used 25°C (room temperature). In some cases, incubation at 4°C induces cell detachment and heavy non-specific labeling via unknown causes
4. A pilot study is needed to determine the appropriate concentration of the EMARS probe in every experiment. If you have non-specific EMARS probe binding, the EMARS probe buffer containing a small amount of a protein component (e.g., bovine serum albumin) can be used to avoid non-specific binding to the cell surface. There is, however, a possibility that the protein components will affect the mass spectrometry analysis

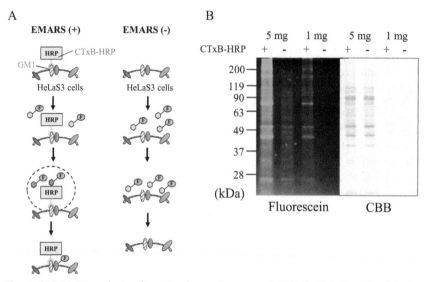

Fig. 4 SDS-PAGE analysis of proximal proteins around GM1 in HeLaS3 cells. (A) The scheme of EMARS-based membrane proximity labeling using CTxB-HRP. HeLaS3 cells were treated with CTxB-HRP and fluorescein-tyramide. The EMARS reaction does not occur in untreated HeLaS3 cells. (B) The EMARS products from CTxB-HRP-treated (+) and untreated (−) cell lysates were prepared and subjected to 10% SDS-PAGE at total protein concentrations of 1 and 5 µg, respectively. The proteins labeled with fluorescein were detected with a fluorescent imager (left column), and Coomassie brilliant blue (CBB) staining was performed for the loading control (right column).

5. Excess reagents cause cell damage and non-specific reactions. A preferable concentration of each reagent, especially H_2O_2, is needed

6. Tyramide-fluorescein can be synthesized by yourself using the following procedure:

   (a) Dissolve 100 mg NHS-fluorescein with 5 mL DMAA

   (b) Dissolve 100 mg tyramine with 5 mL DMAA

   (c) Mix above the NHS-fluorescein and tyramide solutions (total 10 mL).

   (d) Incubate the mixture at 25°C overnight with rotation

   (e) Apply to the 5 mL AG50W–X8 cation exchange resin to remove excess tyramide

   (f) Apply the aliquot of the pass-through fraction to thin-layer chromatography (TLC) to investigate whether the excess tyramide is removed

   (g) Perform TLC (chloroform:methanol:water = 65:25:4) to confirm the removal of excess tyramine (Fig. 5).

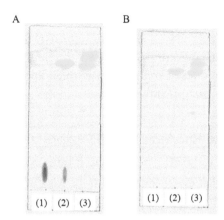

Fig. 5 TLC analysis for fluorescein-tyramide synthesis. An aliquot of the reaction mixture, NHS-fluorescein, and the tyramine standard solution before (A) and after (B) AG-50 resin treatment were subjected to TLC analysis (chloroform:methanol: water = 65:25:4). TLC plates were stained with ninhydrin to detect tyramide. The (1) tyramide standard, (2) sample mixture, and (3) NHS-fluorescein standard. The excess tyramide could be removed after AG-50 resin treatment.

    (h) After confirming the removal of tyramine, dispense the mixture solution and store at −80°C
7. The final products of EMARS can be applied to several subsequent experiments. Therefore, you should select the solvent and volume of the final buffer according to subsequent experiments
8. Both direct detection with a fluorescent imager and a western blot with an anti-fluorescein antibody can be used to identify fluorescein-labeled proteins

By the way, tyramide-biotin and -fluorescein are commercially available at, e.g., https://www.rndsystems.com/products/fluorescein-tyramide_6456 and https://www.sigmaaldrich.com/JP/en/product/sigma/sml2135

3. Intracellular proximity labeling

    Unlike the EMARS method, APEX and BioID are used for intracellular (extracellular free space) protein–protein interaction analysis. In particular, BioID, the technology using biotin ligase (BirA), has recently been used in several studies. BioID can attach biotins to proximal proteins around BirA (a distance of approximately 10 nm from BirA). TurboID and AirID with improved sensitivity have also been developed.

This section describes the protocol for identifying intracellular interacted RhoA proteins, a small GTPase protein, using AirID technology. Although Section 4 describes the proteomics of labeled proteins by proximity labeling, this section also introduces sample purification and peptide preparation for biotinylated protein mass spectrometry.

3.1 Preparing the RhoA-BirA (AirID) expression cell line

In order to perform the AirID experiment, it is necessary to prepare a BirA (AirID) expression vector capable of expressing a fusion protein of BirA with the protein to be analyzed. This section describes how to construct RhoA and BirA fusion protein expression vectors and develop the RhoA-BirA (AirID) expression cell line using synthetic DNA and the Flp-In™ System.

3.1.1 Key resources table

Reagent or Resource	Source	Identifier
Biological Samples		
E. coli. DH5alpha (Competent Quick DH5a)	TOYOBO	Cat#DNA-913F
Chemicals, Peptides, and Recombinant Proteins		
PrimeSTAR HS DNA Polymerase	Takara Bio	Cat#R010A
Ampicillin Sodium Salt, Animal-Free	Nacalai tesque	Cat#19769-64
LB agar	Sigma-Aldrich	Cat#L2897
LB medium	Sigma-Aldrich	Cat#L3022
Dulbecco's modified eagle medium (DMEM)	Nacalai tesque	Cat#08456-36
Fetal bovine serum (FBS)	EQUITECH-bio	Cat#SFBM30-0500
Hygromycin B solution (50 mg/mL)	Wako	Cat#084-07681
Critical Commercial Assays		
In-Fusion HD Cloning Kit	Takara Bio	Cat#639648
Experimental Models: Cell Lines		
Flp-In-293 Cell Line	Thermo Fisher Scientific	Cat#R75007

—cont'd

Reagent or Resource	Source	Identifier
Oligonucleotides		
Primer 1: taaacttaagcttggatgggcaagcccatc	This paper	N/A
Primer 2: gatgggcttgcccatccaagcttaagttta	This paper	N/A
Primer 3: agcctcagatctgctggtggtggtggttctggtggtg gtggttctatggctgccatccgg	This paper	N/A
Primer 4: ccggatggcagccatagaaccaccaccaccagaac caccaccaccagcagatctgaggct	This paper	N/A
Primer 5: tgccttgtcttgtgagctcgagtctagagg	This paper	N/A
Primer 6: cctctagactcgagctcacaagacaaggca	This paper	N/A
Recombinant DNA		
V5-AirID (999 bp; Synthetic DNA)	Integrated DNA Technologies	This paper (see below)
pcDNA5/FRT/TO Vector Kit	Thermo Fisher Scientific	Cat#V652020
RHOA plasmid	addgene	Plasmide#73231
pOG44 plasmid	Thermo Fisher Scientific	Cat#V652020

3.1.2 Materials and equipment
- V5–AirID (999 bp; Synthetic DNA)
- pcDNA5/FRT/TO Vector Kit
- RHOA plasmid
- pOG44 plasmid
- PCR primer (No. 1 to 6)
- PrimeSTAR HS DNA Polymerase
- Ampicillin sodium salt
- Hygromycin B solution (50 mg/mL)
- In-Fusion HD Cloning Kit

- *E. coli* DH5alpha
- LB agar-ampicillin plate
- LB-ampicillin medium
- Flp-In-293 cell Line
- DMEM with 10% FBS

3.1.3 Step-by-step method details (Fig. 6)

1. Synthesize V5-AirID DNA, which has the AirID and V5 tag sequences (see Section 3.1.4)
2. Amplify the V5-AirID sequence using PCR with primers 3 and 4
3. Amplify the pcDNA5/FRT/TO sequence using PCR with primers 1 and 2
4. Amplify the RhoA sequence using PCR with primers 5 and 6
5. Combine and clone these three DNA fragments using the In-Fusion HD Cloning Kit
6. Transform *E. coli* DH5alpha with the cloning products
7. Culture the transformed *E. coli* on the LB agar-ampicillin plate
8. Harvest colonies and amplify *E. coli* on the LB-ampicillin medium
9. Purify the plasmid containing the above three sequences (pcDNA5/FRT/TO-V5-AirID-RhoA)
10. Transfect pcDNA5/FRT/TO-V5-AirID-RhoA and pOG44 into Flp-In-293 cells
11. Culture the transfected cells in DMEM with 10% FBS
12. Add Hygromycin B solution (final concentration: 100 µg/mL medium) to select the transformed cells

3.1.4 Notes

1. The sequence of AirID BirA is based on the paper below:
   Kido et al. AirID, a novel proximity biotinylation enzyme, for analysis of protein–protein interactions (Kido et al., 2020)
2. The V5 tag sequence is atgggcaagcccatccccaaccccctgctgggcctggacagcacc
3. The linker sequence between RhoA and AirID is ggtggtggtggttctgg tggtggtggttct (Gly4-Ser-Gly4-Ser)
4. The pcDNA5/FRT/TO-V5-AirID-RhoA plasmid map:

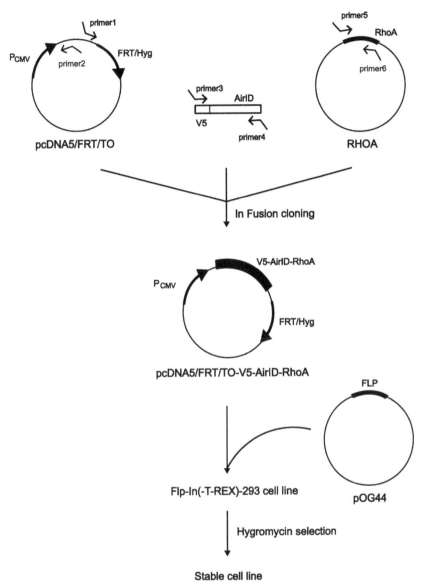

Fig. 6 Flow chart of V5-AirID-expressing stable cell lines using In-Fusion and Flp-In system. The pcDNA5/FRT/TO-based plasmid was constructed using the AirID biotin ligase and V5 tag sequences. The constructed plasmid was transfected to Flp-In 293 cells typically used for the In-fusion system. Green fluorescent protein (GFP)-expressing Flp-In-293 cells were generated by the same method as V5-AirID-RhoA-expressing cells.

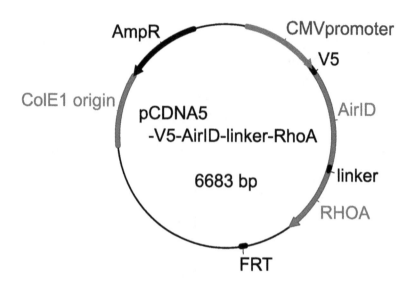

3.2 Proximity labeling using AirID technology

Here, we introduce the AirID experiment procedure using the 293 cells expressing the V5-BirA-RhoA protein prepared above. The proteins located in the RhoA proximity area of the cell may be biotinylated (Fig. 7). The STEP-BY-STEP protocols describe AirID labeling, biotinylated protein purification, and the peptide recovery step after trypsin digestion.

3.2.1 Key resources table

Reagent or Resource	Source	Identifier
Antibodies and Affinity beads		
Rabbit anti-V5 affinity purified antibody	Bethyl labolatories	Cat#A190–120A
Anti-rabbit IgG HRP conjugate	Promega corp.	Cat#401B
Dynabeads M-280 streptavidin	Thermo fisher scientific	Cat#11205D
Chemicals		
Tris(hydroxymethyl)aminomethane	Nacalai tesque	Cat#35406–75
Dithiothreitol (DTT)	Nacalai tesque	Cat#14112–94
Iodoacetamide	Nacalai tesque	Cat#19302–54
Nonidet P-40	Nacalai tesque	Cat#25223–04

—cont'd

Reagent or Resource	Source	Identifier
Protease Inhibitor Cocktail	Nacalai tesque	Cat#25955-24
Phosphatase Inhibitor Cocktail	Nacalai tesque	Cat#07575-51
D-Biotin	Nacalai tesque	Cat#04822-04
Tris(2-carboxyethyl) phosphine hydrochloride (TCEP)	Hampton research	Cat#HR2-651
Trypsin Gold, Mass Spectrometry Grade	Promega	Cat#V5280
25% Trifluoroacetic acid (TFA)	Wako	Cat#201-10781
Cell Lines		
Flp-in-293 cell line	Thermo fisher scientific	Cat#R75007

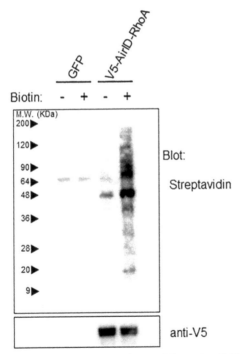

Fig. 7 Detecting the biotinylated proteins obtained from the AirID experiment using AirID-RhoA-expressing cells. The V5-AirID-RhoA or GFP (negative control) transfected 293 cells were treated with or without biotin (biotin triggers AirID biotinylation). Each whole cell lysate (5% v/v) was subjected to a 10% SDS-PAGE gel. (Lower column) V5-tagged proteins were detected with a polyclonal anti-V5 antibody (1:6000), followed by goat anti-rabbit IgG-HRP (1:10,000). (Upper column) Biotin-tagged proteins were detected with streptavidin-peroxidase (1:6000).

3.2.2 Materials and equipment
- PBS (pH 7.2–7.4)
- Protease/phosphatase inhibitor cocktail
- Radioimmunoprecipitation assay (RIPA) buffer (150 mM NaCl, 50 mM Tris–HCl (pH 8.0), 1% NP-40, 0.5% sodium deoxycholate, and 0.1% SDS)
- DTT
- Sonicator (XL-2000 MISONIX)
- Dynabeads streptavidin magnetic beads
- 1 M KCl
- 0.1 M Na_2CO_3
- 2 M Urea/10 mM Tris–HCl (pH 8.0)
- 10 mM Tris–HCl (pH 8.0)
- 20 mM NH_4HCO_3
- 50 mM NH_4HCO_3
- 8 M Urea/50 mM NH_4HCO_3
- 0.5 M TCEP
- 0.5 M Iodoacetamide
- Trypsin, mass spectrometry grade
- 25% Trifluoroacetic acid (TFA)
- Acetonitrile
- Magnetic stand

3.2.3 Step-by-step method details
1. Expand the cell lines expressing GFPs or V5–BirA–RhoA in 6 cm dishes until they reach 70% confluency
2. Add 6 μL of 50 mM D-biotin in 6 mL of medium and incubate for 6 h
3. Wash the cells twice with PBS
4. Add 1 mL of RIPA buffer containing a protease/phosphatase inhibitor cocktail and 1 mM DTT
5. Transfer the samples to new tubes and incubate on ice for 15 min
6. Sonicate the samples for 4×20 s with a 40 s interval between each sonication
7. Centrifuge at $15,000 \times g$ for 15 min at 4 °C
8. Equilibrate the Dynabeads (50–100 μL wet volume per sample) with 1 mL of RIPA buffer and the pellet with the magnetic stand
9. Mix 900 μL of the supernatant (step 7) with the equilibrated Dynabeads (step 8) and rotate at 4°C for 1 h or overnight
10. Apply a magnetic field to pull the beads to the side of the tube. Carefully pipette to remove the supernatant

11. Wash with RIPA buffer without SDS via gentle mixing
12. Apply the magnetic field, then remove the supernatant
13. Repeat this step
14. Wash once with 1 M KCl, 0.1 M Na_2CO_3, and 2 M Urea/10 mM Tris–HCl (pH 8.0)
15. Wash three times with 10 mM Tris–HCl (pH 8.0)
16. After the final wash, carefully remove the wash buffer
17. Resuspend with 50 μM D-biotin in 20 mM NH_4HCO_3
18. Incubate at 37°C for 5 min
19. Wash the beads three times with 20 mM NH_4HCO_3
20. Check the pH of the samples using pH paper (this should be around pH 8.0). If needed, adjust the pH with 200 mM NH_4HCO_3
21. Resuspend with 50 μL of 8 M Urea/50 mM NH_4HCO_3
22. Add 1 μL of 0.5 M TCEP
23. Incubate at 37°C for 60 min
24. Add 2 μL of 0.5 M iodoacetamide
25. Incubate at room temperature in the dark for 30 min
26. Add 350 μL of 50 mM NH_4HCO_3
27. Add 1 μL MS-grade trypsin
28. Incubate at 37 °C with gentle shaking overnight
29. Apply a magnetic field, then collect the supernatant in a new tube
30. Add 50 μL of 50 mM NH_4HCO_3 to the magnetic beads and wash via gentle mixing
31. Apply the magnetic field, collect the supernatant, and combine it with the step 29 tube

3.2.4 Notes
1. In each step, samples can be stored at −80°C for up to 3 weeks
2. Incubation temperature and time might depend on the trypsin manufacturer

4. Proteomics for proximity labeling

Proximity labeling is often used preferentially. However, since it is also necessary to identify the labeled protein after proximity labeling experiments, the name "proximity proteomics" may be more appropriate. This section describes how to identify EMARS products (fluorescein-labeled proteins) via proteomics using mass spectrometry.

4.1 Purification of EMARS products

4.1.1 Key resources table

Reagent or Resource	Source	Identifier
Antibodies		
Anti-fluorescein antibody	SouthernBiotech	Cat# 6400-01
Chemicals, Peptides, and Recombinant Proteins		
NHS-Activated Sepharose 4 Fast Flow	Cytiva	Cat#17-0906-01
Ethanolamine	Nacalai tesque	Cat#23405-55

4.1.2 Materials and equipment
- Anti-fluorescein antibody-conjugated Sepharose (see Section 4.1.4)
- Chloroform/methanol (2:1 v/v)
- NP-40 lysis buffer (20 mM Tris–HCl (pH 7.4), 150 mM NaCl, 5 mM EDTA, 1% NP-40, 10% glycerol, and a protease inhibitor cocktail)

4.1.3 Step-by-step method details
1. Membrane pellets (microsome fraction) after EMARS (see Section 2.2.3)
2. Wash the pellet once with 500 μL of 100 mM Tris–HCl (pH 7.5) and discard the supernatant
3. Add 500 μL of chloroform/methanol (2:1 v/v) to the microsome pellet and mix gently
4. Add 500 μL of distilled water and mix gently
5. Centrifuge at $20{,}000 \times g$ for 2 min at room temperature
6. Remove the supernatant containing the chloroform/methanol solution and distilled water (you can see pellets in the intermediate layer)
7. Wash the pellets with 50% ethanol solution twice
8. Evaporate the pellets for 15 min to completely remove the solution
9. Add 100 μL of 1% SDS-100 mM Tris–HCl (pH 7.4) and heat at 100°C for 10 min
10. Centrifuge at $20{,}000 \times g$ for 15 min at room temperature
11. Transfer the supernatant to a new tube
12. Add 400 μL of the NP-40 lysis buffer and mix
13. Add 100 μL of anti-fluorescein antibody-conjugated Sepharose

14. Rotate the tube at 4 °C overnight
15. Centrifuge at $800 \times g$ for 5 min at 4°C and discard the supernatant
16. Wash the Sepharose resin five times with the NP-40 lysis buffer, twice with 0.5 M NaCl in PBS, and with 50 mM NH_4HCO_3 buffer

4.1.4 Notes

1. Anti-fluorescein antibody-conjugated Sepharose can be used if there is a commercially available one
2. You can make anti-fluorescein antibody-conjugated Sepharose using the following procedure
   (a) Be sure to bring NHS-Activated Sepharose 4 Fast Flow to room temperature before use
   (b) Add 5 mL of coupling buffer (0.2 M $NaHCO_3$ and 0.5 M NaCl (pH 8.3)) to a 15 mL tube
   (c) Add NHS-Activated Sepharose (appropriate to make 2 mL of beads (net))
   (d) Centrifuge at $800 \times g$ for 5 min. Remove the supernatant
   (e) Wash with 5 mL coupling buffer and discard the supernatant (repeated four times)
   (f) Add 5 mL coupling buffer and anti-fluorescein antibody (1 mg)
   (g) Rotate the tube at room temperature overnight
   (h) Centrifuge at $800 \times g$ for 5 min. Remove the supernatant
   (i) Add 5 mL blocking buffer (0.5 M ethanolamine and 0.5 M NaCl (pH 8.3))
   (j) Mix and then remove the supernatant
   (k) Add 5 mL blocking buffer and rotate the tube at room temperature for 1 h
   (l) Centrifuge at $800 \times g$ for 5 min and remove the supernatant
   (m) Wash with 5 mL PBS (repeat three times)
   (n) Add PBS to make 10 mL total (a 20% slurry)
   (o) Stock slurry at 4 °C
3. The anti-fluorescein antibody can be purchased from Southern Biotech or Rockland. Pay attention to the binding capacity when using an anti-fluorescein antibody from other companies. It should be noted that many commercially available anti-fluorescein antibodies have low binding capacities, making them unsuitable for purifying the fluorescein-labeled proteins described in this section

4.2 Peptide preparation and mass spectrometry analysis

The purified labeled proteins are eluted from anti-fluorescein antibody-conjugated Sepharose, trypsinized, and identified by shotgun proteomics using mass spectrometry. Here, we introduce an analysis method using electrospray ionization mass spectrometry (nano LC-ESI-MS/MS: LTQ Orbitrap).

4.2.1 Key resources table

Reagent or Resource	Source	Identifier
Biological Samples		
Trypsin Gold MS grade	Promega	Cat#V5280
Chemicals, Peptides, and Recombinant Proteins		
HRP-conjugated Cholera Toxin B Subunit	LIST Biological Lab.	Cat#105
MPEX PTS reagent kit	GL Science	Cat#5010-21360
Dithiothreitol	Nacalai tesque	Cat#14128–04
Iodoacetamide	Wako	Cat#099-05591
SDS-eliminant	ATTO	Cat#AE-1390
Software and Algorithms	.	
Proteome Discoverer ver. 2.4	Thermo Fisher Scientific	RRID:SCR_014477
Other		
GL-Tip SDB	GL Science	Cat#7820-11200
LTQ Orbitrap XL	Thermo Fisher Scientific	N/A

4.2.2 Materials and equipment
- Elution reagent (1:1 v/v mixture of Solution B in the MPEX PTS reagent kit and 1% SDS-Tris–HCl (pH 7.4))
- SDS
- DTT
- Iodoacetamide (IAA)
- Trypsin solution (1 µg/µL in 50 mM acetic acid)

- Ethyl acetate
- TFA
- Acetonitrile (ACN)
- GL-Tip SDB
- LTQ Orbitrap XL
- Proteome Discoverer ver. 2.4

4.2.3 Step-by-step method details

1. Add 40 μL of the elution reagent (see Section 4.2.4) to the washed Sepharose resin and mix
2. Heat at 100 °C for 10 min (mix after 5 min, and continue heating for another 5 min)
3. Transfer the supernatant to a new tube and add 20 μL of 10 mM DTT/ 50 mM NH$_4$HCO$_3$ solution
4. Incubate at 50 °C for 30 min
5. Add 20 μL of 100 mM IAA/50 mM NH$_4$HCO$_3$
6. Incubate at 37°C for 1 h in the dark
7. Add 5 μL SDS-eliminant solution (see Section 4.2.4)
8. Incubate on ice for 30 min
9. Centrifuge at 13,000 × g for 10 min at 4 °C
10. Transfer the supernatant to a new tube and add 270 μL of 50 mM NH$_4$HCO$_3$
11. Add 2 μL of 20 mM CaCl$_2$ and trypsin solution (1 μg/μL).
12. Incubate at 37°C overnight (see Section 4.2.4)
13. Add 350 μL of ethyl acetate and 3.5 μL TFA to the trypsinized sample solutions and mix gently
14. Centrifuge at 20,000 × g for 1 min at room temperature
15. Discard the upper layer
16. Repeat steps 13–15 five times (see Section 4.2.4)
17. Evaporate the sample for 2 h to completely remove the ethyl acetate
18. Add 200 μL of 5% ACN/0.1% TFA and mix gently
19. Desalt the sample using a GL-Tip SDB (see Section 4.2.4)
20. Evaporate the eluate from the desalting cartridge for 10–15 min to remove the acetonitrile (see Section 4.2.4)
21. Dissolve peptides with the appropriate volume of 2% ACN/0.1% TFA
22. Apply shotgun proteomics to the sample using nano LC-ESI-MS/MS: LTQ Orbitrap XL (see Section 4.2.4)
23. Analyze the data using Proteome Discoverer ver. 2.4 or higher (Table 1)

Table 1 The list of identified proximal proteins around GM1.

Accession No.	Name
sp | P14384 | CBPM_HUMAN	Carboxypeptidase M
sp | Q6YHK3 | CD109_HUMAN	CD109 antigen
sp | Q5ZPR3 | CD276_HUMAN	CD276 antigen
sp | P16070 | CD44_HUMAN	CD44 antigen
sp | P13987 | CD59_HUMAN	CD59 glycoprotein
sp | Q6UVK1 | CSPG4_HUMAN	Chondroitin sulfate proteoglycan 4
sp | P08174 | DAF_HUMAN	Complement decay-accelerating factor
sp | Q14126 | DSG2_HUMAN	Desmoglein-2
sp | Q14118 | DAG1_HUMAN	Dystroglycan
sp | P17813 | EGLN_HUMAN	Endoglin
sp | Q9UNN8 | EPCR_HUMAN	Endothelial protein C receptor
sp | P15328 | FOLR1_HUMAN	Folate receptor alpha
sp | P17931 | LEG3_HUMAN	Galectin-3
sp | P04406 | G3P_HUMAN	GAPDH
sp | P04899 | GNAI2_HUMAN	G(i), alpha-2 subunit
sp | P01891 | 1A68_HUMAN	HLA class I, A-68 alpha chain
sp | P01591 | IGJ_HUMAN	Immunoglobulin J chain
sp | P08648 | ITA5_HUMAN	Integrin alpha-5
sp | P05556 | ITB1_HUMAN	Integrin beta-1
sp | O75145 | LIPA3_HUMAN	Liprin-alpha-3

The trypsinized peptides from fluorescein-labeled proximal proteins by EMARS (see Section 2.2) were applied to nano LC-ESI-MS/MS. The raw data from MS system was subsequently analyzed using Proteome Discoverer. The identified proteins were arranged in descending order of identification score calculated by Proteome Discoverer, and the top 20 candidate proteins are listed.

4.2.4 Notes

1. A detergent, such as SDS, should be used when eluting the fluorescein-labeled protein from the washed Sepharose resin. However, SDS should be removed before trypsinization. In addition, since this analysis often contains membrane proteins, a strong detergent is recommended for the elution process. A specific detergent called MPEX, which can be

removed easily by extraction using an organic solvent, is used together with SDS to adapt to these difficult conditions

2. "SDS-eliminat" is a reagent that can appropriately precipitate SDS and easily remove it. Other SDS removal methods, including a similar reagent, can be used

3. In this section, proteins are digested by typical tryptic digestion, but other enzymes are also available

4. It should be noted that a single treatment with ethyl acetate leaves MPEX and adversely affects the mass spectrometry analysis. Multiple extraction steps are required

5. Other desalting cartridges can be used in the desalting step. For detailed usage of each cartridge, follow the instructions for each cartridge

6. If the sample is completely dried up at the evaporation stage, some peptides may be adsorbed on the tube surface and cannot be eluted

7. We finally analyzed the sample by shotgun proteomics, but other methods, such as in-gel digestion, can also be used

8. The mass spectrometry system parameters and search parameters we used are below

Parameters

Peak list-generating software and release version	Proteome Discoverer ver. 2.4 or higher
Search engine and release version	MASCOT ver. 2.5.1 or higher
Sequence database searched	Swiss Prot
Specificity of all proteases used to generate peptides	Trypsin (Trypsin Gold, Mass Spectrometry Grade)
Number of missed and/or non-specific cleavages permitted	2
Fixed modifications (including residue specificity)	Carbamidomethyl (C)
Variable modifications (including residue specificity)	Oxidation (M), Deamidated (NQ)
Mass tolerance for precursor ions	10 ppm
Mass tolerance for fragment ions	0.8 Da
Threshold score/Expectation value for accepting individual spectra	<0.05

5. Summary and conclusions

In this chapter, we focused on proximity labeling (proximity proteomics), a powerful tool for protein–protein interaction analysis, and explained how to perform it in (cell) membrane or intracellular regions, followed by proteomics analysis of labeled products.

Membrane proximity labeling was first developed to analyze molecular clusters (lipid rafts) on cell membranes. EMARS reactions do not require a genetic engineering technique and only need HRP-labeled recognition molecules (HRP-labeled antibodies), widely used in typical biological research. Therefore, various research samples, such as pathological tissues derived from patients, which are impossible to be genetically modified, are available (EMARS by HRP fusion protein expression using gene recombination is also possible (Miyagawa-Yamaguchi et al., 2014, 2015). We successfully used acute brain slices of mice in an EMARS reaction (Kotani et al., 2018). Recently, an EMARS and membrane proximity labeling method using biotin ligase in BioID was being developed. It can advantageously perform protein–protein interaction analysis in the (cell) membrane and intracellular region. In addition, novel technologies using fluorophore and photochemistry (Sjöstedt et al., 2020) enable spatiotemporal proximity labeling like the optogenetics used in neurochemistry research (Miesenböck, 2009).

Regarding intracellular proximity labeling, many researchers utilize the technology using APEX, a peroxidase used like HRP in EMARS, and BioID (and its advanced technology), which started from cloning the *E. coli* biotin ligase (Cronan, 1990). The technological progress of BioID using biotin ligase is especially remarkable. These methods have led to important research results in life science research for molecular and cell biology and are expected to make further contributions in future life science explorations.

Furthermore, proximity labeling is not only used as a tool for protein–protein interaction research but also for research purposes not previously envisioned. We are currently carrying out drug discovery research using EMARS to identify cancer cell-specific membrane "bi-molecules" (called BiCATs) and utilize them as a novel cancer antigen (Kotani et al., 2019). Likewise, Kaneda et al. (2021) are investigating novel tumor markers to enhance early lung cancer diagnoses using an extracellular vesicle-specific "bi-molecule" on the extracellular vesicle membrane (called BiEV) secreted from cancer cells. A system that can contribute to the virus receptor analysis of anticipated pandemic viruses by establishing a method for identifying virus receptors is also being developed using EMARS. The analysis of the virus co-receptor of SARS-CoV-2, which has been globally rampant, has been performed using this system (Kotani, Nakano, & Kuwahara, 2022).

In this way, proximity labeling (proximity proteomics) is expected to be applied to the evolution of the method and various research fields other than protein–protein interaction analysis, and its future progress will be watched with interest.

Acknowledgments

We thank Kochi University's experimental training equipment facility and the Saitama Medical University Biomedical Research Center for providing general technical assistance. This work was supported by Grants-in-aid for Scientific Research in Japan (No. JP19790060, JP22790071, JP24590082, JP15K07941, No. JP18K06663, and JP21K06562 to N.K.), Mizutani Foundation Research Grants (to N.K.), Japan Science and Technology grants (to N.K.), The Vehicle Racing Commemorative Foundation research grants (to N.K.), and TERUMO Life Science Foundation research grants (to N.K.). We would like to thank Editage (www.editage.com) for English language editing.

Conflicts of interest

The authors have no competing or financial interests to declare.

References

Branon, T. C., Bosch, J. A., Sanchez, A. D., Udeshi, N. D., Svinkina, T., Carr, S. A., et al. (2018). Efficient proximity labeling in living cells and organisms with TurboID. *Nature Biotechnology, 36*(9), 880–898. https://doi.org/10.1038/NBT.4201.

Brown, D. A., & London, E. (1998). Functions of lipid rafts in biological membranes. *Annual Review of Cell and Developmental Biology, 14*, 111–136. https://doi.org/10.1146/annurev. cellbio.14.1.111.

Brown, D. A., & Rose, J. K. (1992). Sorting of GPI-anchored proteins to glycolipid-enriched membrane subdomains during transport to the apical cell surface. *Cell, 68*(3), 533–544. https://doi.org/10.1016/0092-8674(92)90189-J.

Chang, L., Chen, Y. J., Fan, C. Y., Tang, C. J., Chen, Y. H., Low, P. Y., et al. (2017). Identification of siglec ligands using a proximity labeling method. *Journal of Proteome Research, 16*(10), 3929–3941. https://doi.org/10.1021/ACS.JPROTEOME.7B00625/ SUPPL_FILE/PR7B00625_SI_002.ZIP.

Cronan, J. E. (1990). Biotination of proteins in vivo. A post-translational modification to label, purify, and study proteins. *Journal of Biological Chemistry, 265*(18), 10327–10333. https://doi.org/10.1016/S0021-9258(18)86949-6.

Esaki, N., Ohkawa, Y., Hashimoto, N., Tsuda, Y., Ohmi, Y., Bhuiyan, R. H., et al. (2018). ASC amino acid transporter 2, defined by enzyme-mediated activation of radical sources, enhances malignancy of GD2-positive small-cell lung cancer. *Cancer Science, 109*(1). https://doi.org/10.1111/cas.13448.

Frieden, C. (1971). Protein-protein interaction and enzymatic activity. *Annual Review of Biochemistry, 40*, 653–696. https://doi.org/10.1146/ANNUREV.BI.40.070171.003253.

Hashimoto, N., Hamamura, K., Kotani, N., Furukawa, K., Kaneko, K., & Honke, K. (2012). Proteomic analysis of ganglioside-associated membrane molecules: Substantial basis for molecular clustering. *Proteomics, 12*(21), 3154–3163. https://doi.org/10.1002/ pmic.201200279.

Honke, K., & Kotani, N. (2011). The enzyme-mediated activation of radical source reaction: A new approach to identify partners of a given molecule in membrane microdomains. *Journal of Neurochemistry*, *116*(5). https://doi.org/10.1111/j.1471-4159.2010.07027.x.

Honke, K., & Kotani, N. (2012). Identification of cell-surface molecular interactions under living conditions by using the enzyme-mediated activation of radical sources (EMARS) method. *Sensors (Basel, Switzerland)*, *12*(12). https://doi.org/10.3390/s121216037.

Imamaki, R., Ogawa, K., Kizuka, Y., Komi, Y., Kojima, S., Kotani, N., et al. (2018). Glycosylation controls cooperative PECAM-VEGFR2-β3 integrin functions at the endothelial surface for tumor angiogenesis. *Oncogene*, *37*(31), 4287–4299. https://doi.org/10.1038/s41388-018-0271-7.

Ishiura, Y., Kotani, N., Yamashita, R., Yamamoto, H., Kozutsumi, Y., & Honke, K. (2010). Anomalous expression of Thy1 (CD90) in B-cell lymphoma cells and proliferation inhibition by anti-Thy1 antibody treatment. *Biochemical and Biophysical Research Communications*, *396*(2), 329–334. https://doi.org/10.1016/j.bbrc.2010.04.092.

Ito, Y., Honda, A., & Igarashi, M. (2018). Glycoprotein M6a as a signaling transducer in neuronal lipid rafts. *Neuroscience Research*, *128*, 19–24. https://doi.org/10.1016/J.NEURES.2017.11.002.

Iwamaru, Y., Kitani, H., Okada, H., Takenouchi, T., Shimizu, Y., Imamura, M., et al. (2016). Proximity of SCG10 and prion protein in membrane rafts. *Journal of Neurochemistry*, *136*(6), 1204–1218. https://doi.org/10.1111/JNC.13488.

Jiang, S., Kotani, N., Ohnishi, T., Miyagawa-Yamguchi, A., Tsuda, M., Yamashita, R., et al. (2012). A proteomics approach to the cell-surface interactome using the enzyme-mediated activation of radical sources reaction. *Proteomics*, *12*(1), 54–62. https://doi.org/10.1002/pmic.201100551.

Kaneda, H., Ida, Y., Kuwahara, R., Sato, I., Nakano, T., Tokuda, H., et al. (2021). Proximity proteomics has potential for extracellular vesicle identification. *Journal of Proteome Research*, *20*(7). https://doi.org/10.1021/acs.jproteome.1c00149.

Kaneko, K., Ohkawa, Y., Hashimoto, N., Ohmi, Y., Kotani, N., Honke, K., et al. (2016). Neogenin, defined as a GD3-associated molecule by enzyme-mediated activation of radical sources, confers malignant properties via intracytoplasmic domain in melanoma cells. *Journal of Biological Chemistry*, *291*(32), 16630–16643. https://doi.org/10.1074/jbc.M115.708834.

Kido, K., Yamanaka, S., Nakano, S., Motani, K., Shinohara, S., Nozawa, A., et al. (2020). Airid, a novel proximity biotinylation enzyme, for analysis of protein–protein interactions. *eLife*, *9*. https://doi.org/10.7554/ELIFE.54983.

Killinger, B. A., Marshall, L. L., Chatterjee, D., Chu, Y., Bras, J., Guerreiro, R., et al. (2022). In situ proximity labeling identifies Lewy pathology molecular interactions in the human brain. *Proceedings of the National Academy of Sciences of the United States of America*, *119*(5). https://doi.org/10.1073/PNAS.2114405119.

Kotani, N., Gu, J., Isaji, T., Udaka, K., Taniguchi, N., & Honke, K. (2008). Biochemical visualization of cell surface molecular clustering in living cells. *Proceedings of the National Academy of Sciences*, *105*(21), 7405–7409. https://doi.org/10.1073/pnas.0710346105.

Kotani, N., Ishiura, Y., Yamashita, R., Ohnishi, T., & Honke, K. (2012). Fibroblast growth factor receptor 3 (FGFR3) associated with the CD20 antigen regulates the rituximab-induced proliferation inhibition in B-cell lymphoma cells. *Journal of Biological Chemistry*, *287*(44), 37109–37118. https://doi.org/10.1074/jbc.M112.404178.

Kotani, N., Nakano, T., Ida, Y., Ito, R., Hashizume, M., Yamaguchi, A., et al. (2018). Analysis of lipid raft molecules in the living brain slices. *Neurochemistry International*, *119*, 140–150. https://doi.org/10.1016/j.neuint.2017.08.012.

Kotani, N., Nakano, T., & Kuwahara, R. (2022). Screening of candidate host cell membrane proteins involved in SARS-CoV-2 entry. *BioRxiv.* https://doi.org/10.1101/2020.09.09. 289488.

Kotani, N., Yamaguchi, A., Ohnishi, T., Kuwahara, R., Nakano, T., Nakano, Y., et al. (2019). Proximity proteomics identifies cancer cell membrane cis-molecular complex as a potential cancer target. *Cancer Science, 110*(8), 2607–2619. https://doi.org/10. 1111/cas.14108.

Lam, S. S., Martell, J. D., Kamer, K. J., Deerinck, T. J., Ellisman, M. H., Mootha, V. K., et al. (2015). Directed evolution of APEX2 for electron microscopy and proximity labeling. *Nature Methods, 12*(1), 51–54. https://doi.org/10.1038/NMETH.3179.

Lingwood, D., & Simons, K. (2010). Lipid rafts as a membrane-organizing principle. *Science, 327*(5961), 46–50. https://doi.org/10.1126/science.1174621.

Lutter, L. C., & Kurland, C. G. (1975). Chemical determination of protein neighbourhoods in a cellular organelle. *Molecular and Cellular Biochemistry, 7*(2), 105–116. https://doi.org/ 10.1007/BF01792077.

Martell, J. D., Deerinck, T. J., Sancak, Y., Poulos, T. L., Mootha, V. K., Sosinsky, G. E., et al. (2012). Engineered ascorbate peroxidase as a genetically encoded reporter for electron microscopy. *Nature Biotechnology, 30*(11), 1143–1148. https://doi.org/10.1038/NBT.2375.

Miesenböck, G. (2009). The optogenetic catechism. *Science (New York, N.Y.), 326*(5951), 395–399. https://doi.org/10.1126/SCIENCE.1174520.

Miyagawa-Yamaguchi, A., Kotani, N., & Honke, K. (2014). Expressed glycosylphosphatidylinositol-anchored horseradish peroxidase identifies co-clustering molecules in individual lipid raft domains. *PLoS One, 9*(3). https://doi.org/10.1371/ journal.pone.0093054.

Miyagawa-Yamaguchi, A., Kotani, N., & Honke, K. (2015). Each GPI-anchored protein species forms a specific lipid raft depending on its GPI attachment signal. *Glycoconjugate Journal, 32*(7), 531–540. https://doi.org/10.1007/s10719-015-9595-5.

Ohkawa, Y., Momota, H., Kato, A., Hashimoto, N., Tsuda, Y., Kotani, N., et al. (2015). Ganglioside GD3 enhances invasiveness of gliomas by forming a complex with platelet-derived growth factor receptor α and yes kinase. *The Journal of Biological Chemistry, 290*(26), 16043–16058. https://doi.org/10.1074/JBC.M114.635755.

Qin, W., Cho, K. F., Cavanagh, P. E., & Ting, A. Y. (2021). Deciphering molecular inter-actions by proximity labeling. *Nature Methods, 18*(2), 133–143. https://doi.org/10.1038/ s41592-020-01010-5.

Roux, K. J., Kim, D. I., Raida, M., & Burke, B. (2012). A promiscuous biotin ligase fusion protein identifies proximal and interacting proteins in mammalian cells. *The Journal of Cell Biology, 196*(6), 801–810. https://doi.org/10.1083/JCB.201112098.

Simons, K., & Toomre, D. (2000). Lipid rafts and signal transduction. *Nature Reviews Molecular Cell Biology, 1*(1), 31–39. https://doi.org/10.1038/35036052.

Sjöstedt, E., Zhong, W., Fagerberg, L., Karlsson, M., Mitsios, N., Adori, C., et al. (2020). Microenvironment mapping via Dexter energy transfer on immune cells. *Science (New York, N.Y.), 367*(6482). https://doi.org/10.1126/SCIENCE.AAY4106.

Stelzl, U., & Wanker, E. E. (2006). The value of high quality protein-protein interaction networks for systems biology. *Current Opinion in Chemical Biology, 10*(6), 551–558. https://doi.org/10.1016/J.CBPA.2006.10.005.

Vereb, G., Szöllosi, J., Matkó, J., Nagy, P., Farkas, T., Vigh, L., et al. (2003). Dynamic, yet structured: The cell membrane three decades after the singer-Nicolson model. *Proceedings of the National Academy of Sciences of the United States of America, 100*(14), 8053–8058. http://www.ncbi.nlm.nih.gov/entrez/query.fcgi?cmd=Retrieve&db=PubMed&dopt= Citation&list_uids=12832616.

Virkamäki, A., Ueki, K., & Kahn, C. R. (1999). Protein-protein interaction in insulin signaling and the molecular mechanisms of insulin resistance. *The Journal of Clinical Investigation*, *103*(7), 931–943. https://doi.org/10.1172/JCI6609.

Wodak, S. J., Vlasblom, J., Turinsky, A. L., & Pu, S. (2013). Protein-protein interaction networks: The puzzling riches. *Current Opinion in Structural Biology*, *23*(6), 941–953. https://doi.org/10.1016/J.SBI.2013.08.002.

Yamashita, R., Kotani, N., Ishiura, Y., Higashiyama, S., & Honke, K. (2011). Spatiotemporally-regulated interaction between β1 integrin and ErbB4 that is involved in fibronectin-dependent cell migration. *The Journal of Biochemistry*, *149*(3), 347–355. https://doi.org/10.1093/jb/mvq148.

CHAPTER SIX

Looking at LPMO reactions through the lens of the HRP/Amplex Red assay

Anton A. Stepnov and Vincent G.H. Eijsink*

Faculty of Chemistry, Biotechnology and Food Science, NMBU—Norwegian University of Life Sciences, Ås, Norway
*Corresponding author: e-mail address: vincent.eijsink@nmbu.no

Contents

Abstract

Lytic polysaccharide monooxygenases (LPMOs) are unique redox enzymes capable of disrupting the crystalline surfaces of industry-relevant recalcitrant polysaccharides, such as chitin and cellulose. Historically, LPMOs were thought to be slow enzymes relying on O_2 as the co-substrate, but it is now clear that these enzymes prefer H_2O_2, allowing for fast depolymerization of polysaccharides through a peroxygenase reaction. Thus, quantifying H_2O_2 in LPMO reaction set-ups is of a great interest. The horseradish peroxidase (HRP)/Amplex Red (AR) assay is one of the most popular and accessible tools for measuring hydrogen peroxide. This assay has been used in various types of biological and biochemical studies, including LPMO research, but suffers from pitfalls that need to be accounted for. In this Chapter, we discuss this method and its use for assessing the often rate-limiting *in situ* formation of H_2O_2 in LPMO reactions. We show that, after accounting for multiple potential side reactions, quantitative data on H_2O_2 production obtained with the HRP/Amplex Red assay provide useful clues for understanding the catalytic activity of LPMOs, including the impact of reductants and transition metal ions.

Methods in Enzymology, Volume 679
ISSN 0076-6879
https://doi.org/10.1016/bs.mie.2022.08.049

163

1. Introduction

Lytic polysaccharide monooxygenases (LPMOs) are mono-copper enzymes that catalyze oxidative depolymerization of complex carbohydrate substrates, such as chitin or cellulose, in the presence of electron donors (Forsberg et al., 2011; Phillips, Beeson, Cate, & Marletta, 2011; Quinlan et al., 2011; Vaaje-Kolstad et al., 2010). LPMOs are known for a unique ability to act directly on crystalline surfaces of their substrates rather than on isolated and amorphous polysaccharide chains, thus exposing these substrates to the action of canonical hydrolytic enzymes (Harris et al., 2010; Vaaje-Kolstad et al., 2010; Vaaje-Kolstad, Horn, van Aalten, Synstad, & Eijsink, 2005). The contribution of LPMOs to the efficiency of enzymatic biomass conversion has attracted significant attention from industry. New generations of commercial cellulolytic cocktails contain LPMOs alongside other enzymes to improve biomass solubilization (Cannella, Hsieh, Felby, & Jorgensen, 2012; Chylenski et al., 2017). Currently, LPMOs populate eight families in the Carbohydrate-active Enzyme database (Cantarel et al., 2009), referred to as AA9, AA10, AA11 and AA13–17.

Despite more than a decade of ongoing research, the mechanistic aspects of LPMO catalysis are yet to be described in full detail. Historically, LPMO reactions were believed to involve molecular oxygen as a co-substrate (hence the name "monooxygenase"; Horn, Vaaje-Kolstad, Westereng, & Eijsink, 2012), but in 2017 it was discovered that these enzymes prefer hydrogen peroxide (Bissaro et al., 2017). The H_2O_2-dependent (i.e., peroxygenase) LPMO reactions are fast (Bissaro et al., 2017; Filandr et al., 2020; Hedison et al., 2021; Jones, Transue, Meier, Kelemen, & Solomon, 2020; Kuusk et al., 2018) compared to monooxygenase reactions that are generally 2–3 orders of magnitude slower. Next to enhancing catalytic rates, the addition of H_2O_2 to LPMO reactions may promote enzyme inactivation in case the dosage is too high (Bissaro et al., 2017; Kadic, Varnai, Eijsink, Horn, & Liden, 2021; Muller, Chylenski, Bissaro, Eijsink, & Horn, 2018). Notwithstanding the efficiency of hydrogen peroxide as a co-substrate, standard aerobic conditions (i.e., reactions with no added H_2O_2) are commonly used in the LPMO field, both in research laboratories and in industry. In such set-ups, the LPMO reaction is driven by a reductant that delivers the electrons needed for a monooxygenase reaction to occur (R-H + O_2 + $2e^-$ + $2H^+$ → R-OH + H_2O), or for in $situ$ generation of H_2O_2, which is then used in a peroxygenase reaction (R-H + H_2O_2 → R-OH + H_2O). Importantly, the apparent O_2-dependent LPMO activity observed under these conditions may in fact reflect a peroxygenase reaction that is limited by *in situ* production of H_2O_2 (Bissaro et al., 2017; Bissaro, Varnai, Rohr, & Eijsink, 2018). It is well established that hydrogen peroxide can be formed as the result of the oxidase activity of LPMOs, sometimes referred to as the "uncoupled reaction", which is more prone to happen in the absence of substrate (Kittl, Kracher, Burgstaller, Haltrich, & Ludwig, 2012). Furthermore, all commonly used low molecular weight LPMO reductants (such as ascorbic acid, Wilson, Beezer, & Mitchell, 1995; gallic acid, Akagawa, Shigemitsu, & Suyama, 2003; reduced glutathione, Kachur, Koch, & Biaglow, 1998; DTT, Kachur, Held, Koch, & Biaglow, 1997 and L-cysteine, Kachur, Koch, & Biaglow, 1999) are capable of reducing molecular oxygen, leading to generation of H_2O_2. These abiotic reactions are sometimes referred to as "reductant auto-oxidation." All in all, given the role of hydrogen peroxide as the preferred and efficient LPMO co-substrate, determining the rates of *in situ* H_2O_2 generation in LPMO reactions is important to get a better understanding of factors affecting enzyme performance and stability under typical aerobic conditions.

Here, we discuss general approaches to assessing hydrogen peroxide generation in LPMO reactions using the HRP/Amplex Red assay originally introduced to the LPMO field by Kittl et al. (2012). We show that, after properly dealing with potential pitfalls, HRP/Amplex Red assay data may provide valuable clues for understanding LPMO catalysis. Furthermore, such data may help explain the variations in substrate degradation efficiency that are observed when comparing various enzyme-reductant combinations.

Next, we address the growing amount of evidence showing that low and varying amounts of free copper ions that may be present in LPMO reactions can promote the accumulation of excessive amounts of hydrogen peroxide, which affects the reproducibility of kinetic data and may trigger rapid enzyme inactivation. We conclude the Chapter by presenting a modified version of the standard HRP/Amplex Red protocol (Kittl et al., 2012), including procedures that can be used to detect and control free copper contamination in LPMO samples.

2. The HRP/Amplex Red assay: Basic principles and pitfalls

The HRP/Amplex Red assay is a rapid quantitative method commonly used to detect hydrogen peroxide in living organisms (Karakuzu, Cruz, Liu, & Garsin, 2019), cell cultures (Maddalena et al., 2017), cell extracts (Chakraborty et al., 2016) and various enzymatic reactions

(Mishin, Gray, Heck, Laskin, & Laskin, 2010; Zhou & Panchuk-Voloshina, 1997). The first application of the HRP/Amplex Red assay in the LPMO field dates back to 2012, when it was employed by Kittl et al. to demonstrate that LPMOs possess an oxidase side activity leading to the generation of H_2O_2 when the enzymes are supplied with reductant and molecular oxygen (Kittl et al., 2012). The authors suggested that monitoring enzyme-dependent hydrogen peroxide production can be used to screen for active LPMOs, as a faster and cheaper alternative to standard LPMO assays, which involve carbohydrate substrates and complex product detection techniques based on chromatography or mass spectrometry (Eijsink et al., 2019). This new approach was quickly adopted by the research community, making the HRP/Amplex Red assay part of the LPMO characterization toolkit.

Quantification of H_2O_2 in LPMO reactions became particularly important in 2017, when Bissaro et al. showed that hydrogen peroxide is the preferred co-substrate of these enzymes (Bissaro et al., 2017). This crucial discovery revealed that LPMO can not only produce H_2O_2 but also consume it. There is a large and constantly growing amount of HRP/Amplex Red assay data obtained with various LPMOs at various reaction conditions (as discussed in detail below). Prior to discussing these results and defining preferred reaction set-ups, it is important to address basic principles and pitfalls of the HRP/Amplex Red assay.

The HRP/Amplex Red method (Fig. 1) is based on H_2O_2-dependent single-electron oxidation of 10-acetyl-3,7-dihydroxyphenoxazine (Amplex Red) by horseradish peroxidase (HRP). The resulting Amplex Red radicals are not stable and two such radicals will engage in a dismutation reaction leading to the regeneration of one Amplex Red molecule and formation of the highly absorbing and highly fluorescent reporter molecule, resorufin. Note that HRP is thought to carry out two consecutive one-electron oxidations of Amplex Red to the Amplex Red radical (Gorris & Walt, 2009; Fig. 1) per one consumed hydrogen peroxide molecule, hence the formation of resorufin proceeds with a 1:1 ratio relative to the H_2O_2 that is present in the reaction. The HRP/Amplex Red assay allows detection of micromolar concentrations of hydrogen peroxide when using absorbance for detection of resorufin formation, whereas submicromolar concentrations ($\geq 0.06\,\mu M$) can be detected when using fluorescence spectroscopy instead of absorption spectroscopy (Mishin et al., 2010).

As originally proposed by Kittl et al. (2012), H_2O_2 generation by LPMOs is usually measured by supplying an enzyme solution with HRP

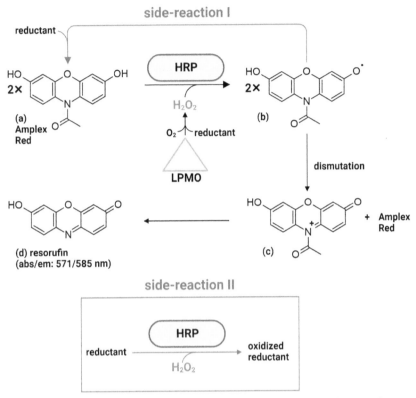

Fig. 1 Schematic representation of the HRP/Amplex Red assay carried out in the presence of an LPMO and a low-molecular-weight reducing compound. One H_2O_2 molecule produced by the reduced LPMO is consumed by horseradish peroxidase (HRP) in a reaction that leads to conversion of two Amplex Red molecules (A) to corresponding radicals (B). The Amplex Red radicals then undergo dismutation resulting in the regeneration of an Amplex Red molecule and formation of a hypothetical unstable intermediate (C) that spontaneously converts into a highly absorbing and highly fluorescent reporter molecule, resorufin (D). Several LPMO reductants are known to repress the HRP/Amplex Red signal, leading to underestimation of H_2O_2. Two side-reactions that are thought to cause this effect in reactions with ascorbic acid are shown with red arrows. Next to activating LPMOs, ascorbic acid is capable of reducing Amplex Red radicals back to Amplex Red ("side-reaction I"), thus suppressing formation of resorufin. Furthermore, HRP may oxidize ascorbic acid instead of Amplex Red, consuming H_2O_2 in a way that is not coupled to generation of the reporter molecule ("side-reaction II"). The mechanisms and stoichiometry of Amplex Red conversion to resorufin shown in this figure are based on the data discussed in Debski et al. (2016) and Gorris and Walt (2009). The figure was created using BioRender.com.

and Amplex Red. The reaction is initiated by the addition of an electron donor (reductant) to activate the LPMO and drive its oxidase activity. The hydrogen peroxide production rate is then assessed by following accumulation of resorufin over time. Importantly, multiple side reactions, discussed below and shown in part in Fig. 1, may interfere with this method.

It is commonly assumed that under the conditions of the HRP/Amplex Red assay LPMOs will only produce, but not consume H_2O_2, since LPMO substrates are not present in these reactions. However, LPMOs are capable of (rather slow) futile H_2O_2 turnover, which will eventually lead to auto-catalytic damage to the active site (Kuusk et al., 2018; Kuusk & Valjamae, 2021). HRP is known for its high affinity for hydrogen peroxide and competition experiments have shown full inhibition of reactions with bacterial and fungal LPMOs when the enzymes were combined at approximately 1:1 or 1:2 (LPMO:HRP) molar ratios (Bissaro et al., 2017; Rieder, Stepnov, Sorlie, & Eijsink, 2021). In the absence of a polysaccharide substrate the LPMO will compete much less efficiently for H_2O_2, meaning that underestimation of H_2O_2 levels due to (futile) LPMO-catalyzed turnover of H_2O_2 is not likely. Nevertheless, competition for H_2O_2 between HRP and LPMOs in the HRP/Amplex Red assay cannot be fully excluded, if low amounts of HRP are used. Of note, the rates of futile H_2O_2 turnover, which, so far, have hardly been studied, may vary between LPMOs.

Another potential cause of underestimation of H_2O_2 are the electron donors (reductants) used to drive the LPMO reaction. It has been demonstrated that the most commonly used reductant, ascorbic acid, interferes with the HRP/Amplex Red assay in a manner that leads to a signal repression, even at concentrations that are well below the commonly used 1 mM (e.g., 50 μM; Bissaro et al., 2020; Stepnov et al., 2021). The mechanism of this interference is not well described and may involve multiple pathways. The most obvious explanation is that ascorbic acid is capable of reducing Amplex Red radicals back to Amplex Red, thus preventing the formation of resorufin (Rodrigues & Gomes, 2010) (see Fig. 1: side-reaction I). To further complicate things, HRP may catalyze direct peroxidation of ascorbic acid (Mehlhorn, Lelandais, Korth, & Foyer, 1996) (Fig. 1: side-reaction II), consuming H_2O_2 in a way that is not coupled to generation of the reporter molecule. Repression of the HRP/Amplex Red assay signal is not a unique property of ascorbic acid. Other popular LPMO reductants, such as gallic acid, L-cysteine and reduced glutathione interfere as well, resulting in different but significant degrees of H_2O_2 underestimation (Fig. 2). These undesirable effects can to some extent be accounted for by supplying the

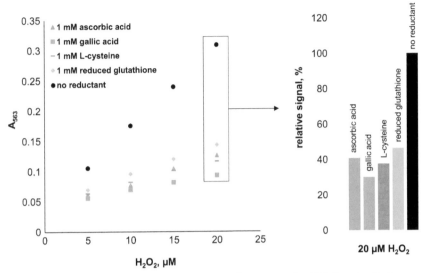

Fig. 2 Interference of commonly used LPMO reductants in the HRP/Amplex Red assay. The figure shows the resorufin signal obtained by absorption spectroscopy ($\lambda = 563$ nm) in the presence or absence of 1 mM reducing compounds with various amounts of H_2O_2. All experiments were carried out in 50 mM sodium phosphate buffer, pH 6.0, at 30 °C. Reaction mixtures contained 5 U/mL HRP and 100 µM Amplex Red. Note that the degree of interference may be pH-dependent as reductant properties may be pH-dependent.

H_2O_2 solutions used for making standard curves with appropriate reductants, as described in Section 6. Note that such a correction is not perfect since it does not account for the fact that the reductant concentration (i.e., the degree of signal repression) may significantly change over time during the HRP/Amplex Red assay. Importantly, there is evidence suggesting that HRP/Amplex Red assays with gallic acid are particularly prone to underestimation of hydrogen peroxide due to the propensity of at least some LPMOs to engage in H_2O_2-consuming side-reactions involving this particular reductant that are not well understood (Rieder, Stepnov, et al., 2021; Stepnov et al., 2021) and likely relate to LPMO reactions with syringol (2,6-dimethoxyphenol) described by Breslmayr et al. (2018).

Next to interfering with the HRP/Amplex Red assay signal, commonly used low-molecular-weight LPMO reductants (such as ascorbic acid, Wilson et al., 1995; gallic acid, Akagawa et al., 2003 and L-cysteine, Kachur et al., 1999) will interact with molecular oxygen, which leads to enzyme-independent production of H_2O_2. Control assay reactions lacking LPMOs should always be carried out to assess this background level of hydrogen peroxide generation, which can be significant depending on the experimental conditions, as discussed further in Sections 3 and 4.

Last but not least, Amplex Red is notorious for its photosensitivity (Summers, Zhao, Ganini, & Mason, 2013). Prolonged exposure of Amplex Red solutions to ambient light will result in an increased background signal due to H_2O_2-independent formation of resorufin. Thus, light exposure should be minimized.

All in all, the HRP/Amplex Red assay is a relatively simple and sensitive tool for detection of H_2O_2 in LPMO reactions. At the same time, this method is prone to multiple side-reactions that make quantitative interpretation of the data non-straightforward.

3. Understanding variations in LPMO performance by measuring *in situ* H_2O_2 generation

Since the initial discovery of the oxidase activity of family AA9 LPMOs from *Neurospora crassa* by Kittl et al. (2012), a large amount of HRP/Amplex Red assay data has become available for several LPMOs, incubated at various reaction conditions (see Table 1, for a brief overview). From these experiments, it is now clear that the ability to produce hydrogen peroxide in the presence of molecular oxygen is a common LPMO feature, found across multiple LPMO types, including members of the recently discovered AA14, AA15 and AA16 families. Looking at LPMO substrates, oxidase activity has been demonstrated for LPMOs acting on cellulose (Hegnar et al., 2019; Kittl et al., 2012), cello-oligomers (Rieder, Stepnov, et al., 2021), starch (Rieder, Ebner, et al., 2021), hemicellulose (Couturier et al., 2018), chitin and chito-oligomers (Rieder, Petrovic, et al., 2021). Interestingly, it has been shown that *Pf*CopC, a copper chaperone from *Pseudomonas fluorescens* which possesses an LPMO-like copper-binding site, albeit with an extra copper ligand that occupies the presumed oxygen binding site, has no ability to generate H_2O_2 (Brander et al., 2020).

Despite almost a decade of research, the exact mechanism of LPMO-dependent hydrogen peroxide formation is yet to be established. It is possible that LPMOs catalyze single-electron reduction of molecular oxygen to superoxide, which then leaves the active site to be converted into H_2O_2 in a reaction with reductants or through dismutation (Kjaergaard et al., 2014). Alternatively, hydrogen peroxide can be directly generated by two-electron reduction of O_2 in the enzyme active site without release of intermediates into solution (Caldararu, Oksanen, Ryde, & Hedegard, 2019; Wang, Walton, & Rovira, 2019).

Table 1 Rates of LPMO-dependent H_2O_2 production determined by using the HRP/Amplex Red assay in the presence of commonly used low-molecular-weight reductants.

CAZy family	LPMO name	Reaction conditions	Reaction rate	Reference
AA9	GtLPMO9B	1 μM LPMO, 30 μM reductant (ascorbic acid or gallic acid), 0.5 U/mL HRP, 50 μM AR in 50 mM Bis–Tris–HCl buffer, pH 6.5	$\approx 0.0015\,s^{-1}$ [a] (ascorbic acid) $\approx 0.0001\,s^{-1}$ [a] (gallic acid)	Hegnar et al. (2019): Fig. 3C, Fig. 4C
	LsLPMO9A	0.5 μM LPMO, 100 μM ascorbic acid, 1 U/mL HRP, 50 μM AR in 40 mM MES buffer, pH 6.6	$\approx 0.0039\,s^{-1}$ [a]	Brander et al. (2021): Fig. 2A
	NcLPMO9C, LsLPMO9A	1 μM LPMO, 1 mM reductant (ascorbic acid, gallic acid or L-cysteine), 0.025 mg/mL HRP, 100 μM AR in 50 mM Bis–Tris–HCl buffer, pH 6.5 at 30°C	NcLPMO9C: $0.017\,s^{-1}$; $0.002\,s^{-1}$; $0.019\,s^{-1}$ (ascorbic acid; gallic acid; L-cysteine)[b] LsLPMO9A: $0.006\,s^{-1}$; $0.002\,s^{-1}$; $0.018\,s^{-1}$ (ascorbic acid; gallic acid; L-cysteine)[b]	Rieder, Stepnov, et al. (2021): Table 1, Table 2
	TaLPMO9A	1 μM LPMO, 50 μM ascorbic acid, 5 U/mL HRP, 100 μM AR in 50 mM Bis–Tris–HCl buffer, pH 6.5 at 40°C	$\approx 0.016\,s^{-1}$	Petrovic et al., 2018
	TaLPMO9A	1 μM LPMO, 100 μM ascorbic acid; HRP and AR concentrations not provided	$\approx 0.014\,s^{-1}$ [a]	Brander et al., 2020: Fig. 6A

Continued

Table 1 Rates of LPMO-dependent H_2O_2 production determined by using the HRP/Amplex Red assay in the presence of commonly used low-molecular-weight reductants.—cont'd

CAZy family	LPMO name	Reaction conditions	Reaction rate	Reference
AA10	BlLPMO10A	5 µM LPMO, 50 µM ascorbic acid, 5 U/mL HRP, 100 µM AR in 20 mM Bis–Tris–HCl buffer, pH 6.0 at 22 °C	$\approx 0.01\,s^{-1}$ [a]	Mutahir et al., 2018: Fig. S3
	CfLPMO10	2 µM LPMO, 25 µM ascorbic acid, 0.2 U/mL HRP, 50 µM AR in 50 mM sodium phosphate buffer, pH 6.0	$\approx 0.0007\,s^{-1}$ [a]	Branch et al., 2021: Fig. 4A
	ScLPMO10C	1 µM LPMO, 1 mM ascorbic acid, 5 U/mL HRP, 100 µM AR in 50 mM sodium phosphate buffer, pH 6.0 at 30 °C	$0.0037\,s^{-1}$ [b]	Stepnov, Eijsink, and Forsberg (2022): Fig. 1A
	SmLPMO10A	1 µM LPMO, 50 µM/250 µM/1 mM ascorbic acid, 0.55 µM HRP, 100 µM AR in 50 mM Bis–Tris–HCl buffer, pH 6.5 at 37 °C	$0.001\,s^{-1}$; $0.002\,s^{-1}$; $0.001\,s^{-1}$ (50 µM; 250 µM; 1 mM ascorbic acid) [b]	Rieder, Petrovic, Valjamae, Eijsink, and Sorlie (2021): Table 2
	SscLPMO10B	0.5 µM LPMO, 1 mM reductant (ascorbic acid, gallic acid or L-cysteine), 5 U/mL HRP, 100 µM AR in 50 mM sodium phosphate buffer, pH 6.0 at 30 °C	$0.0052\,s^{-1}$; $0.012\,s^{-1}$; $0.043\,s^{-1}$ (ascorbic acid; gallic acid; L-cysteine) [b]	Stepnov, Christensen, et al. (2022): Fig. 6
	Tma12	9.8 µM LPMO, 27 µM ascorbic acid, 6.4 U/mL HRP, 45 µM AR in 90 mM sodium phosphate buffer, pH 6.0 at 22 °C	$\approx 0.0016\,s^{-1}$ [a]	Yadav, Archana, Singh, and Vasudev (2019): Table 2

AA11	*Af*LPMO11B	1 µM LPMO, 50 µM/250 µM/1 mM ascorbic acid, 0.55 µM HRP, 100 µM AR in 50 mM Bis–Tris–HCl buffer, pH 6.5 at 37°C	$0.017\,s^{-1}$; $0.091\,s^{-1}$; $0.183\,s^{-1}$ (50 µM; 250 µM; 1 mM ascorbic acid)[b]	Rieder, Petrovic, et al. (2021): Table 2
AA13	*Ao*LPMO13	3 µM LPMO, 50 µM ascorbic acid, 0.025 mg/mL HRP, 100 µM AR in 50 mM Bis–Tris–HCl buffer, pH 6.5 at 30°C	$\approx 0.011\,s^{-1}$ [a,b]	Rieder, Ebner, Glieder, and Sorlie (2021): Fig. 7A
	*Mt*LPMO13, *Nt*LPMO13	1 µM LPMO, 2 mM ascorbic acid, 1.3 µM HRP, 100 µM AR in 50 mM MOPS buffer, pH 7.0	H_2O_2 generation detected; estimation of rate not possible	Vu et al., 2019: Fig. S5
AA14	*Pc*LPMO14A, *Pc*LPMO14B	0.2–4 µM LPMO, 50 µM reductant (ascorbic acid, gallic acid, L-cysteine or epigallocatechin gallate), 7.1 U/mL HRP, 50 µM AR in 50 mM sodium phosphate buffer, pH 6.0 at 30°C	H_2O_2 generation detected; estimation of rate not possible	Couturier et al., 2018: Supplementary Table 2
AA15	*Cg*LPMO15A	1 µM LPMO, 50 µM ascorbic acid, 7 U/mL HRP, 50 µM AR at 30°C	$\approx 0.002\,s^{-1}$ [a,b]	Franco Cairo, Almeida, Damasio, Garcia, and Squina (2022): Fig. 4D
AA16	*Aa*LPMO16	10 µM LPMO, 50 µM reductant (ascorbic acid or L-cysteine), 7.1 U/mL HRP, 50 µM AR in 50 mM citrate–phosphate buffer, pH 6.0 at 30°C	H_2O_2 generation detected; estimation of rate not possible	Filiatrault-Chastel et al. (2019): Fig. S6

[a]The reaction rate was roughly estimated using progress curves or single time-point data presented in the paper.
[b]The data was corrected to account for reductant-dependent H_2O_2 underestimation.
AR, Amplex Red.

The list of published experiments in Table 1 shows large variation in the reaction conditions used to study LPMO-dependent H_2O_2 generation. For instance, the HRP concentrations and reductant concentrations used in these experiments may vary by more than 10-fold, which complicates comparison of the data. It is worth noting that up to this day most researchers use the HRP/Amplex Red assay in a semi-quantitative fashion as a simple tool to demonstrate that LPMOs are present in their enzyme preparations and contain a redox-active copper center. Hence, the repression of resorufin formation in the presence of reducing compounds is typically not accounted for (Table 1). On another note, the reported numbers are sometimes not corrected for LPMO-independent H_2O_2 generation (i.e., reductant auto-oxidation) or it is not clear whether such correction was performed. Finally, some of the published H_2O_2 production experiments did not involve saturation of the LPMOs with copper, meaning that the concentration of catalytically competent enzyme may have been overestimated. Given all the above, one should be careful in drawing generalized conclusions from the literature, e.g., when comparing the rates of H_2O_2 generation across various LPMO families. However, based on a limited number of studies that take reductant interference into account, use properly copper-saturated LPMOs, and assess hydrogen peroxide production by various LPMOs and/or LPMO-reductant combinations under the same reaction conditions, a few inferences can be made:

- The rates of hydrogen peroxide production by LPMOs, even LPMOs from the same CAZy family, may differ considerably under identical reaction conditions (Rieder, Petrovic, et al., 2021; Rieder, Stepnov, et al., 2021).
- The rate of LPMO-dependent H_2O_2 generation depends on the type of reductant used to fuel the reaction (Rieder, Stepnov, et al., 2021; Stepnov et al., 2021; Stepnov, Christensen, et al., 2022).
- The rate of LPMO-dependent H_2O_2 generation depends on the concentration of the reducing compound that is driving the reaction (Rieder, Petrovic, et al., 2021).
- The rate of LPMO-independent H_2O_2 generation obviously also depends on the type of reductant (Stepnov, Christensen, et al., 2022) and reductant concentration (Stepnov et al., 2021). The relative magnitude of LPMO-dependent and LPMO-independent H_2O_2 generation will vary between LPMO-reductant combinations (Stepnov, Christensen, et al., 2022).

Of note, while this has been rarely addressed in the LPMO literature (Hegnar et al., 2019), reductant behavior will be dependent on the pH and so will production of hydrogen peroxide.

While interpretation and comparative analysis of the data provided in Table 1 are difficult due to the variation in reaction conditions, it is worth noting that the observed rates of H_2O_2 production generally are very low, mostly falling between 0.1 and 1 per minute. This correlates well with reported rates for aerobic LPMO reactions (Bissaro, Varnai, et al., 2018), lending support to the claims made by some that O_2-dependent LPMO catalysis observed at standard aerobic conditions does not reflect mono-oxygenase activity but a peroxygenase reaction limited by hydrogen perox-ide generated *in situ* (Bissaro et al., 2017; Bissaro, Varnai, et al., 2018; Hegnar et al., 2019; Stepnov et al., 2021).

Despite the pitfalls of the HRP/Amplex Red assay discussed above, studying the H_2O_2 production rates of various LPMOs or LPMO-reductant combinations under comparable and well-controlled conditions may pro-vide valuable clues for understanding enzyme performance, especially since the discovery that *in situ* H_2O_2 production may be a key determinant of LPMO activity. Given this latter premise, it is reasonable to expect a corre-lation between H_2O_2 production rates estimated by the HRP/Amplex Red assay and LPMO catalytic rates observed in reactions with carbohydrate substrates (using the same pH and reductant). Data for LPMO-dependent H_2O_2 production are commonly obtained using low reductant concentra-tions ($\leq 50\,\mu M$; Table 1), whereas the majority of LPMO reactions with car-bohydrate substrates are carried out using high reductant concentrations ($\geq 1\,mM$) to ensure that catalysis is not limited a by lack of reducing equiv-alents. Needless to say, the quantitative comparison of H_2O_2 generation and substrate degradation data will only make sense in case the same reductant concentration is used in both types of experiments. On a side note, one should expect non-enzymatic H_2O_2 production (i.e., reductant auto-oxidation) to become more significant at higher reductant concentrations, as previously observed (Stepnov et al., 2021). Importantly, the rates of hydrogen peroxide production obtained in reactions containing LPMOs should not be corrected for this "background" abiotic H_2O_2 formation when matching HRP/Amplex Red assay data with substrate degradation data, since non-enzymatic generation of the co-substrate contributes to fueling LPMO activity.

According to a recent study assessing *in situ* H_2O_2 generation in reactions with a bacterial LPMO in the presence of three commonly used reductants (ascorbic acid, gallic acid and L-cysteine) at standard aerobic conditions, the hydrogen peroxide formation rates observed by the HRP/Amplex Red assay, indeed, correlate with the rates of cellulose depolymerization (Stepnov, Christensen, et al., 2022; Fig. 3). These results support the idea that LPMOs operate as strict peroxygenases even when reaction mixtures are not supplied with exogenous H_2O_2, and explain why enzyme activity is strongly dependent on the type of reductant, as previously demonstrated in a seminal paper by Kracher et al., who studied a large variety of electron donors (Kracher et al., 2016). Importantly, these results also show that, instead of just being a method for detecting LPMO activity, the HRP/Amplex Red assay can actually be used to get a meaningful impression of how well an LPMO-reductant combination may work for polysaccharide degradation.

The experiment depicted in Fig. 3 does not show a 1:1 stoichiometry between hydrogen peroxide formation and generation of oxidized LPMO products released from cellulose, which may seem surprising considering

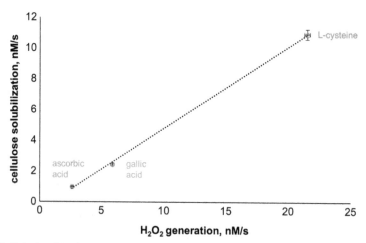

Fig. 3 Relationship between *in situ* H_2O_2 generation and cellulose solubilization observed in LPMO reactions under standard aerobic conditions in the presence of three different electron donors. The experiments were carried out in 50 mM sodium-phosphate buffer, pH 6.0 supplied with 1 mM reductant and 0.5 μM *Ssc*LPMO10B, at 30 °C. The cellulose solubilization rates were assessed using 1% (*w/v*) Avicel, whereas H_2O_2 generation rates were estimated by the HRP/Amplex Red assay in the presence of 5 U/mL HRP and 100 μM Amplex Red. The HRP/Amplex Red signal was corrected for the reductant-dependent underestimation of H_2O_2. Note that the H_2O_2 generation rates reflect overall hydrogen peroxide formation by both enzymatic and non-enzymatic reactions. Error bars indicate standard deviations between triplicates. The figure is based on a previously published dataset (Stepnov, Christensen, et al., 2022).

the proposed reaction mechanism ($R-H + H_2O_2 \rightarrow R-OH + H_2O$) (Bissaro et al., 2017). One explanation is that measuring the release of soluble products does not account for a potentially large fraction (typically 10–50%; Courtade, Forsberg, Heggset, Eijsink, & Aachmann, 2018) of oxidized sites that remain in the insoluble substrate, leading to underestimation of LPMO activity on cellulose.

Another plausible explanation for this discrepancy relates to the impact of substrate-binding on the LPMO oxidase activity. Binding to substrate shields the LPMO active site from the solvent (Bissaro, Isaksen, Vaaje-Kolstad, Eijsink, & Rohr, 2018; Brander et al., 2021; Frandsen et al., 2016), likely preventing the activation of copper by low-molecular-weight reductants. In other words, it is likely that in the presence of bulky polysaccharide substrates H_2O_2 generation is dominated by unbound LPMO species, meaning that the rate of enzymatic H_2O_2 generation will depend on the substrate concentration and LPMO affinity towards the substrate. Thus, LPMO-dependent H_2O_2 generation will be lower in the presence of substrate compared to the absence of substrate. The repressing effect of substrates on hydrogen peroxide production by LPMOs is supported by recently published dose-response experiments with a bacterial LPMO and ascorbic acid (Stepnov et al., 2021), showing that cellulose degradation rates display low dependency on the LPMO dose in a wide range of enzyme concentrations (0.1–4 µM), despite a strong response to the LPMO dose in H_2O_2 production as assessed by the HRP/Amplex Red assay. This shows that, for this LPMO, in the presence of substrate, H_2O_2 production through oxidase activity becomes of minor importance relative to H_2O_2 production resulting from reductant auto-oxidation.

On a side note, the studies alluded to above differ from earlier studies that have firmly established that adding LPMO substrates to HRP/Amplex Red assay reactions leads to complete repression of apparent H_2O_2 production (Hangasky, Iavarone, & Marletta, 2018; Isaksen et al., 2014; Kittl et al., 2012) and thus may be taken to support the notion that the presence of substrate limits, and even abolishes H_2O_2 production. Importantly, the repression of H_2O_2 production in these early studies likely reflects productive consumption of hydrogen peroxide by the LPMOs, rather than substrate-binding effects. Finally, it is important to realize that non-enzymatic routes for H_2O_2 generation, which, as mentioned above, may play a significant role in reactions with LPMO substrates, are sensitive to reaction conditions, such as pH (Pant, Ozkasikci, Furtauer, & Reinelt, 2019; Wilson et al., 1995) and the concentration of transition metal ions (Bissaro, Kommedal, Rohr, & Eijsink, 2020; Stepnov et al., 2021).

4. The effect of free copper ions on LPMO reactions

Although LPMOs are known to be able to utilize multiple types of electron donors, including protein redox partners (Garajova et al., 2016; Kracher et al., 2016; Phillips et al., 2011; Tan et al., 2015), low-molecular-weight reductants are typically employed in experimental set-ups, due to their affordability and ease of handling. However, the use of these seemingly simple compounds involves surprisingly complex and sometimes unpredictable chemistry.

Next to being the most popular LPMO reductant, ascorbic acid is well known for its ability to interact with Cu(II) ions in a manner that leads to dramatically increased auto-oxidation (i.e., hydrogen peroxide production) rates (Buettner & Jurkiewicz, 1996; Zhou et al., 2016). Some researchers suggest that the reaction between ascorbic acid and molecular oxygen always involves trace amounts of copper present in solution and, thus, is not an "auto-oxidation" in a strict sense (Wilson et al., 1995). It has been shown that the addition of micromolar amounts of free copper ions to LPMO reactions fueled by 1 mM ascorbic acid triggers massive *in situ* hydrogen peroxide generation, which, in turn, causes a boost in the enzymatic oxidation of cellulose accompanied by rapid enzyme inactivation (Stepnov et al., 2021) (Fig. 4).

Copper-mediated reactions with O_2 are not unique to ascorbic acid. Other commonly used LPMO reductants, such as DTT (Kachur et al., 1997), L-cysteine (Kachur et al., 1999) and reduced glutathione (Kachur et al., 1998) display similar behavior. Interestingly, the auto-oxidation of gallic acid is virtually unaffected by the presence of copper, as indeed observed in LPMO studies (Stepnov et al., 2021), which is likely explained by the propensity of this reductant to chelate Cu(II) ions instead of reducing them (Severino, Goodman, Reichenauer, & Pirker, 2011).

Importantly, there is evidence suggesting that some published LPMO activity data suffer from copper contamination, most likely in the enzyme stock solutions. When reviewing datasets previously published by our group for *Sc*LPMO10C (e.g., Bissaro et al., 2016; Jensen et al., 2019), we noted a more than 10-fold difference in cellulose oxidation rates derived from studies that were using comparable experimental conditions, with ascorbic acid as a reductant (Stepnov et al., 2021). Indeed, a subsequent analysis of the copper content of various *Sc*LPMO10C stock solutions revealed that a sample showing unusually high activity in cellulose degradation experiments contained $1.7 \pm 0.3 \mu M$ total copper per $1 \mu M$ enzyme, according to

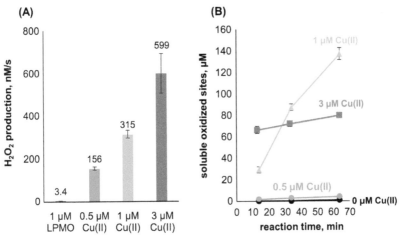

Fig. 4 Effect of free copper ions on hydrogen peroxide generation and LPMO activity. Panel (A) shows the rates of H_2O_2 generation obtained in the HRP/Amplex Red assay for reactions with A10_07 (a close homolog of ScLPMO10C) or with various amounts of Cu(II)SO$_4$, whereas panel (B) shows the progress curves for cellulose solubilization (1% w/v Avicel) by 1 μM AA10_07 in the presence or absence of free copper. Note that H_2O_2 production in the reaction with 1 μM LPMO is negligible compared to the reaction with just 0.5 μM free copper. Also note that substrate oxidation in the reaction lacking free copper (black curve in panel B) is remarkably slow ($\approx 0.06\,\text{min}^{-1}$) (Stepnov et al., 2021). The red progress curve in panel (B) is typical for a reaction with "too much" H_2O_2: the reaction proceeds very fast but at 10 min, the first measuring point, the enzyme is almost completely inactivated. All experiments were carried out at 30 °C in 50 mM sodium phosphate buffer, pH 6.0, using 1 mM ascorbic acid as an electron donor. The reactions featured in panel A contained 5 U/mL HRP and 100 μM Amplex Red. Error bars indicate standard deviations between triplicates. The figure is based on previously published datasets (Stepnov et al., 2021; Stepnov, Eijsink, et al., 2022). Nb. While the impact of copper ions is evident from this figure, the difference between 0.5 and 1 μM copper in panel B is remarkable and possibly due to some copper binding capacity in the substrate or enzyme. See Stepnov et al. (2021) for further discussion.

inductively coupled plasma mass spectrometry (ICP-MS) analysis. After desalting this enzyme batch, its activity became comparable to that of other enzyme batches (see Stepnov et al., 2021 for more data and discussion and see below for how to detect copper contamination without the need for ICP-MS).

There are several potential ways for free copper to enter LPMO reactions. First of all, most LPMO purification protocols involve a copper-saturation procedure followed by desalting. During copper-saturation, LPMO solutions are typically co-incubated with a slight (1:2 or 1:3) molar excess of copper ions to ensure full loading of the metal-binding site. Variation in the efficiency of subsequent desalting procedures may explain differences in the free copper content between enzyme samples. It cannot be excluded that the carry-over of residual free copper into LPMO preparations is facilitated by secondary low-affinity metal binding sites, as observed in some X-ray structures (Borisova et al., 2015). To make things even more complex, it has been recently shown that LPMO inactivation, which is an autocatalytic process that leads to damage of (copper-binding) active site residues (Bissaro et al., 2017), may trigger the release of copper from the damaged active site into solution (Stepnov, Eijsink, et al., 2022). One possible consequence is that free copper ions may accumulate in enzyme stock solutions that are stored for a long time and may suffer from slow inactivation, as observed previously (Kadowaki et al., 2018). Another consequence, addressed in detail in a recent study (Stepnov, Eijsink, et al., 2022), is that copper may accumulate in standard LPMO reactions, due to slow (Kuusk & Valjamae, 2021), but in many cases inevitable enzyme inactivation. Fig. 4 shows that even relatively low amounts of released copper may significantly affect the level of *in situ* H_2O_2 production. In fact, it has been suggested that under some conditions, enzyme inactivation may become a self-reinforcing process (more free copper $\rightarrow$ more H_2O_2 $\rightarrow$ even more enzyme inactivation and free copper (Stepnov, Eijsink, et al., 2022). A final potential source of free copper concerns the substrate; there is evidence suggesting that some commercial-grade LPMO substrates contain significant amounts of copper (or other transition metals) that cause H_2O_2 generation in the presence of copper-sensitive reductants (Rieder, Stepnov, et al., 2021).

These copper effects can be mitigated by using gallic acid as an electron donor (Stepnov et al., 2021), however side-reactions involving LPMOs and this reductant make it a poor choice for the study of *in situ* H_2O_2 generation by the HRP/Amplex Red assay (see Section 2). Preparing copper free reactants and controlling the copper content of enzyme and substrate preparations by ICP-MS seem to be the most straightforward solutions to handling the free copper problem, although issues related to LPMO inactivation will remain. Since ICP-MS requires expensive equipment and can be rather time consuming, an alternative method has been developed in which free copper ions in LPMO preparations are indirectly observed using the HRP/Amplex Red assay, which, as shown in Fig. 4, is sufficiently sensitive to detect sub-micromolar amounts of copper ions. To do so, protein-free samples must be produced by filtering LPMO stock solutions through low molecular weight cut-off (MWCO; e.g., 3 kDa) micro-centrifuge tube filters. These filtrates will contain the same amount of free copper as the original LPMO preparations and can be used to set up HRP/Amplex Red assay reactions, as further discussed in the protocol below.

5. Concluding remarks

All in all, despite the fact that the HRP/Amplex Red assay does not fully reflect true LPMO turnover conditions (e.g., due to substrate binding effects), this assay allows quick assessment of the H_2O_2 generating potential of LPMO-reductant combinations, which is relevant for predicting and understanding enzyme performance in substrate degradation experiments. Therefore, the HRP/Amplex Red assay can be used as an easily accessible tool for optimizing LPMO reactions, such as pre-screening of buffers and enzyme-reductant combinations. As alluded to above, the data obtained in the presence of different reductants (or different concentrations of the same reductant) should always be corrected for underestimation of H_2O_2 due to side reactions involving the electron donor.

Because of its simplicity, including its microtiter plate format, it is tempting to employ the HRP/Amplex Red assay in LPMO engineering campaigns. While the assay indeed may be used for detection of LPMO activity (in rather clean reaction set-ups), it should be obvious from the above that its pitfalls will be particularly problematic when screening non-purified LPMO variants. Furthermore, although typical laboratory reaction set-ups show clear and scientifically interesting correlations between (*in situ*) H_2O_2 production measured by the HRP/Amplex Red assay and LPMO activity on real, insoluble substrates, it is not at all certain that an LPMO variant found to be "more active" after screening with the HRP/Amplex Red assay also works better in a real application. One reason is the H_2O_2-stimulated auto-catalytic inactivation of LPMOs, alluded to above. It is vital that H_2O_2 generation experiments are followed by analysis of LPMO-dependent substrate oxidation, before any claims are made. On a general note, when screening for LPMO variants, we would recommend using real substrates combined with high-throughput mass spectrometry-based product detection, which, admittedly, is considerably more challenging and expensive compared to using the Amplex Red/HRP assay (Forsberg, Stepnov, Naerdal, Klinkenberg, & Eijsink, 2020; Jensen et al., 2019).

There are several limitations to the HRP/Amplex Red assay that follow from the copper-sensitivity of common LPMO reductants, such as ascorbic acid. One should always control the free copper content of LPMO preparations used in H_2O_2 generation experiments, particularly when analyzing samples with low degree of purity. Obviously, the use of cell extracts or secretomes in HRP/Amplex Red experiments should be avoided, because of the potential presence of transition metal ions and/or other HRP substrates and/or other H_2O_2 generating or consuming enzymes. Last but not least, given the evidence showing that copper may leave the active site of damaged LPMOs (Stepnov, Eijsink, et al., 2022), the application of the HRP/Amplex Red assay for measuring residual LPMO activity in various types of inactivation experiments (such as experiments for assessing thermal stability) cannot be recommended when using copper-sensitive reductants such ascorbic acid, reduced glutathione, L-cysteine and DTT.

The final part of this Chapter, Section 6, describes a modified version of the standard HRP/Amplex Red assay protocol (Kittl et al., 2012) that accounts for reductant-dependent signal repression and the potential contamination of LPMO samples with free copper ions.

6. Microtiter plate-based protocol for measuring *in situ* H_2O_2 generation in LPMO reactions

The following protocol relies on detection of resorufin absorbance instead of fluorescence, to make it compatible with most of types of plate readers and microtiter plates. 1 mM ascorbic acid is used in the assay to match the reaction conditions of typical LPMO experiments. The final reaction volume of the assay is 100 µL per well. The final concentrations of ascorbic acid, HRP, Amplex Red and DMSO amount to 1 mM, 5 U/mL, 100 µM and 1% (*v*/v), respectively. We recommend using LPMOs at 1–3 µM concentration to achieve the best results.

6.1 Materials and equipment

1. Amplex Red reagent (e.g., ThermoFisher Scientific, product number A12222)
2. DMSO (e.g., Sigma, product number 34869-2.5L)
3. Hydrogen peroxide (e.g., VWR, product number 23619.297)
4. L-ascorbic acid (e.g., Sigma, product number A5960-100G)
5. Low-binding 3 kDa MWCO micro-centrifugal filters (e.g., VWR, product number 82031–346)
6. Lyophilized horseradish peroxidase type II (e.g., Sigma, product number P8250-100KU)
7. Microplate reader capable of continuously recording absorption at 563 nm and plate shaking
8. Transparent 96-well microtiter plates (e.g., ThermoFisher Scientific, product number 167008)

9. Water of Milli-Q grade or equivalent and in some cases, indicated below, preferably even cleaner metal-free water (e.g., TraceSELECT water, Honeywell, product number 95305-10 L)
10. 0.5 M sodium phosphate (NaPi) buffer, pH 6.0

6.2 Before you begin

1. Prepare a 10 mM solution of ascorbic acid and a 100 μM solution of hydrogen peroxide in Milli-Q water. Make sure to follow safety procedures in case concentrated H_2O_2 is used to produce 100 μM solution. Ideally, highly pure metal-free water should be used instead of Milli-Q water to ensure the long-term stability of reductant and hydrogen peroxide. Filter-sterilize the ascorbic acid solution through a 0.22 μm syringe filter. Aliquot the reductant and H_2O_2 solutions, store at −20 °C away from light, and use only once after thawing.
2. Prepare a 10 mM solution of Amplex Red in DMSO. Aliquot the Amplex Red solution, store at −20 °C away from light, and use only once after thawing.
3. Prepare a 100 U/mL solution of horseradish peroxidase type II (HRP) by diluting the appropriate amount of lyophilized enzyme in Milli-Q water. Filter-sterilize the solution using a 0.22 μm syringe filter and store at +4 °C.

6.3 Step-by-step method details

1. Prepare a protein-free control sample by filtering an appropriate amount of the LPMO stock solution through a 3 kDa MWCO micro-centrifugal filter. Discard the retentate and collect the flow-through. The protein-free flow-through will contain the same amount of free copper as the original (unfiltered) LPMO solution. Use this filtrate in control reactions, using a sample volume identical to the volume of LPMO stock solution used in reactions containing the enzyme (see step 4). Store the protein-free control samples in microcentrifuge tubes at room temperature, or at +4 °C in case of a long-term storage.
2. Prepare an HRP/Amplex Red pre-mix by combining the following components (the volumes refer to one single reaction):

10 mM Amplex Red	1 μL
100 U/mL HRP	5 μL
0.5 M NaPi buffer	5 μL
Milli-Q H_2O	39 μL

Note: the simplest experiment (one sample, one replicate) will require at least eight independent reactions due to the need for control reactions and reactions with H_2O_2 standards. Make sure to prepare the HRP/Amplex Red pre-mix in slight excess (plan for 3–5 extra reactions). Store the reagent in a dark place at room temperature for no longer than one hour.

3. **(a)** Prepare standard solutions by mixing the following components in a transparent 96–well microtiter plate:

Standard sample	Milli-Q H_2O (µL)	0.5 M NaPi buffer (µL)	100 µM H_2O_2 (µL)	10 mM ascorbic acid (µL)
0 µM H_2O_2	35	5	–	10
5 µM H_2O_2	30	5	5	10
10 µM H_2O_2	25	5	10	10
15 µM H_2O_2	20	5	15	10
20 µM H_2O_2	15	5	20	10

    (b) Quickly supply the wells containing standard samples with 50 µL of the HRP/Amplex Red pre-mix and immediately transfer the plate to the plate reader (at room temperature). Mix the solutions by shaking the plate for 30 s at 600 RPM and then measure the absorption at 563 nm. Plot the calibration curve and use it only for data obtained with the same batch of HRP/Amplex Red pre-mix.

Note: the resorufin signal will slowly increase in standard samples due to the auto-oxidation of ascorbic acid. Hence, it is important to minimize the time used to prepare and measure the standard samples. At pH 6.0, the signal drift will be negligible if the procedure is completed within 5–10 min.

4. **(a)** Load the microtiter plate with ≤ 35 µL of LPMO solution. Per well, add 5 µL of 0.5 M NaPi buffer, pH 6.0, and the appropriate amount of Milli-Q water to make the total volume 40 µL.

Note: be aware of possible buffer/pH effects that may emerge if the volume of LPMO solution is high and this solution is buffered.

    (b) Prepare control samples as in step 4a, now substituting the LPMO with the same volume of Milli-Q water (control reaction A) or protein-free ultrafiltrate obtained in step 1 (control reaction B).

    (c) Add 50 µL of the HRP/Amplex Red pre-mix to each well. Avoid formation of air bubbles.

(d) Pre-incubate the plate inside the plate reader for 5 min to reach the desired temperature (usually 30–40 °C). Skip the pre-incubation step if the experiment is to be carried out at room temperature.

(e) Initiate the reactions by adding 10 μL of 10 mM ascorbic acid to each well using a multi-channel pipette. Mix the solutions by shaking the plate for 30 s at 600 RPM and then measure absorption at 563 nm every 10–30 s for at least 90 min.

Note: shorter assay durations may be possible in cases where determination of the (typically low) rate of ascorbic acid auto-oxidation (control reaction A) is not required.

6.4 Quantification and analysis

Calculate the rates of hydrogen peroxide generation in the reactions using the linear parts of the resorufin accumulation curves and the standard curve obtained in step 3. Note that the shape of the resorufin accumulation curves will depend on the H_2O_2 generation rate. Fast reactions (such as reactions with free copper) will slow down and eventually cease due to depletion of Amplex Red.

The rate of control reaction A will reflect the impact of ascorbic acid auto-oxidation on overall H_2O_2 production. If the rate of control reaction B is significantly higher compared to control reaction A, the enzyme sample likely contained free copper and desalting of the LPMO stock solution to remove free (i.e., unbound) copper is advisable.

References

Akagawa, M., Shigemitsu, T., & Suyama, K. (2003). Production of hydrogen peroxide by polyphenols and polyphenol-rich beverages under quasi-physiological conditions. *Bioscience, Biotechnology, and Biochemistry, 67*(12), 2632–2640.

Bissaro, B., Forsberg, Z., Ni, Y., Hollmann, F., Vaaje-Kolstad, G., & Eijsink, V. G. H. (2016). Fueling biomass-degrading oxidative enzymes by light-driven water oxidation. *Green Chemistry, 18*(19), 5357–5366.

Bissaro, B., Isaksen, I., Vaaje-Kolstad, G., Eijsink, V. G. H., & Rohr, A. K. (2018). How a lytic polysaccharide monooxygenase binds crystalline chitin. *Biochemistry, 57*(12), 1893–1906.

Bissaro, B., Kommedal, E., Rohr, A. K., & Eijsink, V. G. H. (2020). Controlled depolymerization of cellulose by light-driven lytic polysaccharide oxygenases. *Nature Communications, 11*(1), 890.

Bissaro, B., Varnai, A., Rohr, A. K., & Eijsink, V. G. H. (2018). Oxidoreductases and reactive oxygen species in conversion of lignocellulosic biomass. *Microbiology and Molecular Biology Reviews, 82*(4).

Bissaro, B., et al. (2017). Oxidative cleavage of polysaccharides by monocopper enzymes depends on H_2O_2. *Nature Chemical Biology, 13*(10), 1123–1128.

Bissaro, B., et al. (2020). Molecular mechanism of the chitinolytic peroxygenase reaction. *Proceedings of the National Academy of Sciences of the United States of America*, *117*(3), 1504–1513.

Borisova, A. S., et al. (2015). Structural and functional characterization of a lytic polysaccharide monooxygenase with broad substrate specificity. *The Journal of Biological Chemistry*, *290*(38), 22955–22969.

Branch, J., et al. (2021). C-type cytochrome-initiated reduction of bacterial lytic polysaccharide monooxygenases. *The Biochemical Journal*, *478*(14), 2927–2944.

Brander, S., et al. (2020). Biochemical evidence of both copper chelation and oxygenase activity at the histidine brace. *Scientific Reports*, *10*(1), 16369.

Brander, S., et al. (2021). Scission of glucosidic bonds by a *Lentinus similis* lytic polysaccharide monooxygenases is strictly dependent on H_2O_2 while the oxidation of saccharide products depends on O_2. *ACS Catalysis*, *11*(22), 13848–13859.

Breslmayr, E., et al. (2018). A fast and sensitive activity assay for lytic polysaccharide monooxygenase. *Biotechnology for Biofuels*, *11*, 79.

Buettner, G. R., & Jurkiewicz, B. A. (1996). Catalytic metals, ascorbate and free radicals: combinations to avoid. *Radiation Research*, *145*(5), 532–541.

Caldararu, O., Oksanen, E., Ryde, U., & Hedegard, E. D. (2019). Mechanism of hydrogen peroxide formation by lytic polysaccharide monooxygenase. *Chemical Science*, *10*(2), 576–586.

Cannella, D., Hsieh, C. W., Felby, C., & Jorgensen, H. (2012). Production and effect of aldonic acids during enzymatic hydrolysis of lignocellulose at high dry matter content. *Biotechnology for Biofuels*, *5*(1), 26.

Cantarel, B. L., Coutinho, P. M., Rancurel, C., Bernard, T., Lombard, V., & Henrissat, B. (2009). The carbohydrate-active enzymes database (CAZy): An expert resource for glycogenomics. *Nucleic Acids Research*, *37*(Database issue), D233–D238.

Chakraborty, S., et al. (2016). Quantification of hydrogen peroxide in plant tissues using Amplex Red. *Methods*, *109*, 105–113.

Chylenski, P., et al. (2017). Enzymatic degradation of sulfite-pulped softwoods and the role of LPMOs. *Biotechnology for Biofuels*, *10*, 177.

Courtade, G., Forsberg, Z., Heggset, E. B., Eijsink, V. G. H., & Aachmann, F. L. (2018). The carbohydrate-binding module and linker of a modular lytic polysaccharide monooxygenase promote localized cellulose oxidation. *The Journal of Biological Chemistry*, *293*(34), 13006–13015.

Couturier, M., et al. (2018). Lytic xylan oxidases from wood-decay fungi unlock biomass degradation. *Nature Chemical Biology*, *14*(3), 306–310.

Debski, D., et al. (2016). Mechanism of oxidative conversion of Amplex(R) Red to resorufin: Pulse radiolysis and enzymatic studies. *Free Radical Biology & Medicine*, *95*, 323–332.

Eijsink, V. G. H., et al. (2019). On the functional characterization of lytic polysaccharide monooxygenases (LPMOs). *Biotechnology for Biofuels*, *12*, 58.

Filandr, F., Man, P., Halada, P., Chang, H., Ludwig, R., & Kracher, D. (2020). The H_2O_2-dependent activity of a fungal lytic polysaccharide monooxygenase investigated with a turbidimetric assay. *Biotechnology for Biofuels*, *13*, 37.

Filiatrault-Chastel, C., et al. (2019). AA16, a new lytic polysaccharide monooxygenase family identified in fungal secretomes. *Biotechnology for Biofuels*, *12*, 55.

Forsberg, Z., Stepnov, A. A., Naerdal, G. K., Klinkenberg, G., & Eijsink, V. G. H. (2020). Engineering lytic polysaccharide monooxygenases (LPMOs). *Methods in Enzymology*, *644*, 1–34.

Forsberg, Z., et al. (2011). Cleavage of cellulose by a CBM33 protein. *Protein Science*, *20*(9), 1479–1483.

Franco Cairo, J. P. L., Almeida, D. V., Damasio, A., Garcia, W., & Squina, F. M. (2022). The periplasmic expression and purification of AA15 lytic polysaccharide monooxygenases from insect species in *Escherichia coli*. *Protein Expression and Purification*, *190*, 105994.

Frandsen, K. E., et al. (2016). The molecular basis of polysaccharide cleavage by lytic polysaccharide monooxygenases. *Nature Chemical Biology, 12*(4), 298–303.

Garajova, S., et al. (2016). Single-domain flavoenzymes trigger lytic polysaccharide monooxygenases for oxidative degradation of cellulose. *Scientific Reports, 6,* 28276.

Gorris, H. H., & Walt, D. R. (2009). Mechanistic aspects of horseradish peroxidase elucidated through single-molecule studies. *Journal of the American Chemical Society, 131*(17), 6277–6282.

Hangasky, J. A., Iavarone, A. T., & Marletta, M. A. (2018). Reactivity of O_2 versus H_2O_2 with polysaccharide monooxygenases. *Proceedings of the National Academy of Sciences of the United States of America, 115*(19), 4915–4920.

Harris, P. V., et al. (2010). Stimulation of lignocellulosic biomass hydrolysis by proteins of glycoside hydrolase family 61: structure and function of a large, enigmatic family. *Biochemistry, 49*(15), 3305–3316.

Hedison, T. M., et al. (2021). Insights into the H_2O_2-driven catalytic mechanism of fungal lytic polysaccharide monooxygenases. *The FEBS Journal, 288,* 4115–4128.

Hegnar, O. A., Petrovic, D. M., Bissaro, B., Alfredsen, G., Varnai, A., & Eijsink, V. G. H. (2019). pH-dependent relationship between catalytic activity and hydrogen peroxide production shown via characterization of a lytic polysaccharide monooxygenase from *Gloeophyllum trabeum. Applied and Environmental Microbiology, 85*(5).

Horn, S. J., Vaaje-Kolstad, G., Westereng, B., & Eijsink, V. G. (2012). Novel enzymes for the degradation of cellulose. *Biotechnology for Biofuels, 5*(1), 45.

Isaksen, T., et al. (2014). A C4-oxidizing lytic polysaccharide monooxygenase cleaving both cellulose and cello-oligosaccharides. *The Journal of Biological Chemistry, 289*(5), 2632–2642.

Jensen, M. S., et al. (2019). Engineering chitinolytic activity into a cellulose-active lytic polysaccharide monooxygenase provides insights into substrate specificity. *The Journal of Biological Chemistry, 294*(50), 19349–19364.

Jones, S. M., Transue, W. J., Meier, K. K., Kelemen, B., & Solomon, E. I. (2020). Kinetic analysis of amino acid radicals formed in H_2O_2-driven Cu(I) LPMO reoxidation implicates dominant homolytic reactivity. *Proceedings of the National Academy of Sciences of the United States of America, 117*(22), 11916–11922.

Kachur, A. V., Held, K. D., Koch, C. J., & Biaglow, J. E. (1997). Mechanism of production of hydroxyl radicals in the copper-catalyzed oxidation of dithiothreitol. *Radiation Research, 147*(4).

Kachur, A. V., Koch, C. J., & Biaglow, J. E. (1998). Mechanism of copper-catalyzed oxidation of glutathione. *Free Radical Research, 28*(3), 259–269.

Kachur, A. V., Koch, C. J., & Biaglow, J. E. (1999). Mechanism of copper-catalyzed autoxidation of cysteine. *Free Radical Research, 31*(1), 23–34.

Kadic, A., Varnai, A., Eijsink, V. G. H., Horn, S. J., & Liden, G. (2021). In situ measurements of oxidation-reduction potential and hydrogen peroxide concentration as tools for revealing LPMO inactivation during enzymatic saccharification of cellulose. *Biotechnology for Biofuels, 14*(1), 46.

Kadowaki, M. A. S., et al. (2018). Functional characterization of a lytic polysaccharide monooxygenase from the thermophilic fungus *Myceliophthora thermophila. PLoS One, 13*(8), e0202148.

Karakuzu, O., Cruz, M. R., Liu, Y., & Garsin, D. A. (2019). Amplex Red assay for measuring hydrogen peroxide production from *Caenorhabditis elegans. Bio-Protocol, 9*(21).

Kittl, R., Kracher, D., Burgstaller, D., Haltrich, D., & Ludwig, R. (2012). Production of four *Neurospora crassa* lytic polysaccharide monooxygenases in *Pichia pastoris* monitored by a fluorimetric assay. *Biotechnology for Biofuels, 5*(1), 79.

Kjaergaard, C. H., et al. (2014). Spectroscopic and computational insight into the activation of O2 by the mononuclear Cu center in polysaccharide monooxygenases. *Proceedings of the National Academy of Sciences of the United States of America, 111*(24), 8797–8802.

Kracher, D., et al. (2016). Extracellular electron transfer systems fuel cellulose oxidative degradation. *Science*, *352*(6289), 1098–1101.

Kuusk, S., & Valjamae, P. (2021). Kinetics of H_2O_2-driven catalysis by a lytic polysaccharide monooxygenase from the fungus *Trichoderma reesei*. *The Journal of Biological Chemistry*, *297*(5), 101256.

Kuusk, S., et al. (2018). Kinetics of H2O2-driven degradation of chitin by a bacterial lytic polysaccharide monooxygenase. *The Journal of Biological Chemistry*, *293*(2), 523–531.

Maddalena, L. A., Selim, S. M., Fonseca, J., Messner, H., McGowan, S., & Stuart, J. A. (2017). Hydrogen peroxide production is affected by oxygen levels in mammalian cell culture. *Biochemical and Biophysical Research Communications*, *493*(1), 246–251.

Mehlhorn, H., Lelandais, M., Korth, H. G., & Foyer, C. H. (1996). Ascorbate is the natural substrate for plant peroxidases. *FEBS Letters*, *378*(3), 203–206.

Mishin, V., Gray, J. P., Heck, D. E., Laskin, D. L., & Laskin, J. D. (2010). Application of the Amplex Red/horseradish peroxidase assay to measure hydrogen peroxide generation by recombinant microsomal enzymes. *Free Radical Biology & Medicine*, *48*(11), 1485–1491.

Muller, G., Chylenski, P., Bissaro, B., Eijsink, V. G. H., & Horn, S. J. (2018). The impact of hydrogen peroxide supply on LPMO activity and overall saccharification efficiency of a commercial cellulase cocktail. *Biotechnology for Biofuels*, *11*, 209.

Mutahir, Z., et al. (2018). Characterization and synergistic action of a tetra-modular lytic polysaccharide monooxygenase from *Bacillus cereus*. *FEBS Letters*, *592*(15), 2562–2571.

Pant, A. F., Ozkasikci, D., Furtauer, S., & Reinelt, M. (2019). The effect of deprotonation on the reaction kinetics of an oxygen scavenger based on gallic acid. *Frontiers in Chemistry*, *7*, 680.

Petrovic, D. M., et al. (2018). Methylation of the N-terminal histidine protects a lytic polysaccharide monooxygenase from auto-oxidative inactivation. *Protein Science*, *27*(9), 1636–1650.

Phillips, C. M., Beeson, W. T., Cate, J. H., & Marletta, M. A. (2011). Cellobiose dehydrogenase and a copper-dependent polysaccharide monooxygenase potentiate cellulose degradation by *Neurospora crassa*. *ACS Chemical Biology*, *6*(12), 1399–1406.

Quinlan, R. J., et al. (2011). Insights into the oxidative degradation of cellulose by a copper metalloenzyme that exploits biomass components. *Proceedings of the National Academy of Sciences of the United States of America*, *108*(37), 15079–15084.

Rieder, L., Ebner, K., Glieder, A., & Sorlie, M. (2021). Novel molecular biological tools for the efficient expression of fungal lytic polysaccharide monooxygenases in *Pichia pastoris*. *Biotechnology for Biofuels*, *14*(1), 122.

Rieder, L., Petrovic, D., Valjamae, P., Eijsink, V. G. H., & Sorlie, M. (2021). Kinetic characterization of a putatively chitin-active LPMO reveals a preference for soluble substrates and absence of monooxygenase activity. *ACS Catalysis*, *11*(18), 11685–11695.

Rieder, L., Stepnov, A. A., Sorlie, M., & Eijsink, V. G. H. (2021). Fast and specific peroxygenase reactions catalyzed by fungal mono-copper enzymes. *Biochemistry*, *60*(47), 3633–3643.

Rodrigues, J. V., & Gomes, C. M. (2010). Enhanced superoxide and hydrogen peroxide detection in biological assays. *Free Radical Biology & Medicine*, *49*(1), 61–66.

Severino, J. F., Goodman, B. A., Reichenauer, T. G., & Pirker, K. F. (2011). Is there a redox reaction between Cu(II) and gallic acid? *Free Radical Research*, *45*(2), 115–124.

Stepnov, A. A., Christensen, I. A., Forsberg, Z., Aachmann, F. L., Courtade, G., & Eijsink, V. G. H. (2022). The impact of reductants on the catalytic efficiency of a lytic polysaccharide monooxygenase and the special role of dehydroascorbic acid. *FEBS Letters*, *596*(1), 53–70.

Stepnov, A. A., Eijsink, V. G. H., & Forsberg, Z. (2022). Enhanced in situ H_2O_2 production explains synergy between an LPMO with a cellulose-binding domain and a single-domain LPMO. *Scientific Reports*, *12*(1), 6129.

Stepnov, A. A., et al. (2021). Unraveling the roles of the reductant and free copper ions in LPMO kinetics. *Biotechnology for Biofuels, 14*(1), 28.

Summers, F. A., Zhao, B., Ganini, D., & Mason, R. P. (2013). Photooxidation of Amplex Red to resorufin: implications of exposing the Amplex Red assay to light. *Methods in Enzymology, 526*, 1–17.

Tan, T. C., et al. (2015). Structural basis for cellobiose dehydrogenase action during oxidative cellulose degradation. *Nature Communications, 6*, 7542.

Vaaje-Kolstad, G., Horn, S. J., van Aalten, D. M., Synstad, B., & Eijsink, V. G. (2005). The non-catalytic chitin-binding protein CBP21 from *Serratia marcescens* is essential for chitin degradation. *The Journal of Biological Chemistry, 280*(31), 28492–28497.

Vaaje-Kolstad, G., et al. (2010). An oxidative enzyme boosting the enzymatic conversion of recalcitrant polysaccharides. *Science, 330*(6001), 219–222.

Vu, V. V., et al. (2019). Substrate selectivity in starch polysaccharide monooxygenases. *The Journal of Biological Chemistry, 294*(32), 12157–12166.

Wang, B., Walton, P. H., & Rovira, C. (2019). Molecular mechanisms of oxygen activation and hydrogen peroxide formation in lytic polysaccharide monooxygenases. *ACS Catalysis, 9*(6), 4958–4969.

Wilson, R. J., Beezer, A. E., & Mitchell, J. C. (1995). A kinetic study of the oxidation of L-ascorbic acid (vitamin C) in solution using an isothermal microcalorimeter. *Thermochimica Acta, 264*, 27–40.

Yadav, S. K., Archana, S. R., Singh, P. K., & Vasudev, P. G. (2019). Insecticidal fern protein Tma12 is possibly a lytic polysaccharide monooxygenase. *Planta, 249*(6), 1987–1996.

Zhou, M., & Panchuk-Voloshina, N. (1997). A one-step fluorometric method for the continuous measurement of monoamine oxidase activity. *Analytical Biochemistry, 253*(2), 169–174.

Zhou, P., et al. (2016). Generation of hydrogen peroxide and hydroxyl radical resulting from oxygen-dependent oxidation of l-ascorbic acid via copper redox-catalyzed reactions. *RSC Advances, 6*(45), 38541–38547. https://doi.org/10.1039/C6RA02843H.

CHAPTER SEVEN

Bioinformatic prediction and experimental validation of RiPP recognition elements

Kyle E. Shelton[a,b] (iD) **and Douglas A. Mitchell**[a,b,c,*] (iD)

[a]Department of Chemistry, University of Illinois at Urbana-Champaign, Urbana, IL, United States
[b]Carl R. Woese Institute for Genomic Biology, University of Illinois at Urbana-Champaign, Urbana, IL, United States
[c]Department of Microbiology, University of Illinois at Urbana-Champaign, Urbana, IL, United States
*Corresponding author: e-mail address: douglasm@illinois.edu

Contents

Methods in Enzymology, Volume 679
ISSN 0076-6879
https://doi.org/10.1016/bs.mie.2022.08.050

Abstract

Ribosomally synthesized and post-translationally modified peptides (RiPPs) are a family of natural products for which discovery efforts have rapidly grown over the past decade. There are currently 38 known RiPP classes encoded by prokaryotes. Half of the prokaryotic RiPP classes include a protein domain called the RiPP Recognition Element (RRE) for successful installation of post-translational modifications on a RiPP precursor peptide. In most cases, the RRE domain binds to the N-terminal "leader" region of the precursor peptide, facilitating enzymatic modification of the C-terminal "core" region. The prevalence of the RRE domain renders it a theoretically useful bioinformatic handle for class-independent RiPP discovery; however, first-in-class RiPPs have yet to be isolated and experimentally characterized using an RRE-centric strategy. Moreover, with most known RRE domains engaging their cognate precursor peptide(s) with high specificity and nanomolar affinity, evaluation of the residue-specific interactions that govern RRE: substrate complexation is a necessary first step to leveraging the RRE domain for various bioengineering applications. This chapter details protocols for developing custom bioinformatic models to predict and annotate RRE domains in a class-specific manner. Next, we outline methods for experimental validation of precursor peptide binding using fluorescence polarization binding assays and in vitro enzyme activity assays. We anticipate the methods herein will guide and enhance future critical analyses of the RRE domain, eventually enabling its future use as a customizable tool for molecular biology.

1. Introduction

Ribosomally synthesized and post-translationally modified peptides (RiPPs) are a rapidly growing family of natural products (Montalbán-López et al., 2021). RiPPs are categorized into molecular classes, which are usually defined by the presence of a defining post-translational modification (PTM) in the mature product (Arnison et al., 2013). RiPPs exhibit a range of desirable biological functions, such as antibacterial, antiviral, and antifungal activities. Additional RiPPs have been shown to act as redox cofactors, exhibit anticancer activities, and display antinociceptive properties (e.g., the FDA-approved conotoxin ziconotide) (Guerrero-Garzón et al., 2020; Hegemann & Süssmuth, 2020; Tocchetti et al., 2021; Walker et al., 2022).

To this end, natural product researchers have a vested interest not only in discovering new RiPPs with clinically useful properties, but also RiPPs with unexploited biological targets and mechanisms of action. Historically, natural product discovery was approached primarily through phenotypic screening of organisms known to be prolific natural product producers, such as soil-dwelling actinobacteria (Dias, Urban, & Roessner, 2012; Katz & Baltz, 2016; Worthen, 2007). However, after many decades of intense screening campaigns by industrial and academic groups, the low-hanging fruit have been picked. Indeed, untargeted methods that rely on readily cultured bacteria now suffer from a high rediscovery rate. In other words, the likelihood of finding a known compound far outweighs the likelihood of discovering a new natural product (Kloosterman, Medema, & van Wezel, 2021).

1.1 Challenges of using bioinformatic genome mining for RiPP discovery

Over the past decade, discovery efforts in the RiPP field have focused to a greater extent on bioinformatics methods, as well as methods to elicit production from cryptic biosynthetic gene clusters (BGCs)—those that are not expressed by the host under standard laboratory growth conditions. New algorithms (e.g., RiPPER and RiPPMiner) incorporate improved algorithms for high-confidence prediction and de-replication of RiPP BGCs, taking advantage of the wealth of publicly available genomic data (de los Santos, 2019; Santos-Aberturas et al., 2019). BGCs mined using these bioinformatic tools are often expressed in a heterologous host, obviating the need to obtain and culture the native host organism, many of which are not readily cultivatable or may only produce a target RiPP under specific growth conditions (Ahmed et al., 2020; Myronovskyi et al., 2018). Likewise, elicitor screening has been carried out to induce expression of cryptic BGCs in the native host, which may only produce detectable titers of product under specific culturing conditions, such as nutrient depletion or other forms of stress (Abdelmohsen et al., 2015; Moon, Xu, Zhang, & Seyedsayamdost, 2019; Pettit, 2011; Pimentel-Elardo et al., 2015). These methods are required for reconstitution of pathways mined from metagenomic data, where the precise host is not usually known. In all, these methods allow for interrogation of lesser-studied bacterial (and archaeal) genera for natural product production and obviate the need for a native host that can be cultured, thus expanding the breadth of RiPP biosynthetic space that can be accessed (Robinson, Piel, & Sunagawa, 2021).

There are several advantages to using bioinformatics-guided approaches to prioritize BGCs with a high probability of producing novel RiPP compounds. The majority of RiPP BGCs are highly compact and require <10 kb of genomic space. For example, most characterized ranthipeptide and streptide BGCs are roughly 3 kb in size, comprising a precursor peptide, one PTM enzyme, and a transporter (Precord, Mahanta, & Mitchell, 2019; Schramma, Bushin, & Seyedsayamdost, 2015). However, some RiPP classes with numerous canonical PTMs (e.g., thiopeptides) can range upwards of ~30 kb (Yu et al., 2009). In many RiPP BGCs, the requisite genes are encoded in a monocistronic operon. Thus, analysis of gene direction can sometimes assist in identification of BGC boundaries and possible multienzyme complexes acting on a previously unrecognized precursor peptide.

Given these parameters, homology-based searching of known RiPP PTM enzymes, paired with analysis of co-occurring protein domains, can be used to generate a list of homologous BGCs to known RiPP biosynthetic pathways, where the class-defining PTMs are characterized (Kloosterman, Shelton, van Wezel, Medema, & Mitchell, 2020; Walker et al., 2020). However, mining for RiPP BGCs by using class-defining enzymes as handles for homology-based searching has intrinsic limitations: many PTM-installing enzymes are unique to specific RiPP classes, and thus are less useful for class-independent discovery efforts. Furthermore, many RiPP enzymes are homologous to other cellular machinery, which can lead to high false-positive rates. For example, lasso peptides cyclases evolved from asparagine synthetase enzymes (DiCaprio, Firouzbakht, Hudson, & Mitchell, 2019; Tietz et al., 2017). Nevertheless, homology-based searching of known PTM enzymes has shown promise in discovering new hybrid RiPP classes that have co-opted canonical PTMs from other RiPP pathways, such as the recently discovered nocathioamides and streptamidines (Fig. 1) (Russell, Vior, Hems, Lacret, & Truman, 2021; Saad et al., 2021).

RiPP discovery can also be viewed from the perspective of substrate precursor peptides. RiPP biosynthesis starts with a ribosomal precursor peptide, typically ~50 amino acids in length (Hudson & Mitchell, 2018). This feature of RiPP BGCs allows, in many cases, for high-confidence prediction of final product structures, based on the sequence of the precursor peptide combined with knowledge about the biosynthetic enzymes encoded nearby. Unfortunately, the small size of precursor peptides means they are largely missed by automated gene finders, which makes homology-based searching using precursor peptides unfeasible. Our group previously addressed this challenge with the development of RODEO, a bioinformatic tool for prediction and annotation of genomic regions with numerous modules for class-specific precursor peptide prediction (Georgiou, Dommaraju, Guo, Mast, & Mitchell, 2020; Oberg, Precord, Mitchell, & Gerlt, 2022; Ramesh et al., 2021; Tietz et al., 2017).

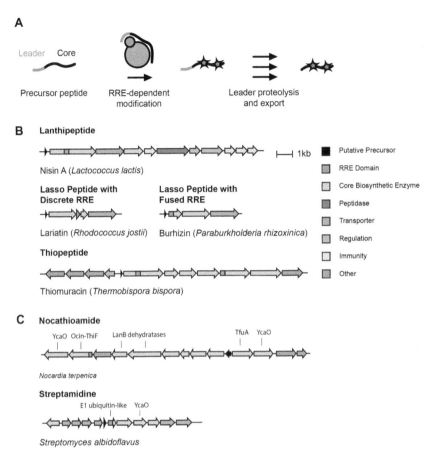

Fig. 1 Function and genomic context of the RRE domain. (A) RiPP recognition elements (RREs) bind the leader peptide region, allowing a fused or complexed enzyme to act on the core peptide residues. (B) Representative examples of fused and discrete RRE domains in RiPP biosynthesis. As shown, RRE domains can exist as N-terminal, C-terminal, or internal domain fusions to other proteins. (C) Two recently discovered hybrid RiPP classes that are RRE-dependent. Nocathioamide BGCs contain an RRE required for thiazole installation, analogous to thiopeptide BGCs. Streptamidine BGCs contain an RRE required for azoline installation, analogous to the cyclodehydratase in linear azole-containing peptide (LAP) biosynthesis. *Figure adapted from Kloosterman, A.M., Shelton, K.E., van Wezel, G.P., Medema, M.H., Mitchell, D.A., 2020. RRE-Finder: A genome-mining tool for class-independent RiPP discovery. mSystems, 5, e00267-20. https://doi.org/10.1128/mSystems.00267-20.*

By using RODEO and related tools, researchers can prioritize putative RiPP BGCs that contain unique precursor peptide sequences or unprecedented enzymes encoded as genomic neighbors, reducing overall rates of rediscovery (Kloosterman et al., 2021). However, determination of exact

BGC boundaries can be complicated by large numbers of PTM enzymes, the presence of tailoring modification enzymes, or distally encoded elements. In particular, in some cases RiPP precursor peptides have been shown to be distally encoded from related PTM-installing enzymes (Haft, 2009; Harris et al., 2020; Li et al., 2010). This phenomenon is likely under-represented in the current literature because RiPP mining bioinformatic tools typically only annotate potential precursors within the local genomic space.

1.2 Initial genomic identification and experimental validation of the RRE domain

While homology-based searching of characterized RiPP PTM enzymes has shown great success in discovery of new compounds of known RiPP classes, this inherently limits the likelihood of novelty in bioactivity (Bushin, Clark, Pelczer, & Seyedsayamdost, 2018; Hudson et al., 2019; Oberg et al., 2022; Schwalen, Hudson, Kille, & Mitchell, 2018; Tietz et al., 2017; Walker et al., 2020). For example, although thiopeptides are a molecularly diverse RiPP class, many thiopeptides inhibit protein translation through either inhibition of elongation factor Tu or the 50S ribosomal subunit (Chan & Burrows, 2021). Class-independent discovery of RiPPs requires an approach to be agnostic toward specific enzymes or PTMs, as these features are usually unique to one or several RiPP classes.

One solution to the challenge of class-independent RiPP discovery employs the use of the RiPP recognition element, or RRE, as a bioinfor-matic handle for novel RiPP BGCs. Initially discovered in 2015, the RRE domain serves to recognize and bind precursor peptides (Fig. 1) (Burkhart, Hudson, Dunbar, & Mitchell, 2015). Although the RRE domain was not formally defined until this time, several crystal structures of leader peptide-bound RRE domains had been solved (e.g., NisB from nisin biosynthesis and LynD from cyanobactin biosynthesis) (Koehnke et al., 2015; Ortega et al., 2014). In addition, the necessity of the RRE domain for precursor peptide processing had been recognized, such as in the streptolysin S biosynthetic pathway (Mitchell et al., 2009).

RREs can exist in RiPP BGCs either as discretely encoded proteins ~80–90 amino acids long, or as fusions to other biosynthetic proteins (Fig. 2) (Kloosterman et al., 2020). RRE domains generally bind their cognate leader peptides with nanomolar affinity. This strong and specific binding is, in all structurally characterized cases, driven by interactions between the third alpha helix and third beta strand of the RRE, and a short motif within the leader peptide, herein called the recognition sequence (Fig. 3) (Chekan et al., 2019).

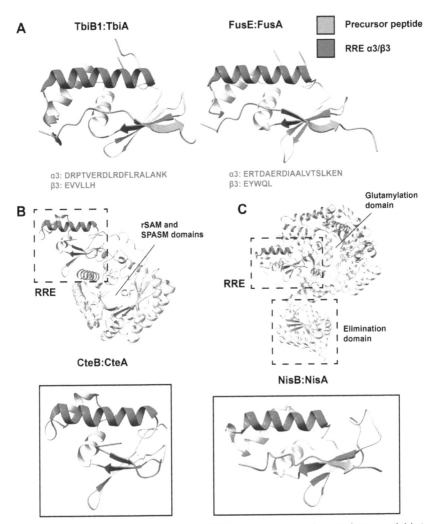

A TbiB1:TbiA

α3: DRPTVERDLRDFLRALANK
β3: EVVLLH

FusE:FusA

α3: ERTDAERDIAALVTSLKEN
β3: EYWQL

Precursor peptide

RRE α3/β3

B

rSAM and
SPASM domains

RRE

CteB:CteA

C

Glutamylation
domain

RRE

Elimination
domain

NisB:NisA

Fig. 2 Representative crystal structures of RRE-precursor peptide complexes available in the PDB. (A) Structures of two discretely encoded RRE proteins in lasso peptide biosynthetic gene clusters (PDB codes 5V1V and 6JX3). TbiA: precursor peptide for therbactin (Chekan, Ongpipattanakul, & Nair, 2019). FusA: precursor peptide for fusilassin (Alfi et al., 2022). (B) An N-terminal RRE fusion to a radical SAM-SPASM enzyme involved in biosynthesis of the ranthipeptide thermocellin (Grove et al., 2017; Hudson et al., 2019). (C) An internal RRE fusion to a dehydratase protein involved in biosynthesis of the lanthipeptide nisin (Repka, Chekan, Nair, & van der Donk, 2017).

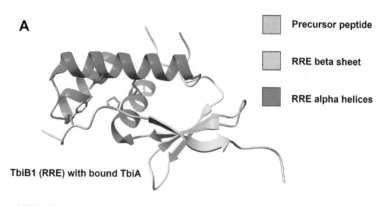

A

☐ Precursor peptide

☐ RRE beta sheet

■ RRE alpha helices

TbiB1 (RRE) with bound TbiA

MEKKKYTAPQLAKVGEFKEATG–WYTAEWGLELIFVFPRFI

Leader Peptide (LP)   Core Peptide (CP)

B

NisB (RRE) with bound NisA

MSTKDFNLDLVSVSKKDSGASPR–ITSISLCTPGCKTGALMG . . .

Leader Peptide (LP)   Core Peptide (CP)

C

TbiB1-TbiA   NisB-NisA

– – Y-(X)$_2$-P-X-L – –   – – F-N-L-D – –

Fig. 3 Leader peptide recognition sequences observed in crystal structures of RREs bound to their cognate leader peptides. (A) The lasso peptide therbactin uses a YxxP leader peptide motif as the recognition sequence, typical of discrete RRE:leader peptide interactions in the lasso peptide class. The TbiB1 RRE binds its cognate precursor

(Continued)

Recognition sequences are highly conserved motifs within RiPP classes, for example, the YxxP motif of lasso peptides, the FNLD motif of lanthipeptides, or the FxxxB (B, branched chain amino acid) motif in cytolysins (DiCaprio et al., 2019; Mitchell et al., 2009; van der Donk & Nair, 2014). The experimental determination of key residues, both in the RRE and the precursor peptide, is especially important for engineering applications. For example, new-to-nature RiPPs can be produced by concatenating the recognition sequence for multiple RRE domains on one leader peptide (Burkhart, Kakkar, Hudson, van der Donk, & Mitchell, 2017).

As of the most recent survey of RRE domains, roughly half of prokaryotic RiPP classes encode one or more RRE domains in the BGC, presumably for substrate recognition and processing although that has not been experimental validated in every case (Kloosterman et al., 2020). The role of the RRE domain in precursor peptide engagement (i.e., substrate recognition) has been biochemically determined for several RiPP classes, including lanthipeptides, lasso peptides, thiopeptides, linear azole-containing peptides, and bottromycins, among others (Burkhart et al., 2015; DiCaprio et al., 2019; Dunbar, Tietz, Cox, Burkhart, & Mitchell, 2015; Hegemann & van der Donk, 2018; Mavaro et al., 2011; Melby, Dunbar, Trinh, & Mitchell, 2012; Schwalen et al., 2017). Taken wholistically, the RRE is a common motif spanning a range of known RiPP classes and, although it is not ubiquitous to RiPP biosynthetic pathways, it has been proposed as a potentially useful bioinformatic handle for first-in-class RiPP discovery (Fig. 1).

Fig. 3—Cont'd   peptides with a K_D of 266 nM (Chekan et al., 2019). (B) The lanthipeptide nisin employs an FNLD recognition sequence, typical of class I lanthipeptides. NisB binds the unmodified NisA precursor peptide with a K_D of 1.05 µM, an example of a lower-affinity RRE interaction (Bothwell et al., 2019; Mavaro et al., 2011). (C) In general, RRE:leader peptide interactions are driven by hydrophobic packing interactions at the interface of the third alpha helix of the RRE and the leader peptide. The exact residues responsible for recognition are dependent on the class of RiPP being studied. *Parts of figure adapted from Chekan, J.R., Ongpipattanakul, C., Nair, S.K., 2019. Steric complementarity directs sequence promiscuous leader binding in RiPP biosynthesis. Proceedings. National Academy of Sciences. United States of America 116, 24049–24055. https://doi.org/10.1073/pnas.1908364116.*

1.3 The RRE domain as a class-independent discovery tool

Although RRE domains are common to a host of RiPP classes, they are difficult to detect by traditional homology-based searches (e.g., BLAST). This arises owing to RRE domains sharing structural homology but exhibiting high sequence divergence (Fig. 2) (Kloosterman et al., 2020). One strategy for bioinformatically defining and predicting a sequence-diverse protein family is the use of hidden Markov models (HMMs), which are statistical models that correlate each residue position with an amino acid probability score (Eddy, 2004). For example, the Pfam database uses a collection of HMMs to define conserved protein families (Mistry et al., 2021).

Until 2020, the only Pfam HMM that accurately predicted the presence of an RRE domain was the PqqD model (PF05402), which is named after the first RRE domain to be structurally characterized, the PqqD protein in pyrroloquinoline quinone (PQQ) biosynthesis (Evans, Latham, Xia, Klinman, & Wilmot, 2017; Tsai, Yang, Shih, Wang, & Chou, 2009). While PF05402 robustly identifies discretely encoded RRE domains (i.e., small proteins that contain only an RRE domain), it fails to identify most RRE domains in other RiPP classes, particularly in cases where the RRE domain exists as a fusion to a PTM enzyme (Fig. 2) (Kloosterman et al., 2020; Tietz et al., 2017). In short, one bioinformatic model is insufficient to capture the diversity of primary RRE sequences found across RiPP classes.

In 2020, we created a bioinformatic tool, RRE-Finder, which addresses the shortcomings of existing models for RRE prediction in two ways (Kloosterman et al., 2020). Two modes were designed to enable class-independent RiPP discovery as well as class-specific annotation of RRE domains. First, exploratory mode of RRE-Finder uses a truncated version of HHpred (described in the following section), which uses secondary structure prediction and comparison to existing RRE crystal structures to predict RRE domains (Kloosterman et al., 2020; Soding, Biegert, & Lupas, 2005). The accompanying mode, precision mode, uses a library of custom HMMs, each designed to specifically predict RRE domains belonging to a specific RiPP class. This chapter includes protocols and guidelines for building such custom HMMs, including how to curate RRE sequences representative of a subfamily and how to validate the accuracy and precision of resulting models. Although this example is specific to RRE domains, the techniques employed here could ostensibly be applied to annotation of any protein domain that is inconsistently annotated by comparison to protein family databases, such as Pfam and TIGRfam (Fig. 4) (Haft, 2001; Mistry et al., 2021).

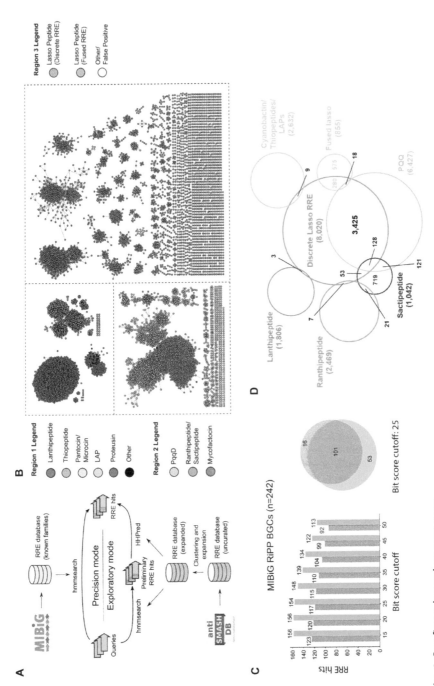

Fig. 4 See figure legend on next page.

Following bioinformatic prioritization of RRE domains, there are several ways to experimentally validate that these RRE are functional, necessary for installation of class-defining PTMs, and which specific residue-level interactions are necessary for RRE:leader peptide binding. Herein, we cover methods for heterologous expression and affinity purification of RRE domains and their cognate precursor peptides. Next, methods are provided for systematic mutagenesis of RRE domains and precursors, paired with fluorescence polarization binding assays, which we have found to be a fast and effective means of assessing which residues are indispensable for the nanomolar affinity observed in most RRE-binding interactions. Finally, we provide methods for in vitro activity assays to determine the functional role of a given RRE in a RiPP biosynthetic pathway. In general, we believe the methods contained herein could be adapted to study any RRE-dependent binding interaction or RiPP biosynthetic pathway, opening the door to discovering and studying novel RRE-dependent RiPP biochemistry.

2. Generation of custom models for RRE prediction and annotation

This section covers strategies for generating custom hidden Markov models (HMMs) for prediction of RRE domains from genomic data. The pipeline for generating custom HMMs (Fig. 5) requires a suite of bioinformatic tools, listed in Section 3.1. Sections 3.2 and 3.3 outline the

Fig. 4 Development and validation of the original RRE-Finder bioinformatic tool. (A) Overall pipeline of RRE-Finder development. The methodology behind precision mode of RRE-Finder is detailed in Section 3 of this chapter. Exploratory mode uses a truncated version of the HHpred pipeline, with further details found in the original publication. (B) The predicted family of RRE-dependent RiPPs retrieved from the UniProtKB database using RRE-Finder in precision mode with a bit score cutoff of 25 (The UniProt Consortium et al., 2021). Predicted RREs are colored based on RiPP class annotation from precision mode HMMs. (C) Validation of RRE-Finder precision and exploratory models using true positive RiPP BGCs from the MIBiG database at a bit score cutoff of 25 (Kautsar et al., 2019). (D) Schematic of HMM predictive overlap in the selected models of RRE-Finder precision mode. Numbers in parentheses indicate the RREs retrieved from UniProtKB in June 2020 using a class-specific HMM at a bit score cutoff of 25. Numbers in overlapping regions indicate that multiple models called this RRE at a greater significance than the bit score threshold. *Figure adapted from Kloosterman, A.M., Shelton, K.E., van Wezel, G.P., Medema, M.H., Mitchell, D.A., 2020. RRE-Finder: A genome-mining tool for class-independent RiPP discovery. mSystems, 5, e00267-20. https://doi.org/10.1128/mSystems.00267-20.*

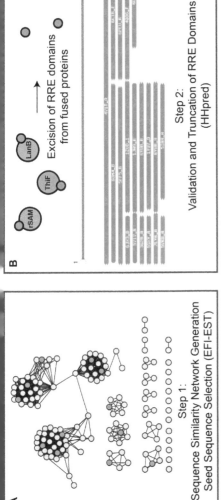

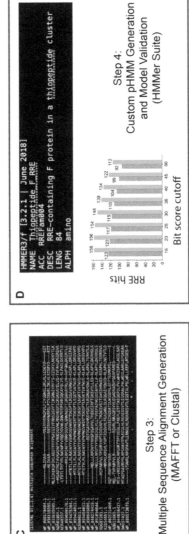

Fig. 5 General pipeline for creating a set of custom HMMs. (A) Datasets of known and predicted RRE-containing proteins are generated and visualized using Cytoscape and the Enzyme Function Initiative Enzyme Similarity Tool (Shannon et al., 2003; Zallot, Oberg, & Gerlt, 2019). From the sequence similarity network, seed sequences are selected. (B) RRE domains fused to other proteins and enzymes are truncated in silico using the HHpred web tool. (C) MAFFT is used to generate a multiple sequence alignment representative of the target RRE family (Katoh & Standley, 2013). (D) HMMER tools and protein database search tools are used to generate and validate the custom HMM (Mistry, Finn, Eddy, Bateman, & Punta, 2013, p. 3).

process for selecting representative protein sequences to define a domain and generating custom HMMs. Where possible, we have indicated options at each stage of this pipeline to use either web tool or downloadable versions of the tools employed. Those without prior bioinformatics experience may wish to use web tools to eliminate user error and save hard drive storage. However, the downloadable versions of these programs will be more suitable for researchers wishing to generate a large library of HMMs and those with command line experience. Creating effective HMMs is an iterative process, and Section 3.4 outlines the process for improvement and validation of custom HMMs (Johnson, Eddy, & Portugaly, 2010). Although this workflow applies specifically to generating custom HMMs for RRE domain identification, the strategy here easily extends to any other protein domain. While certain numbers, such as bit score cutoffs, will change from those outlined below, the general pipeline and guidelines can be applied to any domain that is not well defined by existing databases, such as Pfam or TIGRfam (Haft, 2001; Mistry et al., 2021).

2.1 Software, online tools, and hardware requirements

2.1.1 Hardware

Use of HMMER to generate HMMs requires a Unix-based operating system or simulated environment. If you are a Mac user, you are already using a Unix system and can use the Terminal application for these steps. If you are using a Windows OS, you will need to use a Unix emulator, such as the Mintty console emulator for Cygwin (found at https://mintty.github.io).

Memory is a key consideration for successful manipulation of sequence similarity networks (SSNs) in Cytoscape. Networks with large numbers of edges (lines on the output network that connect homologous proteins) can be difficult to manipulate. The Cytoscape user manual suggests that users have 1 GB of RAM per 150,000 edges. Network sizes are significantly cut down by employing RepNode networks, which are covered in this section. As discussed in the following section, the number of nodes per network at a given RepNode is output as part of the EFI-EST tool.

2.1.2 HMMER suite (Finn et al., 2015)

The HMMer suite tools can be downloaded at www.hmmer.org, where complete installation instructions and usage documentation can be found. This workflow will employ some of the HMMer tools, including HMM scan, HMM build, and HMM press functions. Validation of HMMs will require use of the web tool HMM search function, which can be found at www.ebi.ac.uk/Tools/hmmer/search/hmmsearch.

2.1.3 EFI-EST web tools (Zallot et al., 2019)

The EFI-EST tools, used for generation of sequence similarity networks and genome neighboring networks, are exclusively web tools. They can be found at https://efi.igb.illinois.edu/efi-est.

2.1.4 Cytoscape (Shannon et al., 2003)

Cytoscape must be downloaded for the visualization and manipulation of sequence similarity networks. This open-source program can be downloaded at https://cytoscape.org.

2.1.5 RODEO (Georgiou et al., 2020; Kloosterman et al., 2020; Ramesh et al., 2021; Schwalen et al., 2018; Tietz et al., 2017)

RODEO is an artificial intelligence-driven tool for compilation, categorization, visualization, and annotation of RiPP BGCs, including precursor peptide prediction. The web tool can be accessed at https://rodeo.igb.illinois.edu. For large batch runs (>1000 query sequences), users are encouraged to download the command line version from https://github.com/the-mitchell-lab/rodeo2.

2.1.6 HHpred (Soding et al., 2005)

HHpred predicts protein domain homology using a combination of primary and predicted secondary structure alignment to structures in the Protein Data Bank (PDB; https://www.rcsb.org) (Zardecki, Dutta, Goodsell, Voigt, & Burley, 2016). For most applications, using the web tool will be sufficient: https://toolkit.tuebingen.mpg.de/tools/hhpred. A downloadable version of HHpred may be needed for large-scale analysis, but requires local download of the entire PDB, so this is not recommended for casual users.

2.1.7 MAFFT (optional) (Katoh & Standley, 2013)

MAFFT is a command line-only tool for generating multiple sequence alignments (MSAs), appropriate for either HMM or phylogenetic tree generation. It can be downloaded at https://mafft.cbrc.jp/alignment/software/. Other tools for MSA generation can be used in place of MAFFT and several web tools can serve as substitutes, such as the Clustal Omega tool available at https://www.ebi.ac.uk/Tools/msa/clustalo/.

2.2 Curation of seed sequences for model generation

Timing: 2–3 h

1. Acquire or generate a dataset from which to select sequences to use for hidden Markov model generation. Ideally, this dataset should represent all known members of a given RRE class, e.g., all sactipeptide associated RREs. In cases where a dataset does not exist, one should be generated by using a target RRE as a query for a PSI-BLAST search. Parameters should be set to three iterative rounds of searching with an expect value cutoff of 0.05

2. Generate a sequence similarity network of the target dataset using the EFI-EST tools. Input for the SSN tool can be in the form of a FASTA file with all query sequences or a list of UniProt or GenBank accession IDs. All other settings are kept at default

3. When prompted by email, generate the final SSN by choosing a starting alignment score. The alignment score determines how proteins in the SSN cluster together as a function of their sequence identity. Choose an alignment score that corresponds to clustering of sequences at 40% identity, as determined by the output graphs generated by EFI-EST

4. The output SSN can be downloaded at several different RepNode cutoffs. RepNode networks conflate proteins that share greater identity than a set percentage cutoff. For custom HMM generation, it is recommended to download the RepNode60 network. This ensures that proteins occupying separate nodes on the network do not share greater than 60% sequence identity

5. Use Cytoscape to visualize the sequence similarity network. Data can be imported as an xgmml file. Select the Layouts drop-down menu, then select the "organic" layout under the yFiles tab. In this view, clustered proteins are visualized as circular "nodes" and edges connecting two nodes represents homology more significant than the cutoff specified by the chosen alignment score (Fig. 6)

6. Select 5–20 nodes from the SSN that will comprise the seed sequences for the HMM. See notes below for considerations on choosing a diverse set of seed sequences and selecting an appropriate number of sequences to represent your target RRE family

7. Once nodes are selected, output the selected data to a tab-delimited file (readable either with a text editor or Microsoft Excel). This file contains data useful for a variety of downstream analyses, including phylogeny of the producing organism, sequence length, and matches to families in Pfam or InterPro databases. The column called "shared name" can

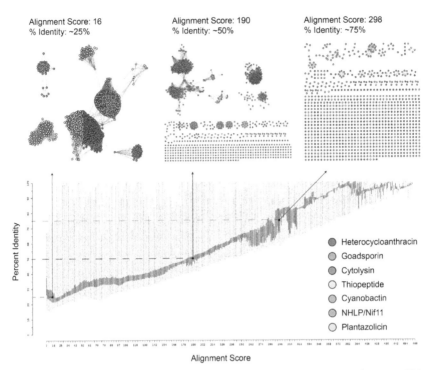

Fig. 6 Selection of an appropriate alignment score for SSN visualization. This network shows a dataset of RRE-containing proteins from various RiPP natural products with azoles/thiazoles in their final structures. As shown, at a low alignment score, RREs belonging to disparate classes of RiPPs cluster together (e.g., thiopeptides and heterocycloanthracins). This is known as underfractionation. Raising the alignment score above a reasonable threshold fragments clusters to an extent that they are not useful for HMM generation. This is known as overfractionation.

be copied and pasted into the InterPro protein retrieval tool (https://www.ebi.ac.uk/interpro/search/sequence/) to generate a FASTA file of target RRE seed sequences. This FASTA file can be either be used as direct input for MSA generation, as covered by the following section, or can be truncated to contain only RRE domains (highly recommended if you are dealing with RRE-domain fusions to larger enzymatic domains)

8. (Optional) Truncate seed sequences using HHpred to contain only the residues comprising the RRE domain. Each seed sequence should be individually submitted to the HHpred web tool (https://toolkit. tuebingen.mpg.de/tools/hhpred). Output from this tool shows alignment of the query protein to proteins in the Protein Data Bank (https://www.rcsb.org), using a combination of primary and (predicted)

secondary structure homology (Fig. 5). The most common PDB accessions identified for RRE domains are listed in Table 1. Navigate to the
alignment section for the top-scoring RRE hit to identify the query
protein residues corresponding to the RRE domain. Your FASTA file
from the previous step should be modified to contain only the RRE
domain residues before generating a multiple sequence alignment

Notes

1. There is no minimum or maximum number of seed sequences that
   can be incorporated into an HMM; however, we have found that using
   too many sequences can lead to a model that is biased toward known,
   characterized RREs and not useful for exploratory, genome-mining
   applications. For smaller RiPP classes (<1000 predicted members),
   10 diversity-maximized sequences are usually sufficient. For large RiPP
   classes, more sequences should be employed

2. As shown in Fig. 7, sequences should be chosen to represent the natural
   diversity of an RRE family and not bias the model toward commonly
   occurring motifs. If multiple clusters occur in the SSN at the starting
   alignment score (representing clustering at 40% identity), seed sequences
   should be selected to sample from each primary cluster. Singleton nodes
   or clusters with a small number of nodes (<5 nodes) should usually be
   ignored as these are outliers that are not representative of the RRE class
   being targeted

2.3 Custom HMM generation

Timing: <1 h

1. Use MAFFT to generate a multiple sequence alignment for the target
   RRE class. The default L-INS-I alignment option should be used.
   The input for MAFFT should be a FASTA format file of your individual
   RRE seed sequences. The output MSA will be a text-readable file

2. Use the command line terminal to navigate to the folder containing your
   MSA. Use the following command to generate a custom HMM from the
   output MSA:

   % hmmbuild <hmm_file> <msafile>

   The output HMM should be appended with the suffix .hmm for the
   following steps

3. If the output HMM will be used with the RODEO command line tool
   (rather than the web tool), press it into binary form using the following
   HMMER command:

Table 1 List of RRE crystal and NMR structures available in the PDB. LP: leader peptide.

RRE	RiPP class	PDB accession	UniProtKB accession	Precursor bound?	Citation DOI
LynD	Cyanobactin	4V1T	A0YXD2	Yes	https://doi.org/10.1038/nchembio.1841
TruD	Cyanobactin	4BS9	B2KYG8	No	https://doi.org/10.1002/anie.201306302
ThcOx	Cyanobactin	5LQ4	B8HTZ1	No	https://doi.org/10.1107/S2059798316015850
MibB	Lanthipeptide	5EHK	E2IHB7	No	https://doi.org/10.1016/j.chembiol.2015.11.017
NisB	Lanthipeptide	4WD9/6M7Y	P20103	Yes	https://doi.org/10.1038/nature13888
McbB	LAP	6GOS/6GRH	P23184	Yes	https://doi.org/10.1016/j.molcel.2018.11.032
FusB1	Lasso peptide	6JX3	Q47QT5	Yes (LP only)	https://doi.org/10.1021/acschembio.9b00348
TbiB1	Lasso peptide	5V1V	D1CIZ5	Yes (LP only)	https://doi.org/10.1073/pnas.1908364116
MccB	Microcin	6OM4	Q47506	Yes	https://doi.org/10.1039/c8sc03173h
PaaA	Pantocin	5FF5	Q9ZAR3	No	https://doi.org/10.1021/jacs.5b13529
PqqD	PQQ	3G2B/5SXY	Q8P6M8	No	https://doi.org/10.1002/prot.22461 https://doi.org/10.1021/acs.biochem.7b00247
CteB	Ranthipeptide	5WGG	A3DDW1	Yes (LP only)	https://doi.org/10.1021/jacs.7b01283
SkfB	Sactipeptide	6EFN	O31423	No	https://doi.org/10.1074/jbc.RA118.005369
SuiB	Streptide	5V1T	A0A0Z8EWX1	Yes	https://doi.org/10.1073/pnas.1703663114
TbtB	Thiopeptide	6EC7/6EC8	D6Y502	No	https://doi.org/10.1073/pnas.1905240116

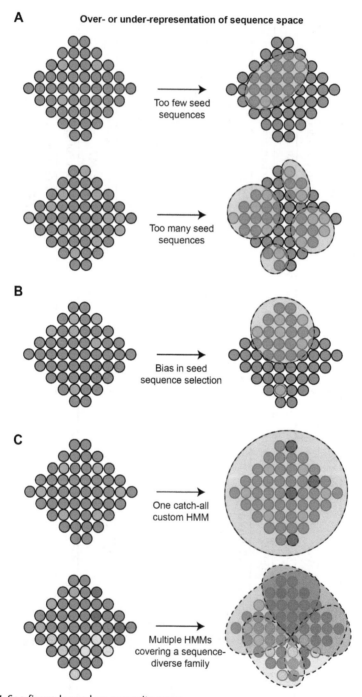

Fig. 7 See figure legend on opposite page.

% hmmpress < hmm_file >

This command should produce four separate files with the same parent name as the original HMM but with different suffixes (h3f, h3i, h3m, h3p).

4. To ensure the HMM is properly annotated by RODEO, use a text editor to change or add the following fields at the top of the HMM:

NAME: Target RRE class (e.g., thiopeptide RRE)

ACC: Identifier of your HMM, helpful if you are creating multiple models (e.g., RREfam004)

DESC: Description of the target protein domain

2.4 Custom HMM model validation

Custom HMMs generated using the method above should be validated either against a trusted dataset of RRE-containing proteins or, in cases where such a dataset does not exist and would be difficult to generate, using the online hmmsearch tool to interrogate the UniProtKB database using the custom HMM. In many cases, model validation may reveal predictive gaps of the custom HMM, necessitating iterative improvement of the HMM by modifying the input seed sequences. The below section outlines a protocol for assessment of model recall and precision, and advice for improvement of HMMs in cases where model recall fails to meet a designated threshold.

Timing: 2–3 h

1. Validate recall of your HMM against the original dataset of RREs generated in Section 3.2 using the hmmsearch function of the HMMer suite. Input for this function includes a FASTA format file containing all RRE sequences comprised by the original dataset and the HMM generated in the previous section. Hmmsearch can be run in the terminal using the following command:

% hmmsearch –tblout <f> -T <x> <hmmfile> <seqdb>

Fig. 7 Implications of seed sequence selection on custom HMM predictive capabilities. (A) Using too few or too many seed sequences for MSA generation can limit HMM scope as shown in this theoretical dataset. Green nodes represent selected seed sequences, while the larger dotted green circles represent homologous sequences called by the resulting HMM. (B) Choosing highly identical seed sequences is not recommended. This problem is largely eliminated by employing RepNode networks. (C) For large or highly diverse RiPP classes, multiple HMMs can be employed to better cover a sequence space. In our dataset, such division into subclasses was carried out for sactipeptides and linear azole-containing peptides, both of which contain sequence-diverse RREs spanning the class.

In the above command, $<$hmmfile$>$ should direct to your custom HMM, while $<$seqdb$>$ directs to the FASTA file of RRE sequences. The output file will be a tab-delimited file, where $<$f$>$ represents your desired output file name. The -T option indicates a bit score significance cutoff. We have found a bit score of 25 to be an appropriate cutoff for high-confidence RRE domains

2. An effective custom HMM should detect $>$90% of the RRE domains in the original dataset at a bit score threshold of 25. If this threshold is not met, follow the steps below to improve the model

3. Evaluate custom HMM specificity through an unbiased query of the UniProtKB protein database. Using the hmmsearch web tool (https://www.ebi.ac.uk/Tools/hmmer/search/hmmsearch), input the custom HMM to query the UniProtKB database. For RRE domains, we have found a bit score threshold of 25 to be ideal for retrieving high confidence, class-specific RRE domains with minimal false positives. Lowering the bit score threshold may be necessary for very large or diverse RiPP classes. Raising the bit score threshold reduces false positive rates and is useful in cases where class-specific identification is important, or model overlap is high (Fig. 5).

4. Use all UniProtKB retrievals at the specific bit score threshold as queries for RODEO (either the web tool or command line tool) to gather the genome neighborhood flanking each predicted RRE. RODEO takes NCBI accession identifiers as input, thus UniProtKB accessions must first be converted to "EMBL/NCBI CDS" accessions using the UniProt mapping tool (https://www.uniprot.org/id-mapping).

3. Cloning, expression, and purification of RRE domains and precursor peptides

Proper expression and purification of RRE domains and their cognate precursor peptides is a critical first step to performing in vitro activity-based assays or binding assays such as fluorescence polarization. Impurities could result in miscalculation of protein concentration, which will decrease the accuracy of K_D and IC_{50} measurements, as outlined in a later section. In general, we have had success expressing codon-optimized RRE and precursor constructs in the E. coli BL21 (DE3)-RIPL strain. RREs are generally small ($\sim$85 amino acids) and fold readily, so the use of chaperone proteins or cold induction temperatures is usually unnecessary when expressing a discrete or excised RRE. We have had success purifying both RREs and their cognate precursor peptides as constructs with N-terminal His6 tags or maltose-binding protein (MBP) tags. The MBP tag is preferable in most cases for two reasons. First, MBP is highly soluble and will help ensure the RRE can be concentrated without risking precipitation. Second, the MBP tag increases the overall size difference between the tagged RRE and precursor peptide, which improves fluorescence polarization signals during binding assays.

3.1 Materials, reagents, and equipment

Equipment

1. Thermo Scientific Sorvall Legend Micro 17
2. Sorvall RC6 Plus floor centrifuge with SS-34 rotor (Thermo Scientific)
3. Ultrasonic cell disruptor (Microson, NY)

Materials

1. EconoSpin silica membrane mini-spin column (Epoch Life Science, TX)
2. Amylose resin (New England BioLabs, MA)
3. 1.5 × 20 cm CrystalCruz chromatography column (Santa Cruz Biotechnology, CA)
4. Amicon Ultra centrifugal filter, 30 kDa molecular weight cutoff (MWCO) (EMD Millipore)

Reagents and Buffers

1. Q5 DNA polymerase (New England BioLabs, MA)
2. Restriction enzymes (New England BioLabs, MA)
3. T4 DNA ligase (New England BioLabs, MA)
4. Protease inhibitor cocktail ($1 \times$ solution contains $2 \mu M$ leupeptin, $2 \mu M$ benzamide HCl, and $2 \mu M$ E64; solution can be prepared at a $100 \times$ concentration and stored at $-80\,°C$ until use)
5. Luria-Bertani (LB) growth medium: 1 L of LB medium contains 10 g tryptone, 10 g NaCl, 5 g yeast extract; supplement with 15 g of agar for growth on solid medium
6. Lysis buffer: 50 mM Tris-HCl at pH 7.5, 500 mM NaCl, 2.5% glycerol (v/v), and 0.1% Triton X-100 (v/v)
7. Wash buffer: 50 mM Tris-HCl, 500 mM NaCl, 2.5% glycerol (v/v), and 0.5 mM tris(2-carboxyethyl)phosphine (TCEP)
8. Elution buffer: 50 mM Tris-HCl at pH 7.5, 150 mM NaCl, 2.5% glycerol (v/v), 10 mM maltose, and 0.5 mM TCEP

9. Protein storage buffer: 50 mM N-2-hydroxyethylpiperazine-N'-2-ethanesulfonic acid (HEPES) at pH 7.5, 300 mM NaCl, 2.5% glycerol (v/v), and 0.5 mM TCEP

10. Phosphate-buffered saline (PBS): 137 mM NaCl, 2.7 mM KCl, 10 mM Na_2HPO_4, and 1.8 mM KH_2PO_4

3.2 Cloning strategies

The *E. coli* DH5α strain was used for plasmid amplification and storage. Genes of interest can either be ordered as codon-optimized constructs directly from companies like GenScript or Twist Bioscience. Alternatively, if the native host strain is available and culturable, the strain can be grown on appropriate medium and genomic DNA can be extracted using a commercial gDNA extraction kit. Genes of interest are then amplified from gDNA using PCR with primers that introduce N- and C-terminal restriction sites appropriate for ligating into an expression vector (in this case a pET28-MBP expression vector). The resulting PCR products and host vectors are digested, purified, and ligated using T4 DNA ligase and following conventional molecular biology protocols. If genes of interest are ordered for custom synthesis, PCR amplification can be carried out directly on the received plasmid and then subcloned into an appropriate expression vector. The resulting constructs were validated through Sanger sequencing.

3.3 Expression protocol for RRE domains

Timing: 2–3 d

1. Transform the pET28-MBP-RRE plasmid into BL21 (DE3) chemically competent cells using standard heat shock procedures

2. Grow cells overnight (~16 h) on LB agar plates containing 50 µg/mL kanamycin at 37 °C. If an expression vector other than pET28 is employed, the antibiotic resistance cassette contained within the plasmid will determine what antibiotic should be added to the growth medium

3. The following day, pick an isolated colony and inoculate a 10 mL starter culture of liquid LB containing 50 µg/mL kanamycin and 34 µg/mL chloramphenicol. Chloramphenicol can be omitted if using a regular BL21 strain of *E. coli* not containing the pACYC-based plasmid for rare codons

4. Incubate the starter culture overnight (16–18 h) at 37 °C with moderate shaking

5. Inoculate 1 L of sterile LB medium with the entire 10 mL starter culture. This larger culture should also contain equal concentrations of kanamycin and chloramphenicol as the starter culture

6. Grow the 1 L culture at 37 °C with moderate shaking until the optical density reaches 0.6. This will take several hours and OD_{600} can be periodically measured using any standard spectrophotometer. Cultures expressing RRE domains should be grown to an OD_{600} of 0.4–0.6, while cultures expressing precursor peptides can be grown to a higher density of 0.8–1.0

7. After the proper optical density is reached, cool the cultures in ice water for 20 min or place in a 4 °C cold room

8. Induce protein expression by adding 0.4 mM isopropyl β-D-1-thiogalactopyranoside (IPTG). For substrate peptides, induce with 1 mM IPTG

9. Incubate cultures overnight (16 h) at 22 °C with shaking set to 220 rpm. Precursor peptides are highly susceptible to proteolytic cleavage, and should only be incubated for 1–2 h at a higher temperature, usually 37 °C

10. Harvest cells by centrifugation at $3000 \times g$ for 10 min. Discard spent media, resuspend cell pellet in 50 mL of chilled PBS, and harvest cells once more by centrifugation at $3000 \times g$ for 10 min. Resuspension of the cell pellet in PBS provides an opportunity to transfer cells to a 50 mL conical tube, which is suitable for storage at −80 °C

11. Discard the supernatant. Flash-freeze cell pellets in liquid nitrogen and store at −80 °C until use

3.4 Purification protocol for RRE domains and precursor peptides

Timing: 6–8 h

1. Thaw cell pellets on ice for 30 min. After thawing, resuspend the cell pellets in pre-chilled lysis buffer. Add in 4 mg/mL of lysozyme, 1 × protease inhibitor cocktail, and 10 mg of phenylmethylsulfonyl fluoride (PMSF) protease inhibitor (only needed for precursor purification, optional for RRE purification)

2. Resuspend cell pellet fully in lysis buffer by using a vortex mixer

3. Lyse cells with an ultrasonic cell disruptor (sonicator) for 30 s increments with a power output of 10–12 W. Alternate sonication intervals

with 10 min periods of gentle rocking at 4 °C until cell pellets are fully broken up and solution appears cloudy. It is easiest to perform this step in a cold room; otherwise, cells should be kept on ice whenever possible

4. Remove insoluble cellular components by centrifugation at $20,000 \times g$ for 1 h at 4 °C

5. Load affinity chromatography columns with amylose resin (10 mL per 1 L of original cell culture) and pre-equilibrate column by running through 20 mL of chilled lysis buffer supplemented with 0.5 mM TCEP

6. Load entire cell lysate supernatant onto the pre-equilibrated column and allow for gravity flow through the column. It is good practice to collect the flow-through in case binding of the analyte to column is not successful

7. Wash the column with 40 mL of wash buffer, then elute target protein/peptide into a clean falcon tube using 30 mL of chilled elution buffer

8. Concentrate the eluent using a 30 kDa MWCO Amicon Ultra centrifugal filter

9. Perform buffer exchange with 30 mL of chilled protein storage buffer prior to determination of final concentration and storage

10. A rough concentration of the eluted protein/peptide can be determined by absorbance at 280 nm. Expected extinction coefficients can be predicted using the ExPASy ProtParam tool available at http://web.expasy.org/protparam. A Bradford or bicinchoninic acid (BCA) colorimetric protein concentration assay should be performed. If an accurate protein concentration is needed, dry weight or quantitative amino acid analysis should be conducted. Protein purity should also be evaluated via Coomassie-stained SDS-PAGE with a molecular weight standard to assess full-length expression

Notes

1. The best results will come from performing the cell lysis and affinity chromatography steps above in a 4 °C cold room. If this is not possible, protein solutions should be kept on ice whenever possible, including directly before and after column chromatography

2. RRE domains are highly soluble and can generally be concentrated to up to 20 mg/mL before there is risk of precipitation

3. If columns become clogged during affinity chromatography, flow can be re-established by gently resuspending the amylose resin using a pipette. This can be done without affecting overall yield

4. The above protocol will generally yield protein purities >95%, as determined by band intensity on an SDS-PAGE gel. If additional purification is required for a particular application, either size-exclusion or ion-exchange chromatography can be employed

4. Fluorescence polarization-binding assays

We have found a straightforward and relatively high-throughput way of determining key binding residues to be a tandem site-directed mutagenesis (SDM) and fluorescence polarization (FP) assay workflow (Burkhart et al., 2015; Zhang et al., 2016). First, residues suspected to contribute to key binding interactions are selected; there are several strategies for selecting target residues, outlined in this section. Then, either the plasmid construct for the MBP-tagged precursor peptide or RRE is subjected to SDM. Variants are then expressed and purified following the protocols in the previous section. Finally, binding parameters are measured using either fluorescence polarization (FP) or competition FP assays.

Because the FP signal is enhanced by a large molecular weight difference between protein and ligand, we have had the most success with assays where the RRE is kept as an MBP-tagged protein, but the leader (or precursor) peptide is not. For this purpose, we employ a pET28-MBP expression plasmid with a TEV-cleavable linker, which is a peptide motif recognized and cleaved by the tobacco etch virus protease (Raran-Kurussi, Cherry, Zhang, & Waugh, 2017). The combination of nanomolar affinity of the RRE for its cognate LP and the large size difference between protein and ligand makes FP an effective and rapid way to assess binding for any RRE-dependent RiPP pathway.

4.1 Materials, reagents, and equipment

Equipment
1. Thermo Scientific Sorvall Legend micro 17 centrifuge
2. Synergy H4 hybrid plate reader (BioTek)
3. Multi-channel pipette, 0.5–10 µL and 20–200 µL

Materials
1. 96-well microplates (Corning)
2. 384-well black polystyrene microplates (Corning)

Reagents
1. Q5 DNA polymerase (New England BioLabs, MA)
2. *Dpn*I restriction enzyme (New England BioLabs, MA)

3. pET28b-MBP-precursor peptide plasmid (constructed as described in Section 4)
4. pET28b-MBP-RRE plasmid (constructed as described in Section 4)
5. Fluorescein isothiocyanate (Sigma-Aldrich)

4.2 Strategy and protocol for site-directed mutagenesis

Timing: 2–3 d

1. Use Q5 DNA polymerase to amplify the pET28-MBP plasmid containing either a target RRE domain or leader peptide. Primers should be designed to anneal to regions both 15 base pairs upstream and downstream of the region targeted for mutagenesis. The PCR cycle protocol is shown in Table 2
2. Following PCR, digest any remaining parental plasmid by adding 1 μL DpnI restriction enzyme and incubating at 37 °C for 2–4 h
3. Transform DH5α chemically-component E. coli cells with the mutated plasmid by standard heat shock procedures. Plate transformed cells onto LB agar plates supplemented with 50 μg/mL kanamycin. Allow colonies to grow at 37 °C overnight (12–16 h)

Table 2 PCR cycling protocol used for site-directed mutagenesis. Annealing temperature of 65 °C as indicated should work for most amplifications using long primers designed as in Section 4.2 but can be adjusted if initial PCR cycles fail.

Step	Temperature (°C)	Duration (min:sec)
1	98	10:00
2	98	0:30
3	65	0:30
4	72	16:00
5	Repeat steps 2–4 for 10 cycles	
6	98	0:30
7	65	0:30
8	72	16:00 (+15 s/cycle)
9	Repeat steps 6–8 for 20 cycles	
10	72	30:00
11	12	Hold until use

4. Choose 3 distinct colonies and use to inoculate cultures of 10 mL LB with 50 µg/mL kanamycin. Allow cultures to grow at 37 °C with moderate agitation for 12–16 h

5. Once cultures have grown to be visibly cloudy ($\sim OD_{600}$ of 2), harvest cells in microcentrifuge tubes ($4000 \times g$ for 10 min). Isolate plasmid from the cell pellets using standard protocol with a miniprep kit

6. Verify the correct sequence of the mutated plasmid using primers designed to flank the target region. When using the pET28-MBP, we employ an MBP-forward primer and a T7-terminator primer with the following sequences:

   MBP F primer: GAGGAAGAGTTGGCGAAAGATCCAGGTA
   T7 R primer: GCTAGTTATTGCTCAGCGG

7. RRE and peptide variants should be expressed and purified using the protocols outlined in the previous section

4.3 FITC-labeling of precursor peptides for use in binding assays

A stock solution of fluorescently labeled precursor peptide must be synthesized for fluorescence polarization (FP) and competition FP experiments in the following sections. Synthetic peptides containing the leader peptide of interest can be ordered from GenScript. For a larger cost, fluorescently labeled peptides can be ordered and this protocol can be skipped entirely. We have had success using fluorescein isothiocyanate (FITC) as a fluorescent tag, due to its high extinction coefficient and compatibility with filters on most commercial plate readers. The example protocol herein uses the thiomuracin leader peptide, but this protocol can be adapted to any leader peptide that can be purified by standard reverse-phase HPLC.

Timing: 3 d

1. Dissolve the synthetic leader peptide in dimethyl sulfoxide (DMSO) to a final concentration of 10 mg/mL

2. Dissolve FITC in 10% aqueous DMSO to a final concentration of 6 mg/mL. Adjust pH of the resulting solution to 9.5 using $NaHCO_3$ and Na_2CO_3

3. Combine 100 µL of dissolved peptide with 300 µL of FITC solution. Allow N-terminal FITC labeling to proceed overnight (12–16 h) at 25 °C. The reaction should be wrapped in foil to prevent photobleaching of the FITC label

4. Quench the reaction using 50 mM of NH_4Cl, allowing the quenched reaction to sit in darkness for at least 1 h

5. Evaluate labeling by MALDI-TOF MS using the methods in Section 4. FITC labeling should result in an overall mass shift of 389.4 Da relative to the mass of the unmodified leader peptide

6. Purify labeled peptide using CombiFlash instrument equipped with a reverse-phase C18 column (RediSep Rf 4.3 g). General separation conditions are a 5–90% methanol gradient, with 10 mM NH_4HCO_3 used as the aqueous phase

7. Spot fractions on a stainless-steel target and analyze by MALDI-TOF MS. Combine any fractions containing labeled peptide and concentrate down using a rotary evaporator

8. Re-dissolve labeled peptide in 5% methanol. Remove insoluble components by filtering mixture through a 0.22 μm syringe filter

9. Inject peptide solution onto a Betasil C18 HPLC column (250 × 4.6 mm). Separate at a flow rate of 1 mL/min using a gradient of 20–98% methanol, with 100 mM NH_4HCO_3 as the aqueous phase

10. Analyze fractions by MALDI-TOF MS for presence of labeled peptide. Pool fractions containing the desired peptide and concentrate using rotary evaporation

11. Dissolve dried peptide in binding buffer to a final concentration of 250 nM. This is the stock solution that will be employed for FP and comp. FP protocols below. The concentration of labeled peptide can be estimated based on the absorption at 495 nm, using the extinction coefficient of FITC ($73,000 M^{-1} cm^{-1}$)

Notes

1. Use of fluorescein is advantageous because of its high extinction coefficient, compatibility with excitation/emission filters for most plate readers, and its sequence-independent labeling of the peptide N-terminus. Addition of an N-terminal Gly-Gly linker to the synthesized leader peptide is recommended to spatially separate fluorescein from the RRE and obviate any potential binding interference

2. We have had the most success with a single Gly-Gly linker. Use of a longer linker region allows the bound fluorophore too much flexibility relative to the bound leader peptide

4.4 Fluorescence polarization (FP) binding assay

Timing: 2–4 h

1. Dilute a master solution of the MBP-tagged RRE of choice to an appropriate starting concentration. The starting concentration should be chosen to center the FP binding curve around the predicted K_D of

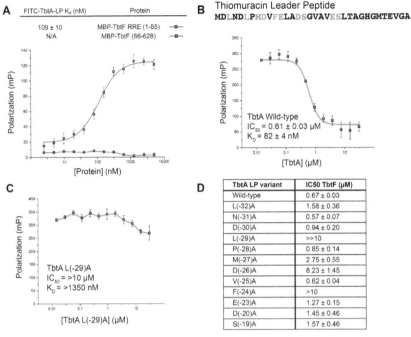

A

FITC-TbtA-LP K_d (nM)	Protein	
109 ± 10	MBP-TbtF RRE (1-85)	▪
N/A	MBP-TbtF (86-628)	■

[Protein] (nM)

B

Thiomuracin Leader Peptide
MDLNDLPMDVFELADSGVAVESLTAGHGMTEVGA

TbtA Wild-type
IC_{50} = 0.61 ± 0.03 μM
K_D = 82 ± 4 nM

[TbtA] (μM)

C

TbtA L(-29)A
IC_{50} = >10 μM
K_D = >1350 nM

[TbtA L(-29)A] (μM)

D

TbtA LP variant	IC50 TbtF (μM)
Wild-type	0.67 ± 0.03
L(-32)A	1.58 ± 0.36
N(-31)A	0.57 ± 0.07
D(-30)A	0.94 ± 0.20
L(-29)A	>>10
P(-28)A	0.85 ± 0.14
M(-27)A	2.75 ± 0.55
D(-26)A	8.23 ± 1.45
V(-25)A	0.62 ± 0.04
F(-24)A	>10
E(-23)A	1.27 ± 0.15
D(-20)A	1.45 ± 0.46
S(-19)A	1.57 ± 0.46

Fig. 8 Analysis of recognition sequence-binding residues in the thiomuracin system via fluorescence polarization assays. (A) Fluorescence polarization binding curves for wild-type thiomuracin leader peptide (TbtA) binding to the RRE of TbtF versus the fused C-terminal portion of the protein. In all cases, error bars represent a standard deviation of the mean ($n = 3$). (B) Competition binding curve for wild-type TbtA leader competing against a FITC-labeled leader peptide. In the leader sequence shown, important binding residues (>2-fold perturbation of binding upon mutation) are highlighted in blue, while critical binding residues (ablation of binding upon mutation) are highlighted in green. (C) Competitive FP binding curve for a low affinity variant of the TbtA leader peptide. This data is representative of a variant with severely impacted binding and an IC_{50} outside of the range of the assay parameters. (D) Summary of competitive FP data for alanine scan of the thiomuracin leader peptide. The negative numbers indicate residue position N-terminal to the leader peptide cleavage site. *Figure adapted from Zhang, Z., Hudson, G.A., Mahanta, N., Tietz, J.I., van der Donk, W.A., Mitchell, D.A., 2016. Biosynthetic timing and substrate specificity for the thiopeptide thiomuracin. Journal of the American Chemical Society 138, 15511–15514. https:/doi.org/10.1021/jacs.6b08987.*

the RRE:LP interaction (~60–100 nM for most RRE:LP binding interactions). For the example outlined in Fig. 8, we have chosen a starting concentration of 3.2 μM

2. Perform 11 consecutive two-fold dilutions of the MBP-tagged RRE in binding buffer. This is most easily done in a 96-well plate

3. Transfer the serially diluted MBP-RRE solutions to a non-binding-surface, 384-well black microplate. To each well, add FITC-labeled precursor peptide to a final concentration of 25 nM, mixing very gently by pipetting to avoid the formation of bubbles. Each unique RRE:LP interaction should be assayed in triplicate
4. Let the binding partners equilibrate for 30 min with shaking at 25 °C. If a microplate shaker is not available, plates can also be equilibrated for 1 h without shaking and will result in the same binding curve
5. Collect FP data using a plate reader with the appropriate emission and excitation filters installed ($\lambda_{ex} = 485$ nm; $\lambda_{em} = 538$ nm for FITC labels). Our data was collected using a Synergy H4 Hybrid plate reader with SoftMax Pro Gen5 software
6. Calculate polarization units at each concentration using the following equation, where P is the polarization, $I_{\parallel}$ is emission fluorescence intensity parallel to excitation, and $I_{\parallel}$ is emission fluorescence intensity perpendicular to excitation. G represents a differential sensitivity correction and is a plate-reader specific parameter

$$P = \frac{I_{\parallel} - G \times I_{\perp}}{I_{\parallel} + G \times I_{\perp}}$$

7. Fit the data using OriginPro9.1 (OriginLab) with a non-linear dose response curve to estimate the dissociation constant (K_D). For most RRE:leader peptide interactions, the K_D will lie in the mid nM range

Notes:

1. An ideal FP binding curve will have a sigmoidal shape, flattening out at both high and low concentrations of the MBP-tagged RRE. These regions represent unbound leader peptide polarization and fully saturated binding, respectively. Not observing these regions in a binding curve indicates that the starting concentration of MBP-tagged RRE should be adjusted to center the curve around the observed dissociation constant. Although many RRE:LP interactions are of nanomolar affinity, there are known examples of diminished affinity RRE interactions (Melby et al., 2012). In these cases, a higher initial concentration of MBP-tagged RRE should be employed

4.5 Competition FP assay

Timing: 2–4 h

1. Competition FP can be used to assess the effect of recognition sequence mutation on leader peptide affinity toward the RRE. The same FITC-labeled wild-type leader peptide used in part 5.4 can be employed here

2. Perform 11 consecutive two-fold dilutions of the MBP-tagged leader peptide variant in binding buffer. This is most easily done in a 96-well plate. As in Section 5.4, the initial concentration of leader peptide variant should be chosen to center the inhibition curve around the expected IC_{50} value. We have found $20 \mu M$ to be a reasonable starting concentration for many RRE:leader peptide interactions

3. Mix each well of serially diluted leader peptide variant with MBP-RRE (to a final concentration of $4 \mu M$) and $3 \mu L$ of 250 nM stock solution of FITC-labeled leader peptide. Transfer mixed solutions to a 384-well black microplate. Let mixtures either equilibrate with shaking for $30 °C$ or without shaking for 1 h at $25 °C$. All assays should be performed in triplicate

4. Fluorescence polarization data can be collected and processed in an identical manner to Section 5.4. When using OriginPro to estimate the IC_{50} value, data should be fit through a non-linear regression analysis

5. K_i, the inhibition constant, can be calculated from the estimated IC_{50} values using the following equation. In this equation, L_{50} is the concentration of FITC-labeled peptide and P_O is the final concentration of MBP-tagged RRE ($3 \mu M$).

$$K_i = \frac{IC_{50}}{1 + \frac{L_{50}}{K_D} + \frac{P_O}{K_D}}$$

5. Enzyme activity assays and mass spectrometric analysis

While the FP-based binding assays described in the previous section are suitable for confirming and quantifying binding of an RRE to its cognate precursor peptide, it can be useful to assess the role of RRE-binding from the broader perspective of RiPP PTMs. RREs themselves do not install PTMs on their cognate precursor, but rather serve as part of larger enzymatic complexes, serving to position the core peptide region in the active site of RiPP PTM-installing enzymes. In cases where the RRE exists as a fusion to a larger enzymatic domain, the relationship between leader peptide binding and related PTMs is more obvious. In cases where the RRE exists as a discrete protein, it may be less apparent which partner enzyme(s) form the components of the PTM-installing enzyme complex. If potential partner enzymes have been identified and are amenable to heterologous expression and purification using the methods outlined in Section 4, we have found in vitro enzymatic activity assays using a "leave one out" approach to be a fast way to assess which PTMs in each RiPP biosynthetic pathway are RRE-dependent. Analysis is carried out using MALDI-TOF mass spectrometry. In the below section, we outline an example activity-based assay for assessing the role of the RRE in thiopeptide biosynthesis (Hudson, Zhang, Tietz, Mitchell, & van der Donk, 2015). While these conditions are a starting point for assessing the RRE's role in novel RiPP biochemistry, some parameters, like enzyme loading, equilibration time, and reaction additives, will necessarily change depending on the type of RiPP chemistry involved. For example, RREs that are associated with radical S-adenosyl methionine enzymes (rSAMs) will be much more difficult to assess through in vitro activity assays, due to rSAM sensitivity to oxygen (Oberg et al., 2022).

5.1 Materials, reagents, and equipment

Equipment
1. Bruker UltrafleXtreme mass spectrometer (Bruker Daltonics)
2. Speedvac concentrator (Optional; Thermo Fisher)

Materials
1. C18 ZipTip pipette tips (Millipore Sigma)

Reagents
1. TEV protease (prepared in-house through Ni-NTA affinity chromatography and stored at −80 °C until use)
2. Synthetase buffer: 50 mM Tris-HCl at pH 7.5, 125 mM NaCl, 20 mM MgCl$_2$, 5 mM DTT, 5 mM ATP
3. High performance liquid chromatography grade (HPLC) acetonitrile (Millipore Sigma)
4. ZipTip solutions: Solution A (100% HPLC-grade acetonitrile), solution B (50% acetonitrile, 50% H$_2$O, 0.1% formic acid), and solution C (H$_2$O with 0.1% formic acid)
5. ProteoMass Peptide MALDI-MS calibration kit (Millipore Sigma)

5.2 Activity assay protocol: Thiopeptide azole/azoline installation

Timing: 6–8 h
1. All reaction components should be expressed and purified separately using the protocols outlined in Section 4. For this activity-based assay, there are four reaction components: the full-length precursor peptide

(TbtA), the RRE-containing protein (TbtF), the azoline-installing YcaO protein (TbtG), and a dehydrogenase (TbtE) that oxidizes azolines to azoles

2. Digest all MBP-tagged components using TEV protease. Add protein components (TbtE/F/G) to a final concentration of 4 μM, peptide to a final concentration of 100 μM, and TEV (1:10 M ratio). Dilute reactions to 50 μL using synthetase buffer. Allow digestion to proceed for 30 min at 25 °C without agitation

3. After TEV digestion is complete, add in 5 mM ATP to initiate azoline/ azole formation. Let the reaction proceed for 5–6 h at 25 °C

4. Desalt the reaction for analysis by MALDI-TOF MS using the standard ZipTip procedure, outlined in the following section

5. (Optional) If your ZipTip is getting clogged in the previous step, larger protein components should be precipitated out of solution by addition of 50% acetonitrile (this will also quench the reaction). Remove precipitated proteins by centrifugation of the reaction mixture (17,000 × g, 10 min, 25 °C) (Fig. 9)

5.3 MALDI-TOF-MS qualitative analysis of modified products

Timing: 30 min

1. Prime a ZipTip by sequentially pipetting up and down gently, moving from solution A to solution C using a 10- or 20 μL micropipette. Avoid air bubble formation during this process

2. Load the reaction mixture onto the pre-equilibrated ZipTip by pipetting up and down 10 times into the reaction supernatant, avoiding disruption of the pelleted protein precipitate

3. De-salt the sample by washing 3 times with 10 μL of solution C. Wash can be discarded as waste

4. Elute target peptides from the ZipTip using 3 μL of 70% aqueous acetonitrile saturated with sinapinic acid. Deposit the eluted peptide onto a stainless-steel MALDI target plate and allow the spot to dry before analysis. Drying can be sped up using a small fan and lightly scraping the target's surface with the pipette tip during elution can encourage crystallization

5. Analyze samples by MALDI-TOF mass spectrometry using a Bruker UltraflexXtreme instrument operating in reflector positive-ion mode. The instrument should be calibrated using the MALDI calibration kit

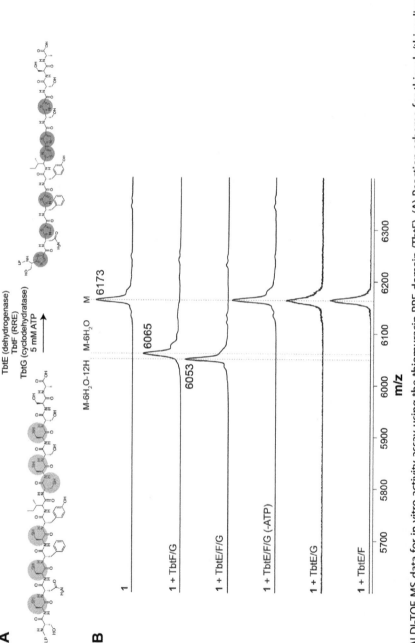

Fig. 9 MALDI-TOF MS data for in vitro activity assay using the thiomuracin RRE domain (TbtF). (A) Reaction scheme for thiazole/thiazoline formation on the thiomuracin precursor peptide. Cysteine residues highlighted in purple are cyclized by TbtG in an RRE-dependent process, and further oxidized by TbtE. (B) MALDI-TOF mass spectra showing the results of leaving individual reaction components out during in vitro activity assays. As shown, the TbtG cyclization reaction is dependent on the RRE (TbtF) as well as ATP. *Figure adapted from Hudson, G.A., Zhang, Z., Tietz, J.I., Mitchell, D.A., van der Donk, W.A., 2015. In vitro biosynthesis of the Core scaffold of the Thiopeptide Thiomuracin. Journal*

prior to data collection. Depending on the expected size of the product (the molecular weight of the unmodified precursor peptide can be obtained easily using the Expasy ProtParam tool at https://web.expasy.org/protparam/), adjust the mass window for detection within your targeted mass range

6. Analyze data using the Bruker FlexAnalysis software (this software is only available for machines running a Windows OS).

5.4 ESI-HR-MS/MS analysis of modified products

Timing: 1 h

(1) Desalt the reaction mixture supernatant using ZipTips and the protocol described above, eluting into 30 μL of 75% aqueous acetonitrile. Repeat this process with 3–5 separate ZipTips, to ensure a concentrated sample of the analyte peptides

(2) Dry down the desalted sample using a Speedvac concentrator set to 25 °C

(3) Redissolve the reaction mixture in 35% aqueous acetonitrile with 0.1% formic acid added. This solution can be directly infused onto the Orbitrap Fusion ESI-MS using a TriVersa Nanomate 100

(4) Before infusing the sample, the ESI-MS should be calibrated and tuned appropriately. Our collected data used the Pierce LTQ Velos ESI Positive Ion Calibration Solution

(5) Data can be analyzed using the Xcalibur software from Thermo Fisher or an equivalent software. Data should be averaged across the time dimension

6. Summary

In this chapter, we have outlined the main challenges and opportunities associated with bioinformatic prediction of RiPP recognition elements (RREs), a prevalent domain in prokaryotic RiPP biosynthesis. RRE domains are highly sequence diverse but display well conserved secondary and tertiary structures. We have provided strategies to customize the prediction and annotation of RRE domains using state-of-the-art bioinformatic tools. In addition, we have reiterated the most useful methods for expression and purification of RRE domains for use in experimental assays, such as in vitro binding and enzyme activity assays. Herein, we have provided methods for targeted mutagenesis of RRE and leader peptide binding residues, and quantification of binding through fluorescence polarization and competition FP assays. Given that RRE domains are present in the majority of known prokaryotic RiPP classes, and numerous examples of novel RRE-dependent natural products have been reported in recent years, we anticipate yet-undiscovered classes of RiPPs will emerge in coming years. The strategies herein will enable natural product researchers to both mine novel gene clusters using the RRE as a bioinformatic handle and confirm function of these RRE domains through binding and enzyme activity assays.

Acknowledgments

This work was supported by the National Institutes of Health (AI144967 to D.A.M.) and the Chemistry-Biology Interface Research Training Program (GM070421 to K.E.S.).

Conflict of interest statement

The authors declare no conflicts of interest.

References

Abdelmohsen, U. R., Grkovic, T., Balasubramanian, S., Kamel, M. S., Quinn, R. J., & Hentschel, U. (2015). Elicitation of secondary metabolism in actinomycetes. *Biotechnology Advances, 33*, 798–811. https://doi.org/10.1016/j.biotechadv.2015.06.003.

Ahmed, Y., Rebets, Y., Estévez, M. R., Zapp, J., Myronovskyi, M., & Luzhetskyy, A. (2020). Engineering of Streptomyces lividans for heterologous expression of secondary metabolite gene clusters. *Microbial Cell Factories, 19*, 5. https://doi.org/10.1186/s12934-020-1277-8.

Alfi, A., Popov, A., Kumar, A., Zhang, K. Y. J., Dubiley, S., Severinov, K., et al. (2022). Cell-free mutant analysis combined with structure prediction of a lasso peptide biosynthetic protease B2. *ACS Synthetic Biology, 11*, 2022–2028. https://doi.org/10.1021/acssynbio.2c00176.

Arnison, P. G., Bibb, M. J., Bierbaum, G., Bowers, A. A., Bugni, T. S., Bulaj, G., et al. (2013). Ribosomally synthesized and post-translationally modified peptide natural products: Overview and recommendations for a universal nomenclature. *Natural Product Reports, 30*, 108–160. https://doi.org/10.1039/C2NP20085F.

Bothwell, I. R., Cogan, D. P., Kim, T., Reinhardt, C. J., van der Donk, W. A., & Nair, S. K. (2019). Characterization of glutamyl-tRNA–dependent dehydratases using nonreactive substrate mimics. *Proceedings. National Academy of Sciences. United States of America, 116*, 17245–17250. https://doi.org/10.1073/pnas.1905240116.

Burkhart, B. J., Hudson, G. A., Dunbar, K. L., & Mitchell, D. A. (2015). A prevalent peptide-binding domain guides ribosomal natural product biosynthesis. *Nature Chemical Biology, 11*, 564–570. https://doi.org/10.1038/nchembio.1856. http://www.nature.com/nchembio/journal/v11/n8/abs/nchembio.1856.html#supplementary-information.

Burkhart, B. J., Kakkar, N., Hudson, G. A., van der Donk, W. A., & Mitchell, D. A. (2017). Chimeric leader peptides for the generation of non-natural hybrid RiPP products. *ACS Central Science, 3*, 629–638. https://doi.org/10.1021/acscentsci.7b00141.

Bushin, L. B., Clark, K. A., Pelczer, I., & Seyedsayamdost, M. R. (2018). Charting an unexplored streptococcal biosynthetic landscape reveals a unique peptide cyclization motif. *Journal of the American Chemical Society, 140*, 17674–17684. https://doi.org/10.1021/jacs.8b10266.

Chan, D. C. K., & Burrows, L. L. (2021). Thiopeptides: Antibiotics with unique chemical structures and diverse biological activities. *The Journal of Antibiotics*, *74*, 161–175. https://doi.org/10.1038/s41429-020-00387-x.

Chekan, J. R., Ongpipattanakul, C., & Nair, S. K. (2019). Steric complementarity directs sequence promiscuous leader binding in RiPP biosynthesis. *Proceedings. National Academy of Sciences. United States of America*, *116*, 24049–24055. https://doi.org/10.1073/pnas.1908364116.

de los Santos, E. L. C. (2019). NeuRiPP: Neural network identification of RiPP precursor peptides. *Scientific Reports*, *9*, 13406. https://doi.org/10.1038/s41598-019-49764-z.

Dias, D. A., Urban, S., & Roessner, U. (2012). A historical overview of natural products in drug discovery. *Metabolites*, *2*, 303–336. https://doi.org/10.3390/metabo2020303.

DiCaprio, A. J., Firouzbakht, A., Hudson, G. A., & Mitchell, D. A. (2019). Enzymatic reconstitution and biosynthetic investigation of the lasso peptide Fusilassin. *Journal of the American Chemical Society*, *141*, 290–297. https://doi.org/10.1021/jacs.8b09928.

Dunbar, K. L., Tietz, J. I., Cox, C. L., Burkhart, B. J., & Mitchell, D. A. (2015). Identification of an auxiliary leader peptide-binding protein required for Azoline formation in ribosomal natural products. *Journal of the American Chemical Society*, *137*, 7672–7677. https://doi.org/10.1021/jacs.5b04682.

Eddy, S. R. (2004). What is a hidden Markov model? *Nature Biotechnology*, *22*, 1315–1316. https://doi.org/10.1038/nbt1004-1315.

Evans, R. L., Latham, J. A., Xia, Y., Klinman, J. P., & Wilmot, C. M. (2017). Nuclear magnetic resonance structure and binding studies of PqqD, a chaperone required in the biosynthesis of the bacterial dehydrogenase cofactor Pyrroloquinoline Quinone. *Biochemistry*, *56*, 2735–2746. https://doi.org/10.1021/acs.biochem.7b00247.

Finn, R. D., Clements, J., Arndt, W., Miller, B. L., Wheeler, T. J., Schreiber, F., et al. (2015). HMMER web server: 2015 update. *Nucleic Acids Research*, *43*, W30–W38. https://doi.org/10.1093/nar/gkv397.

Georgiou, M. A., Dommaraju, S. R., Guo, X., Mast, D. H., & Mitchell, D. A. (2020). Bioinformatic and reactivity-based discovery of Linaridins. *ACS Chemical Biology*, *15*, 2976–2985. https://doi.org/10.1021/acschembio.0c00620.

Grove, T. L., Himes, P. M., Hwang, S., Yumerefendi, H., Bonanno, J. B., Kuhlman, B., et al. (2017). Structural insights into Thioether bond formation in the biosynthesis of Sactipeptides. *Journal of the American Chemical Society*, *139*, 11734–11744. https://doi.org/10.1021/jacs.7b01283.

Guerrero-Garzón, J. F., Madland, E., Zehl, M., Singh, M., Rezaei, S., Aachmann, F. L., et al. (2020). Class IV Lasso Peptides Synergistically Induce Proliferation of Cancer Cells and Sensitize Them to Doxorubicin. *iScience*, *23*, 101785. https://doi.org/10.1016/j.isci.2020.101785.

Haft, D. H. (2001). TIGRFAMs: A protein family resource for the functional identification of proteins. *Nucleic Acids Research*, *29*, 41–43. https://doi.org/10.1093/nar/29.1.41.

Haft, D. H. (2009). A strain-variable bacteriocin in Bacillus anthracis and Bacillus cereus with repeated Cys-Xaa-Xaa motifs. *Biology Direct*, *4*, 15. https://doi.org/10.1186/1745-6150-4-15.

Harris, L. A., Saint-Vincent, P. M. B., Guo, X., Hudson, G. A., DiCaprio, A. J., Zhu, L., et al. (2020). Reactivity-based screening for Citrulline-containing natural products reveals a family of bacterial peptidyl arginine deiminases. *ACS Chemical Biology*, *15*, 3167–3175. https://doi.org/10.1021/acschembio.0c00685.

Hegemann, J. D., & Süssmuth, R. D. (2020). Matters of class: Coming of age of class III and IV lanthipeptides. *RSC Chemical Biology*, *1*, 110–127. https://doi.org/10.1039/D0CB00073F.

Hegemann, J. D., & van der Donk, W. A. (2018). Investigation of substrate recognition and biosynthesis in class IV Lanthipeptide systems. *Journal of the American Chemical Society*, *140*, 5743–5754. https://doi.org/10.1021/jacs.8b01323.

Hudson, G. A., Burkhart, B. J., DiCaprio, A. J., Schwalen, C. J., Kille, B., Pogorelov, T. V., et al. (2019). Bioinformatic mapping of radical S-Adenosylmethionine-dependent Ribosomally synthesized and post-translationally modified peptides identifies new Cα, Cβ, and Cγ-linked Thioether-containing peptides. *Journal of the American Chemical Society, 141*, 8228–8238. https://doi.org/10.1021/jacs.9b01519.

Hudson, G. A., & Mitchell, D. A. (2018). RiPP antibiotics: Biosynthesis and engineering potential. *Current Opinion in Microbiology, 45*, 61–69. https://doi.org/10.1016/j.mib.2018.02.010.

Hudson, G. A., Zhang, Z., Tietz, J. I., Mitchell, D. A., & van der Donk, W. A. (2015). In vitro biosynthesis of the Core scaffold of the Thiopeptide Thiomuracin. *Journal of the American Chemical Society, 137*, 16012–16015. https://doi.org/10.1021/jacs.5b10194.

Johnson, L. S., Eddy, S. R., & Portugaly, E. (2010). Hidden Markov model speed heuristic and iterative HMM search procedure. *BMC Bioinformatics, 11*, 431. https://doi.org/10.1186/1471-2105-11-431.

Katoh, K., & Standley, D. M. (2013). MAFFT multiple sequence alignment software version 7: Improvements in performance and usability. *Molecular Biology and Evolution, 30*, 772–780. https://doi.org/10.1093/molbev/mst010.

Katz, L., & Baltz, R. H. (2016). Natural product discovery: Past, present, and future. *Journal of Industrial Microbiology and Biotechnology, 43*, 155–176. https://doi.org/10.1007/s10295-015-1723-5.

Kautsar, S. A., Blin, K., Shaw, S., Navarro-Muñoz, J. C., Terlouw, B. R., van der Hooft, J. J. J., et al. (2019). MIBiG 2.0: A repository for biosynthetic gene clusters of known function. *Nucleic Acids Research.* https://doi.org/10.1093/nar/gkz882.

Kloosterman, A. M., Medema, M. H., & van Wezel, G. P. (2021). Omics-based strategies to discover novel classes of RiPP natural products. *Current Opinion in Biotechnology, 69*, 60–67. https://doi.org/10.1016/j.copbio.2020.12.008.

Kloosterman, A. M., Shelton, K. E., van Wezel, G. P., Medema, M. H., & Mitchell, D. A. (2020). RRE-Finder: A genome-mining tool for class-independent RiPP discovery. *mSystems, 5.* e00267-20 https://doi.org/10.1128/mSystems.00267-20.

Koehnke, J., Mann, G., Bent, A. F., Ludewig, H., Shirran, S., Botting, C., et al. (2015). Structural analysis of leader peptide binding enables leader-free cyanobactin processing. *Nature Chemical Biology, 11*, 558–563. https://doi.org/10.1038/nchembio.1841.

Li, B., Sher, D., Kelly, L., Shi, Y., Huang, K., Knerr, P. J., et al. (2010). Catalytic promiscuity in the biosynthesis of cyclic peptide secondary metabolites in planktonic marine cyanobacteria. *Proceedings. National Academy of Sciences. United States of America, 107*, 10430–10435. https://doi.org/10.1073/pnas.0913677107.

Mavaro, A., Abts, A., Bakkes, P. J., Moll, G. N., Driessen, A. J. M., Smits, S. H. J., et al. (2011). Substrate recognition and specificity of the NisB protein, the Lantibiotic dehydratase involved in Nisin biosynthesis. *The Journal of Biological Chemistry, 286*, 30552–30560. https://doi.org/10.1074/jbc.M111.263210.

Melby, J. O., Dunbar, K. L., Trinh, N. Q., & Mitchell, D. A. (2012). Selectivity, directionality, and promiscuity in peptide processing from a *Bacillus* sp. Al Hakam Cyclodehydratase. *Journal of the American Chemical Society, 134*, 5309–5316. https://doi.org/10.1021/ja211675n.

Mistry, J., Chuguransky, S., Williams, L., Qureshi, M., Salazar, G. A., Sonnhammer, E. L. L., et al. (2021). Pfam: The protein families database in 2021. *Nucleic Acids Research, 49*, D412–D419. https://doi.org/10.1093/nar/gkaa913.

Mistry, J., Finn, R. D., Eddy, S. R., Bateman, A., & Punta, M. (2013). Challenges in homology search: HMMER3 and convergent evolution of coiled-coil regions. *Nucleic Acids Research, 41*, e121. https://doi.org/10.1093/nar/gkt263.

Mitchell, D. A., Lee, S. W., Pence, M. A., Markley, A. L., Limm, J. D., Nizet, V., et al. (2009). Structural and functional dissection of the heterocyclic peptide Cytotoxin Streptolysin S. *The Journal of Biological Chemistry, 284*, 13004–13012. https://doi.org/10.1074/jbc.M900802200.

Montalbán-López, M., Scott, T. A., Ramesh, S., Rahman, I. R., van Heel, A. J., Viel, J. H., et al. (2021). New developments in RiPP discovery, enzymology and engineering. *Natural Product Reports, 38*, 130–239. https://doi.org/10.1039/D0NP00027B.

Moon, K., Xu, F., Zhang, C., & Seyedsayamdost, M. R. (2019). Bioactivity-HiTES unveils cryptic antibiotics encoded in Actinomycete Bacteria. *ACS Chemical Biology, 14*, 767–774. https://doi.org/10.1021/acschembio.9b00049.

Myronovskyi, M., Rosenkränzer, B., Nadmid, S., Pujic, P., Normand, P., & Luzhetskyy, A. (2018). Generation of a cluster-free Streptomyces albus chassis strains for improved heterologous expression of secondary metabolite clusters. *Metabolic Engineering, 49*, 316–324. https://doi.org/10.1016/j.ymben.2018.09.004.

Oberg, N., Precord, T. W., Mitchell, D. A., & Gerlt, J. A. (2022). RadicalSAM.org: A resource to interpret sequence-function space and discover new radical SAM enzyme chemistry. *ACS Bio Med Chem Au, 2*, 22–35. https://doi.org/10.1021/acsbiomedchemau.1c00048.

Ortega, M. A., Hao, Y., Zhang, Q., Walker, M. C., van der Donk, W. A., & Nair, S. K. (2014). Structure and mechanism of the tRNA-dependent lantibiotic dehydratase NisB. *Nature*. https://doi.org/10.1038/nature13888.

Pettit, R. K. (2011). Small-molecule elicitation of microbial secondary metabolites: Elicitation of microbial secondary metabolites. *Microbial Biotechnology, 4*, 471–478. https://doi.org/10.1111/j.1751-7915.2010.00196.x.

Pimentel-Elardo, S. M., Sørensen, D., Ho, L., Ziko, M., Bueler, S. A., Lu, S., et al. (2015). Activity-independent discovery of secondary metabolites using chemical elicitation and cheminformatic inference. *ACS Chemical Biology, 10*, 2616–2623. https://doi.org/10.1021/acschembio.5b00612.

Precord, T. W., Mahanta, N., & Mitchell, D. A. (2019). Reconstitution and substrate specificity of the thioether-forming radical *S*-adenosylmethionine enzyme in freyrasin biosynthesis. *ACS Chemical Biology, 14*, 1981–1989. https://doi.org/10.1021/acschembio.9b00457.

Ramesh, S., Guo, X., DiCaprio, A. J., De Lio, A. M., Harris, L. A., Kille, B. L., et al. (2021). Bioinformatics-guided expansion and discovery of Graspetides. *ACS Chemical Biology, 16*, 2787–2797. https://doi.org/10.1021/acschembio.1c00672.

Raran-Kurussi, S., Cherry, S., Zhang, D., & Waugh, D. S. (2017). Removal of affinity tags with TEV protease. In N. A. Burgess-Brown (Ed.), *Heterologous gene expression in E. coli: Methods in molecular biology* (pp. 221–230). New York, NY: Springer. https://doi.org/10.1007/978-1-4939-6887-9_14.

Repka, L. M., Chekan, J. R., Nair, S. K., & van der Donk, W. A. (2017). Mechanistic understanding of Lanthipeptide biosynthetic enzymes. *Chemical Reviews, 117*, 5457–5520. https://doi.org/10.1021/acs.chemrev.6b00591.

Robinson, S. L., Piel, J., & Sunagawa, S. (2021). A roadmap for metagenomic enzyme discovery. *Natural Product Reports, 38*, 1994–2023. https://doi.org/10.1039/D1NP00006C.

Russell, A. H., Vior, N. M., Hems, E. S., Lacret, R., & Truman, A. W. (2021). Discovery and characterisation of an amidine-containing ribosomally-synthesised peptide that is widely distributed in nature. *Chemical Science, 12*, 11769–11778. https://doi.org/10.1039/D1SC01456K.

Saad, H., Aziz, S., Gehringer, M., Kramer, M., Straetener, J., Berscheid, A., et al. (2021). Nocathioamides, uncovered by a tunable metabologenomic approach, define a novel class of chimeric lanthipeptides. *Angewandte Chemie, International Edition, 60,* 16472–16479. https://doi.org/10.1002/anie.202102571.

Santos-Aberturas, J., Chandra, G., Frattaruolo, L., Lacret, R., Pham, T. H., Vior, N. M., et al. (2019). Uncovering the unexplored diversity of thioamidated ribosomal peptides in Actinobacteria using the RiPPER genome mining tool. *Nucleic Acids Research, 47,* 4624–4637. https://doi.org/10.1093/nar/gkz192.

Schramma, K. R., Bushin, L. B., & Seyedsayamdost, M. R. (2015). Structure and biosynthesis of a macrocyclic peptide containing an unprecedented lysine-to-tryptophan crosslink. *Nature Chemistry, 7,* 431–437. https://doi.org/10.1038/nchem.2237.

Schwalen, C. J., Hudson, G. A., Kille, B., & Mitchell, D. A. (2018). Bioinformatic expansion and discovery of Thiopeptide antibiotics. *Journal of the American Chemical Society, 140,* 9494–9501. https://doi.org/10.1021/jacs.8b03896.

Schwalen, C. J., Hudson, G. A., Kosol, S., Mahanta, N., Challis, G. L., & Mitchell, D. A. (2017). In vitro biosynthetic studies of bottromycin expand the enzymatic capabilities of the YcaO superfamily. *Journal of the American Chemical Society, 139,* 18154–18157. https://doi.org/10.1021/jacs.7b09899.

Shannon, P., Markiel, A., Ozier, O., Baliga, N. S., Wang, J. T., Ramage, D., et al. (2003). Cytoscape: A software environment for integrated models of biomolecular interaction networks. *Genome Research, 13,* 2498–2504. https://doi.org/10.1101/gr.1239303.

Soding, J., Biegert, A., & Lupas, A. N. (2005). The HHpred interactive server for protein homology detection and structure prediction. *Nucleic Acids Research, 33,* W244–W248. https://doi.org/10.1093/nar/gki408.

The UniProt Consortium, Bateman, A., Martin, M.-J., Orchard, S., Magrane, M., Agivetova, R., et al. (2021). UniProt: The universal protein knowledgebase in 2021. *Nucleic Acids Research, 49,* D480–D489. https://doi.org/10.1093/nar/gkaa1100.

Tietz, J. I., Schwalen, C. J., Patel, P. S., Maxson, T., Blair, P. M., Tai, H.-C., et al. (2017). A new genome-mining tool redefines the lasso peptide biosynthetic landscape. *Nature Chemical Biology, 13,* 470–478. https://doi.org/10.1038/nchembio.2319.

Tocchetti, A., Iorio, M., Hamid, Z., Armirotti, A., Reggiani, A., & Donadio, S. (2021). Understanding the mechanism of action of NAI-112, a lanthipeptide with potent antinociceptive activity. *Molecules, 26,* 6764. https://doi.org/10.3390/molecules26226764.

Tsai, T.-Y., Yang, C.-Y., Shih, H.-L., Wang, A. H.-J., & Chou, S.-H. (2009). *Xanthomonas campestris* PqqD in the pyrroloquinoline quinone biosynthesis operon adopts a novel saddle-like fold that possibly serves as a PQQ carrier. *Proteins, 76,* 1042–1048. https://doi.org/10.1002/prot.22461.

van der Donk, W. A., & Nair, S. K. (2014). Structure and mechanism of lanthipeptide biosynthetic enzymes. *Current Opinion in Structural Biology, 29,* 58–66. https://doi.org/10.1016/j.sbi.2014.09.006.

Walker, M. C., Eslami, S. M., Hetrick, K. J., Ackenhusen, S. E., Mitchell, D. A., & van der Donk, W. A. (2020). Precursor peptide-targeted mining of more than one hundred thousand genomes expands the lanthipeptide natural product family. *BMC Genomics, 21,* 387. https://doi.org/10.1186/s12864-020-06785-7.

Walker, J. A., Hamlish, N., Tytla, A., Brauer, D. D., Francis, M. B., & Schepartz, A. (2022). Redirecting RiPP biosynthetic enzymes to proteins and backbone-modified substrates. *ACS Central Science, 8,* 473–482. https://doi.org/10.1021/acscentsci.1c01577.

Worthen, D. B. (2007). Streptomyces in nature and medicine: The antibiotic makers. *Journal of the History of Medicine and Allied Sciences, 63,* 273–274. https://doi.org/10.1093/jhmas/jrn016.

Yu, Y., Duan, L., Zhang, Q., Liao, R., Ding, Y., Pan, H., et al. (2009). Nosiheptide bio-synthesis featuring a unique indole side ring formation on the characteristic Thiopeptide framework. *ACS Chemical Biology, 4*, 855–864. https://doi.org/10.1021/cb900133x.

Zallot, R., Oberg, N., & Gerlt, J. A. (2019). The EFI web resource for genomic enzymology tools: Leveraging protein, genome, and metagenome databases to discover novel enzymes and metabolic pathways. *Biochemistry, 58*, 4169–4182. https://doi.org/10.1021/acs.biochem.9b00735.

Zardecki, C., Dutta, S., Goodsell, D. S., Voigt, M., & Burley, S. K. (2016). RCSB protein data Bank: A resource for chemical, biochemical, and structural explorations of large and small biomolecules. *Journal of Chemical Education, 93*, 569–575. https://doi.org/10.1021/acs.jchemed.5b00404.

Zhang, Z., Hudson, G. A., Mahanta, N., Tietz, J. I., van der Donk, W. A., & Mitchell, D. A. (2016). Biosynthetic timing and substrate specificity for the thiopeptide thiomuracin. *Journal of the American Chemical Society, 138*, 15511–15514. https://doi.org/10.1021/jacs.6b08987.

CHAPTER EIGHT

The preparation of recombinant arginyltransferase 1 (ATE1) for biophysical characterization

Misti Cartwright, Verna Van, and Aaron T. Smith∗
Department of Chemistry and Biochemistry, University of Maryland Baltimore County, Baltimore, MD, United States
∗Corresponding author: e-mail address: smitha@umbc.edu

Contents

Methods in Enzymology, Volume 679
ISSN 0076-6879
https://doi.org/10.1016/bs.mie.2022.07.036

235

Abstract

Arginyltransferases (ATE1s) are eukaryotic enzymes that catalyze the non-ribosomal, post-translational addition of the amino acid arginine to an acceptor protein. While understudied, post-translation arginylation and ATE1 have major impacts on eukaryotic cellular homeostasis through both degradative and non-degradative effects on the intracellular proteome. Consequently, ATE1-catalyzed arginylation impacts major eukaryotic biological processes including the stress response, cellular motility, cardio-vascular maturation, and even neurological function. Despite this importance, there is a lack of information on the structural and biophysical characteristics of ATE1, prohibiting a comprehensive understanding of the mechanism of this post-translational modification, and hampering efforts to design ATE1-specific therapeutics. To that end, this chapter details a protocol designed for the expression and the purification of ATE1 from *Saccharomyces cerevisiae*, although the approaches described herein should be generally applicable to other eukaryotic ATE1s. The detailed procedures afford high amounts of pure, homogeneous, monodisperse ATE1 suitable for downstream biophysical analyses such as X-ray crystallography, small angle X-ray scattering (SAXS), and cryo-EM techniques.

1. Introduction

Proteins are a versatile group of biological macromolecules that are responsible for undertaking many indispensable functions within the cell. However, when constrained to only the 20 canonical amino acids, not every necessary cellular function may be fulfilled. Thus, the encoding capacity of the genome must be expanded, and such augmentation is achieved by the ability of proteins to be co- and/or post-translationally modified. Post-translational modifications (PTMs) of proteins may be highly varied and include the covalent addition of functional groups and/or reactive moieties, as well as the proteolysis or cleavage of the polypeptide (Conibear, 2020; Walsh, Garneau-Tsodikova, & Gatto, 2005). Many storied PTMs such as phosphorylation, methylation, and acetylation (Fig. 1) have been well-studied over the last several decades, and essential cellular functions have been ascribed to the fidelity of these modifications. For example, phosphorylation is a reversible PTM that is responsible for the addition or removal of a phosphate group on a protein at a hydroxyl group on amino acid side chains. This addition and its removal are both vital for regulating cellular metabolism and respiration (Humphrey, James, & Mann, 2015). As another example, acetylation is a reversible PTM that is characterized by the addition of an acetyl group (typically derived from acetyl CoA) that is attached to the N-terminus of a protein or at the side chain of lysine

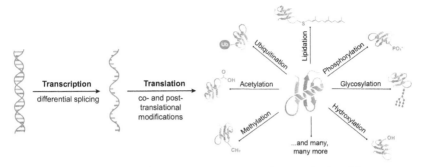

Fig. 1 Post-translational modifications (PTMs) increase the complexity of the proteome. After transcription and subsequent translation, PTMs may alter protein function, oligomerization, activity, and even stabilization, thereby affording an expansion of the encoding capacity of the genome.

(Lys) residues. Acetylation has been shown to be important in gene regulation and transcription whereby the acetyl group neutralizes the positively-charged Lys allowing for the unwinding of DNA from the histone (Drazic, Myklebust, Ree, & Arnesen, 2016; Kumar, Thakur, & Prasad, 2021). Despite our understanding of these examples, several additional and important PTMs exist within the eukaryotic cell but have been historically understudied.

A lesser-studied but essential eukaryotic PTM is arginylation, which is the covalent addition of the amino acid arginine (Arg) to an acceptor protein, catalyzed by the enzyme arginyltransferase 1 (ATE1) (Balzi, Choder, Chen, Varshavsky, & Goffeau, 1990; Saha & Kashina, 2011; Tasaki, Sriram, Park, & Kwon, 2012; Van & Smith, 2020). Typically, Arg is added to an N-terminal aspartic acid (Asp), glutamic acid (Glu), or oxidized cysteine (Cys) residue of a polypeptide through the formation of a conventional peptide bond, extending the length of the polypeptide by one amino acid residue (Saha & Kashina, 2011). However, data have emerged demonstrating that arginylation may also occur at a side chain residue (typically Asp or Glu) through formation of an isopeptide bond rather than a conventional peptide bond (Wang et al., 2014). Regardless of the location of arginylation, this energy-independent process is catalyzed by ATE1s that use the high-energy aminoacylated Arg-tRNA^{Arg} (or fragments thereof) (Avcilar-Kucukgoze et al., 2020; Wang et al., 2011) as the Arg donor (Fig. 2), although the mechanism of this process is poorly understood due to a lack of ATE1 structural and biophysical information. Once arginylated, this PTM can have dramatic effects on protein structure and function.

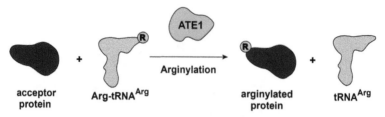

Fig. 2 Cartoon depiction of post-translational arginylation. ATE1 catalyzes the energy-independent transfer of the amino acid Arg (R) from the aminoacylated Arg-tRNA^{Arg} to an acceptor protein to create an arginylated protein.

Arginylation is emerging as a global regulator of eukaryotic cellular homeostasis through its degradative and non-degradative effects on the proteome. Arginylation has a major role within the N-degron pathway, a hierarchal determinant of intracellular protein half-life (Timms & Koren, 2020). The first step of the N-degron pathway is the methionine (Met) aminopeptidase-catalyzed removal of the N-terminal Met, which occurs for at least half of all proteins (Varshavsky, 2011, 2019). This process exposes what was previously the second amino acid as a primary (1°), secondary (2°), or tertiary (3°) destabilizing residue. Tertiary (3°) destabilizing residues are asparagine (Asn), Cys, or glutamine (Gln), and require chemical modifications before arginylation may occur. In animals and plants, Asn and Gln must be deamidated commonly by Ntan1 (Nt^N-amidase) and Ntaq1 (Nt^Q-amidase), respectively. Cys, however, must be oxidized to Cys sulfinic acid via Cys dioxygenases (Varshavsky, 2019); this process may also be catalyzed spuriously by reactive oxygen and reactive nitrogen species (Tasaki et al., 2012). Once a negative charge is imparted on the side chain of the now chemically-modified secondary (2°) destabilizing residue, proteins bearing Asp, Glu, or oxidized Cys can then be arginylated in a tRNA-dependent manner that is catalyzed by ATE1 (Van & Smith, 2020). Once these proteins are tagged with Arg—a primary (1°) destabilizing residue—these modified proteins may then be recognized by N-recognins, ubiquitinated, and degraded in a proteasomal-dependent manner. Examples of proteins arginylated and subsequently degraded in this manner include regulators of G-protein signaling (RGS), such as RGS4, RGS5, RGS16 and (most recently) RGS7 that are important in cardiovascular maturation as well as in the nervous system (Fina et al., 2021; Van & Smith, 2020). However, degradation is not the only fate for proteins and enzymes that have been post-translationally arginylated.

Arginylation was originally thought to be involved in only protein degradation, but recent and important research has shown that proteins may also be stabilized and oligomerize differently once arginylated (Van & Smith, 2020). Examples of some proteins that change their oligomeric state in response to arginylation include β-actin and calreticulin. Regarding calreticulin, arginylation during stress facilitates the production of stress granules, which is an important adaption in promoting cell survival (Carpio et al., 2013; Kashina, 2014). Calreticulin is a monomer before arginylation and a dimer after arginylation, which promotes a change in olig-omerization that then allows the formation of disulfide bridges upon calcium depletion (Carpio et al., 2013). In *ate1* knockouts of amoeba (*Dictyostelium discoideum*), a lack of arginylation resulted in β-actin abnormalities, indicating that actin-dependent processes such as cell adhesion and other cytoskeletal activities such as movement are ATE1-dependent (Batsios et al., 2019; Kashina, 2014). Very recently, studies have even shown that arginylation even stabilizes the human immunodeficiency virus (HIV) core during the uncoating process (Kishimoto et al., 2021). These cellular studies clearly demonstrate that ATE1-catalyzed arginylation affects proteins via non-degradative pathways as well, but the precise mechanism of action is unclear.

Despite its physiological importance, the structural and the biophysical properties of ATE1 are poorly understood. For example, there is currently no known structure of an ATE1 from any organism. Some mechanistic inferences have been garnered via comparison to functional analogs of ATE1 that are found in prokaryotes and are known as leucyl/phenylalanyl (L/F) transferases, which have been structurally characterized. L/F transfer-ases are prokaryotic enzymes that transfer hydrophobic residues such as leu-cine (Leu) and phenylalanine (Phe) to the N-terminus of protein substrates (Suto et al., 2006). Structures of L/F transferases with and without tRNA analogs have been determined, and these structures have been used to sug-gest post-translational arginylation mechanisms (Suto et al., 2006; Van & Smith, 2020; Watanabe et al., 2007); however, these comparisons are no substitute for the structural and biophysical characterizations of a bona fide ATE1. The lack of this information prohibits the design of novel com-pounds to target intracellular arginylation as a therapeutic strategy.

To that end, in this chapter, we describe the recombinant expression and preparation of eukaryotic ATE1 to suitable purity and homogeneity for bio-physical and structural characterizations. While ATE1 is found in virtually all eukaryotes (Jiang et al., 2020), in higher-ordered organisms, the gene may be alternatively spliced, leading to different isoforms and different intra-cellular localizations (Kwon, Kashina, & Varshavsky, 1999). To simplify this difficulty, in this chapter we have focused on the isolation and purification of ATE1 from *Saccharomyces cerevisiae* (*Sc*ATE1), which exists as a single (iso)

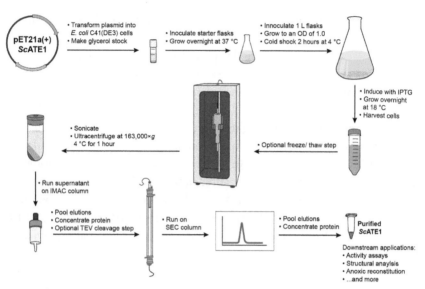

Fig. 3 Schematic cartoon depicting the process of ScATE1 purification described in this protocol.

form, but this protocol may be generally applicable to other eukaryotic ATE1s. We describe in detail a general procedure for the cloning, expression, metal-affinity purification, tag cleavage, and gel filtration of ScATE1 (Fig. 3) that reproducibly yields pure, monomeric, monodisperse ATE1 of sufficient purity for enzymatic assays, small-angle X-ray scattering, and X-ray crystallography, all of which may be used to decipher the enigmatic mechanism of post-translational arginylation.

2. General methods and analysis

Personal protective equipment such as safety glasses, lab coats, and gloves should be used while conducting these experiments. Additionally, adherence to all safety and security measures relevant for a biosafety level-1 (BSL-1) laboratory should be followed.

3. Construct design and protein expression

This section describes the large-scale expression of ScATE1 to suitable yields for downstream biophysical analyses. The expression construct of S. cerevisiae ATE1 (ScATE1) was based on the sequence from S. cerevisiae strain ATCC 204508 (Uniprot identifier P16639). A codon-optimized, synthetically-generated gene containing an additional Tobacco Etch Virus (TEV)-protease cleavage site (ENLYFQS) was subcloned into the pET-21a(+) expression plasmid using the *Nde*I and *Xho*I restriction sites, resulting in an encoded C-terminal (His)$_6$ affinity tag when read in-frame for ease of purification. Experience shows that an affinity tag must not be placed on the N-terminus of ATE1. N-terminally tagged ATE1 generally fails to accumulate as a soluble protein, and any purified, N-terminally tagged protein is generally inactive, presumably due to misfolding of the N-terminal regulatory domain that houses the presence of the ATE1 [Fe—S] cluster (Van et al., 2021).

3.1 Equipment

- Bunsen burner
- *Sc*ATE1 gene subcloned into the pET-21a(+) expression plasmid, ca. 80–100 ng/μL concentration
- BioRad MicroPulser Electroporator
- Sterile 0.1 cm electroporation cuvettes
- Sterilized pipette tips
- Sterilized 1.5 mL microcentrifuge tubes
- Thermomixer Eppendorf Thermomixer C
- Sterile 0.22 μm pore size syringe filters
- Disposable plastic syringes
- Sterile 25 mL serological pipettes
- Thermo Scientific Nanodrop OneC
- Thermo Scientific MaxQ 8000 incubator shaker
- Thermo Scientific Lynx 6000 centrifuge
- Sterile 50 mL conical tubes

3.2 Reagents

- Filter sterilized stock ampicillin at 1000× concentration (0.1 g/mL in distilled H$_2$O)
- Electrocompetent *Escherichia coli* C41 (DE3) expression cells
- Recovery medium for cloning (MilliporeSigma part number CMR0002)
- LB agar plates supplemented with ampicillin (100 μg/mL)
- 3 × 100 mL sterile Luria Broth (LB): 1 g NaCl, 1 g yeast extract, 0.5 g Tris base/Tris(HCl) dissolved in 100 mL distilled H$_2$O and autoclaved

- $12 \times 1\,L$ sterile Luria Broth (LB): $10\,g$ NaCl, $10\,g$ yeast extract, $5\,g$ Tris base/Tris(HCl) dissolved in $1\,L$ distilled H_2O and autoclaved in $2\,L$ baffled, glass trypsinizing flasks
- $12\,mL$ $1\,M$ sterile isopropyl β-D-1-thiogalactopyranoside (IPTG): $2.86\,g$ IPTG dissolved in $12\,mL$ distilled water and filter sterilized
- Resuspension buffer: $100\,mM$ NaCl, $50\,mM$ Tris pH 7.5, 5% (v/v) glycerol

3.3 Procedure

Day 1
1. Thaw plasmid and recovery media.
2. Set thermomixer to $37\,°C$ with shaking of $300\,RPM$.
3. Place tubes of desired electrocompetent *E. coli* cell line on ice for $15\,min$.
4. Add $2\,\mu L$ of plasmid to the cells under the flame of a Bunsen burner.
5. Under flame of a Bunsen burner, mix and transfer the plasmid/cell mixture to a sterile, $0.1\,cm$ electroporation cuvette, ensuring the mixture is pipetted fully into the cuvette slot (see Note 1).
6. Immediately place the covered cuvette in the electroporator and pulse using the "EC1" program.
7. After pulsed, remove the cuvette from the electroporator. Under the flame of a Bunsen burner, add $500\,\mu L$ of cold recovery media to the cuvette and gently pipette up and down to mix (the mixture should appear cloudy).
8. Still under the flame of a Bunsen burner, transfer this mixture to a new, sterile $1.5\,mL$ tube.
9. Place the tube in the Thermomixer and mix for $30\,min$ at $300\,RPM$.
10. Warm an LB ampicillin agar plate at room temperature and label.
11. After $30\,min$, remove the tubes containing the transformed cells from the thermomixer. Under the flame of a Bunsen burner, pipette $80\,\mu L$ onto the LB ampicillin agar plate and spread using a sterilized spreader.
12. Allow colonies to grow overnight by placing in a warm room or $37\,°C$ incubator.

Day 2
1. Under the flame of a Bunsen burner and using a sterile pipet tip, pluck a single colony from the LB ampicillin agar plate and inoculate $3 \times$ labeled, sterile flasks charged with $100\,mL$ of sterile Luria Broth (LB) supplemented with $100\,\mu L$ of $0.1\,g/mL$ filter sterilized ampicillin.

2. Place these flasks into the incubator shaker, and grow the *E. coli* overnight with shaking of 200 RPM at 30 °C.
3. Seal the plate with parafilm and store at 4 °C.

Day 3

1. Under the flame of a Bunsen burner, add 1 mL of 0.1 g/mL filter sterilized ampicillin to each of 12 baffled, glass trypsinizing flasks charged with 1 L of autoclaved LB.
2. Retrieve the 3 × 100 mL culture flasks from the incubator shaker and combine the cultures under the flame of a Bunsen burner.
3. To each of the 12 flasks charged with 1 L of autoclaved LB and supplemented with 100 μg/mL, inoculate with approximately 25 mL of the combined culture from step 2 (see Note 2).
4. Allow the bacteria to grow with shaking of 200 RPM in an incubator shaker set to 37 °C until the cultures reach an optical density (OD) at 600 nm (OD_{600}) of ca. 0.6–0.8.
5. Remove the flasks from the incubator shaker and cold shock the cultures by placing them at 4 °C for approximately 2 h.
6. Approximately 30 min prior to the end of step 5, set an incubator shaker to 18 °C.
7. Once step 5 is finished, to each of the 12 flasks induce expression by adding 1 mL of 1 M filter sterilized IPTG.
8. Allow induction to occur with shaking of 200 RPM in an incubator shaker set to 18 °C overnight (typically 16–20 h).

Day 4

1. Harvest cells by spinning at $5000 \times g$, for 12 min at 4 °C in 4 × 1 L centrifuge bottles and discard the supernatant.
2. Repeat step 1 three times until all cells have been harvested and all supernatant has been discarded.
3. Pour off the remaining supernatant.
4. Resuspend cells in approximately 35 mL of resuspension buffer per 3 L of culture.
5. Transfer cellular resuspension to 50 mL conical tubes, flash-freeze on $N_{2(l)}$ and store at −80 °C (see Note 3).

3.4 Notes

1. If the cell-plasmid mixture does not go into the slot, gently tap the cuvette to allow the mixture to get to the bottom of the cuvette.

2. A glycerol stock can be made from the remaining cells and stored in
   $-80\,^\circ$C for future cell growths.

3. The freezing step is optional. It is possible to go directly to cell lysis and
   purification without the intermediate freezing step.

4. Purification of (His)$_6$-tagged ATE1

This section describes the initial purification of ScATE1 by utilizing
the C-terminal (His)$_6$ tag, which allows for the protein to be purified with
immobilized metal affinity chromatography (IMAC). As previously noted
(vide supra), our experience shows that the location of the ATE1 affinity
tag is important, and properly folded and enzymatically active ATE1 is best
achieved when the affinity tag is present on the C-terminus of the protein.
After cell lysis and centrifugation, the clarified lysate is applied to the
nickel-containing resin, and purification is achieved using a fast protein liquid chromatography (FPLC) instrument, which results in good purity as
determined by sodium dodecyl sulfate-polyacrylamide gel electrophoresis
(SDS-PAGE) and high yields (average of approximately $25\,\mathrm{mg\,L}^{-1}$ culture).

4.1 Equipment

- QSonica Q700 ultrasonic cell disruptor operating at $4\,^\circ$C
- Stainless-steel beaker
- Polycarbonate ultracentrifuge tubes
- Beckman-Coulter Optima XE-90 ultracentrifuge
- ÄKTA™ Pure protein purification system
- HisTrap™ HP 5 mL pre-packed protein purification column
- Amicon® Ultra-15 centrifugal filter unit with a 30 kDa molecular weight
  cut off (MWCO) spin filter
- Eppendorf 5810R centrifuge
- Gel electrophoresis equipment

4.2 Reagents

- Wash buffer: 50 mM Tris pH 8.0, 300 mM NaCl, 10% (v/v) glycerol,
  1 mM tris(2-carboxyethyl)phosphine (see Note 1)
- Elution buffer: 50 mM Tris pH 8.0, 300 mM NaCl, 10% (v/v) glycerol,
  1 mM tris(2-carboxyethyl)phosphine, 300 mM imidazole
- 15% (m/v) SDS-polyacrylamide protein electrophoresis gels
- Solid phenylmethylsulfonyl fluoride (PMSF)

4.3 Procedure

1. If the ATE1-containing cells were frozen in the optional part of Step 3 (vide supra), thaw cells gently to room temperature and transfer to a stainless-steel beaker.

2. To the thawed, homogenized cells, add ca. 50–100 mg of solid PMSF and stir vigorously prior to lysis (see Note 2).

3. Immerse the stainless-steel beaker in an ice-water slurry. Sonicate for ca. 10–12 min at 80% amplitude, alternating 30 s on sonication pulse and 30 s off sonication pulse.

4. Transfer the lysate to polycarbonate ultracentrifuge tubes. Centrifuge at $163,000 \times g$ for 1 h at 4 °C to pellet insoluble cellular debris.

5. After ultracentrifugation, decant the supernatant into a cooled beaker.

6. Prepare your buffers, column, and sample inlet on your FPLC (see Note 3).

7. Pre-wash the column with 5 column volumes (CVs) of a buffer mixture containing 93% (v/v) wash buffer and 7% (v/v) elution buffer.

8. Apply your sample lysate to the column.

9. Wash the column again with 8 CVs of a buffer mixture containing 93% (v/v) wash buffer and 7% (v/v) elution buffer.

10. Elute any non-specifically bound proteins by washing the column with 6 CVs of a buffer mixture containing 90% (v/v) wash buffer and 10% (v/v) elution buffer.

11. Elute ATE1 by washing the column with 6 CVs of 100% elution buffer.

12. Pool together the elution fractions from Step 11.

13. Wash an Amicon Ultra-15 centrifugal filter unit with a 30 kDa MWCO spin filter with distilled H_2O by centrifuging at $5000 \times g$ for 5 min at 4 °C.

14. Discard the H_2O from the tube and apply 14 mL of lysate to the spin concentrator. Spin the concentrator at $5000 \times g$ for ca. 10–20 min at 4 °C (see Note 4).

15. Resuspend the protein gently in the spin filter and add more elution fractions until the protein has been concentrated to ca. 1–1.5 mL (typically ca. 20–40 mg/mL in our hands).

16. Remove the protein from the filter and transfer to a clean, labeled 1.5 mL microcentrifuge tube.

17. To determine the purity of the protein, apply 1–5 µg of purified protein per well to a 15% (m/v) SDS-polyacrylamide gel.

18. Aliquot the protein into fractions of desired sizes (ca. 10–1000 µL), flash-freeze on $N_{2(l)}$ and store at −80 °C (see Note 5).

4.4 Notes

1. In our experience, it is imperative to include some sort of reducing agent in the recombinant preparation of ATE1 in order to prevent protein aggregation. Other reducing agents, such as dithiothreitol (DTT) and/or β-mercaptoethanol (BME) are suitable but will oxidize more rapidly than tris(2-carboxyethyl)phosphine.

2. If proteolysis is observed, other inhibitor cocktails may be substituted for PMSF. However, we have not noted a need for these alternative protease inhibitors in our hands.

3. While an FPLC makes the IMAC purification procedure more automated, it is unnecessary, and a loose resin in a gravity column may be substituted for this portion of the protocol.

4. Depending on protein yield, concentration may be slower or quicker than reported here. This step may need to be optimized for each ATE1 construct.

5. Freezing of the protein is optional at this step. One may go directly to tag cleavage and/or size-exclusion chromatography at this step.

5. Cleavage of (His)$_6$-tagged ATE1

This section describes the removal of the C-terminal (His)$_6$ tag using the Tobacco Etch Virus (TEV) protease cleavage site (ENLYFQS) that was genetically added to the construct (see Section 3). The removal of the C-terminal (His$_6$) tag is optional, but removal of this tag may be useful for downstream applications, although in our hands, we have not noted a difference in the behavior of ATE1 with and without its C-terminal tag. After purification by an affinity column, the buffer of the purified enzyme is modified, and the ATE1 is then incubated with TEV-protease (commercial or house-made). Once cleavage has occurred, the solution is then applied to a nickel-containing resin, and the cleaved protein now elutes during the column-washing step instead of the elution step. The cleavage may be confirmed via Western blotting using an anti-(His)$_6$ antibody (see Note 1).

5.1 Equipment

- ÄKTA™ Pure protein purification system
- HisTrap™ HP 5 mL pre-packed protein purification column
- Amicon® Ultra-15 centrifugal filter unit with a 30 kDa molecular weight cut off (MWCO) spin filter

- Eppendorf 5810R centrifuge
- Gel electrophoresis equipment

5.2 Reagents

- Wash buffer: 50 mM Tris pH 8.0, 300 mM NaCl, 10% (v/v) glycerol, 1 mM tris(2-carboxyethyl)phosphine
- Elution buffer: 50 mM Tris pH 8.0, 300 mM NaCl, 10% (v/v) glycerol, 1 mM tris(2-carboxyethyl)phosphine, 300 mM imidazole
- 15% (m/v) SDS-polyacrylamide protein electrophoresis gels
- Ethylenediaminetetraacetic acid (EDTA)
- Dithiothreitol (DTT)
- Tobacco Etch Virus (TEV) protease (commercial or house-made)

5.3 Procedure

1. Measure the concentration of purified ATE1 from Section 4.
2. Dilute ATE1 protein to a concentration of approximately 1 mg/mL.
3. To the protein solution, add EDTA to a final concentration of 0.5 mM.
4. To the protein solution, add DTT to a final concentration of 100 mM.
5. To the protein solution, add TEV protease to the protein in a 1:100 (w/w) ratio.
6. Incubate the ATE1 and protease solution overnight at 4 °C while constantly rocking (see Note 2).
7. After tag cleavage, buffer exchange the ATE1 and protease mixture into wash buffer using an Amicon Ultra-15 centrifugal filter unit with a 30 kDa MWCO spin filter. Spin the concentrator at $5000 \times g$ for ca. 10–20 min at 4 °C until the mixture is concentrated to ca. 1–2 mL. Dilute the concentrated mixture with fresh wash buffer to approximately 14 mL and centrifuge again until the mixture is concentrated to ca. 1–2 mL. Repeat this process again until you have buffer exchanged your ATE1 and protease mixture 3–4 times.
8. Prepare your buffers, column, and sample inlet on your FPLC.
9. Pre-wash the column with 6 CVs of a buffer mixture containing 93% (v/v) wash buffer and 7% (v/v) elution buffer.
10. Apply your ATE1 and protease sample solution to the column (see Note 3).
11. Wash the column again with 8 CVs of a buffer mixture containing 93% (v/v) wash buffer and 7% (v/v) elution buffer. Make certain to collect these fractions, as your cleaved ATE1 should elute at this stage.

12. Elute any non-specifically bound proteins by washing the column with 6 CVs of a buffer mixture containing 90% (v/v) wash buffer and 10% (v/v) elution buffer (see Note 4).

13. Elute any protein still stuck to the column by washing the column with 6 CVs of 100% elution buffer.

14. Separately pool together the elution fractions from Steps 11, 12, and 13.

15. To determine the location of ATE1, apply 1–5 µg of purified protein per well to a 15% (m/v) SDS-polyacrylamide gel.

16. Once the location of cleaved ATE1 is identified, concentrate the protein to ca. 1–1.5 mL (typically ca. 20–40 mg/mL in our hands).

17. Aliquot the protein into fractions of desired sizes (ca. 10–1000 µL), flash-freeze on $N_{2(l)}$ and store at −80 °C (see Note 5).

5.4 Notes

1. While Western blotting is an easy and convenient method for determining the presence or absence of the $(His)_6$ tag, other methods such as mass spectrometry may be used instead.

2. The temperature and time for incubation of TEV protease with ATE1 may need to be optimized. For some constructs, room (approximately 25 °C) or elevated (approximately 37 °C) temperatures and shorter incubation times may be necessary to accomplish complete cleavage.

3. If you have concentrated your sample to ca. 1–2 mL, you may use a small sample loop to apply your mixture to your column (vide infra). For larger volumes, and external sample line or a superloop may be necessary.

4. Because ATE1s are also Cys-rich proteins, you may experience some modest binding of the cleaved protein to Ni-containing resin even in the absence of the $(His)_6$ tag. We recommend keeping all of your elution fractions and testing them to see whether your cleaved protein is present.

5. Freezing of the protein is optional at this step. One may go directly to size-exclusion chromatography at this step.

6. Size-exclusion purification of ATE1

This section describes the final purification (or "polishing") of ScATE1 via size-exclusion chromatography. This process may be performed with either the $(His)_6$-tagged or the cleaved form of the protein. In both cases, the protein will be both further purified and buffer exchanged into a more suitable downstream buffer, while in the latter case, this process of gel filtration will also remove the small amount of TEV protease present in the cleavage mixture. Moreover, this process affords the separation of differing oligomeric states of *Sc*ATE1 and allows for the removal of any aggregated protein that may have been present during purification and/or created due to spurious oxidation of the O_2-sensitive *Sc*ATE1. Finally, using calibration standards for the column of interest, this process reveals that *Sc*ATE1 is a predominantly monomeric protein (regardless of the presence or absence of the $(His)_6$ tag), which is the first time this observation has been reported, to our knowledge. This procedure is broadly applicable to other eukaryotic ATE1s, and the final purified protein is suitable for structural and biophysical characterizations.

6.1 Equipment

- ÄKTA™ Pure protein purification system
- Eppendorf 5418R microcentrifuge
- 3 mL syringes
- BD PrecisionGlide™ 20 G needle
- ÄKTA™ sample loop, 2.0 mL
- HiLoad™ 16/600 Superdex 200 pg preparative gel filtration column pre-calibrated at 4 °C
- Sterile 0.22 μm bottle top filters for 45 mm media bottles
- Schlenk line
- Schlenk line adapter top for 45 mm media bottles
- Stir plate
- Stir bar
- 1.5 mL microcentrifuge tubes
- Sterile 1.5 mL microcentrifuge tubes containing 0.22 μm filters
- Agilent Cary 60 UV–Visible spectrophotometer
- UV-transparent 1 cm cuvette

6.2 Reagents

- Size-exclusion buffer: 50 mM Tris pH 7.5, 100 mM KCl, 5% (v/v) glycerol, 1 mM dithiothreitol (DTT) (see Note 1)

6.3 Procedure

Day 1
1. Prepare the size-exclusion buffer and filter through a sterile 0.22 μm bottle top filter.

2. Add a stir bar to the buffer, attach the Schlenk line adapter to the bottle, and connect the adapter to the Schlenk line vacuum.

3. With stirring, degas the buffer for at least several hours, typically overnight (see Note 2).

Day 2

1. Pack the outside of the degassed buffer bottle with ice to cool the buffer.

2. If the purified ATE1 were frozen in the optional part of Step 4 or Step 5 (vide supra), thaw the protein (ca. 10–20 mg) gently on ice.

3. Once thawed, filter the protein briefly by centrifuging at $5000 \times g$ for 5 min at 4 °C through a 1.5 mL microcentrifuge tube containing 0.22 μm filter in order to remove any insoluble material.

4. Attach the 2 mL sample loop to the FPLC, connect to a clean 3 mL syringe, and rinse the sample loop several times with size-exclusion buffer.

5. Using a 20 G needle attached to a 3 mL syringe, load the purified ATE1 into the syringe, invert, and remove any bubbles.

6. Remove the needle carefully and attach the syringe to the sample loop to load the purified ATE1 into the sample loop (see Note 3). Once loaded, finish attaching the sample loop to the FPLC.

7. Pre-equilibrate the size-exclusion column with 1.5 CVs of cold, degassed size-exclusion buffer.

8. Apply the sample in the 2 mL sample loop to the column by washing the sample loop with 6 mL of cold, degassed size-exclusion buffer.

9. Elute the protein isocratically from the size-exclusion column by applying another 1.5 CVs cold, degassed size-exclusion buffer. For a column of this size, we typically collect protein in 1 or 2 mL fractions, but the fraction size can be optimized for each construct, depending on its oligomeric homogeneity.

10. Based on the elution volume and standards that were previously calibrated on the size-exclusion column, calculate the estimated oligomeric state of each elution peak. If it is unclear which peak is your protein of interest, apply 1–5 μg of purified protein per well to a 15% (m/v) SDS-polyacrylamide gel to determine which fractions to pool.

11. If you have multiple elution peaks containing ATE1, pool and concentrate each in separate Amicon Ultra-15 centrifugal filter units each containing a 30 kDa MWCO spin filter. Our preparations are typically dominated by a single, monomeric elution peak (Fig. 4), which is generally the only protein form that we pool and concentrate.

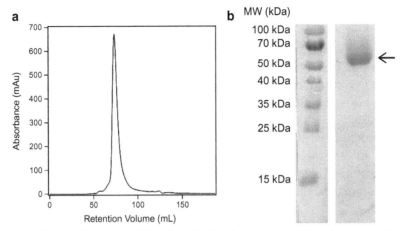

Fig. 4 Following the protocol described in this chapter, recombinantly expressed and purified ScATE1 is homogeneous and monodisperse. (A) Superdex 200 size-exclusion chromatogram of post-IMAC-purified ScATE1. Based on calibration standards, ScATE1 is predominantly monomeric under these conditions. (B) 15% SDS-PAGE analysis of IMAC- and SEC-purified ScATE1. The left lane is the molecular weight (MW) marker, while the right lane is purified ScATE1. The left arrow points to purified ScATE1 with an apparent molecular weight of ≈ 60 kDa.

12. Once concentrated to the desired concentration (typically 10 mg/mL for our purposes), apply 1–5 µg of purified protein per well to a 15% (m/v) SDS–polyacrylamide gel to determine the final purity of the protein (Fig. 4).

13. Estimate the total amount of protein retrieved by measuring the protein absorbance at 280 nm (A_{280}) and calculating the concentration using an estimated molar absorptivity (ε) of 87,700 M^{-1} cm^{-1} for apo ScATE1 (see Note 4).

14. Aliquot the protein into fractions of desired sizes (ca. 10–1000 µL), flash-freeze on $N_{2(l)}$ and store at $-80\,°C$ (see Note 5).

6.4 Notes

1. The size-exclusion buffer composition can require trial-and-error testing. In general, for structural studies, lower salt concentrations are preferred. However, not all ATE1 constructs tolerate lower salt concentrations. In lieu of DTT, tris(2-carboxyethyl)phosphine or BME are acceptable substitutes, but their oxidation times are different than DTT.

2. This step is important to prevent bubbles from disrupting the gel-filtration matrix. The buffer should be degassed until no more bubbles are visibly observed while stirring under vacuum, although the precise length of time may differ based on buffer composition.

3. The exact amount of protein that may be applied to the size-exclusion column will differ from column to column and protein to protein. Generally, we apply no more than 10–20 mg per run for HiLoad™ 16/600 Superdex 200 pg, but we have applied up to 100 mg of protein in some cases with only minimal loss of peak-to-peak resolution.

4. The molar absorptivity presented is calculated theoretically but is generally accurate in our experience if the protein is apo, i.e., lacks the [Fe—S] cluster.

5. Freezing of the protein is optional at this step, and the protein may be used directly for downstream applications.

7. Summary and conclusions

This chapter describes the recombinant expression and preparation of a eukaryotic ATE1 to suitable purity and homogeneity for downstream biophysical and structural characterizations. Understanding the mechanism of ATE1-catalyzed post-translation arginylation is paramount, as it is well known that ATE1 is a critical enzyme for a variety of biological processes including cellular homeostasis, cardiovascular development, and cellular motility. Moreover, recent research has shown that ATE1 is linked to several human diseases, such as cancer and HIV. However, despite this importance, there is currently no published structure of an ATE1 from any organism, and there is very little published research on any biophysical or enzymatic characterization of this important enzyme, which is a major roadblock in efforts to understand the mechanism of this emerging post-translational modification. The protocol detailed in this chapter focuses on efforts to express S. cerevisiae ATE1 (ScATE1) and to purify this ATE1 using chromatographic methods. The data presented here clearly show that this approach delivers large quantities of monomeric and monodisperse ScATE1. Based on size-exclusion data, we also show that ScATE1 behaves monomerically under these conditions, which was inferred but not known previously. This well-behaved protein can be used for downstream biophysical and biochemical studies, which should result in a better understanding of the structure, the mechanism, and the enzymatic properties of ScATE1. Additionally, the procedure outlined in this chapter may be generally applied to other eukaryotic ATE1s allowing for the isolation and purification of enzyme from other organisms, which is currently underway. Ultimately, a greater understanding of the structure and the function of ATE1 as well as the mechanism of post-translation arginylation should lead to further research to allow the targeting of this PTM for therapeutic purposes.

Acknowledgments

This work was supported by NIH-NIGMS grant R35 GM133497 (A.T.S and V.V.) and in part by NIH-NIGMS supplement R35 GM133497-03S1 (M.C.).

References

Avcilar-Kucukgoze, I., Gamper, H., Polte, C., Ignatova, Z., Kraetzner, R., Shtutman, M., et al. (2020). tRNAArg-derived fragments can serve as arginine donors for protein arginylation. *Cell Chemical Biology*, *27*(7), 839–849.e834. https://doi.org/10.1016/j.chembiol.2020.05.013.

Balzi, E., Choder, M., Chen, W. N., Varshavsky, A., & Goffeau, A. (1990). Cloning and functional analysis of the arginyl-tRNA-protein transferase gene ATE1 of Saccharomyces cerevisiae. *Journal of Biological Chemistry*, *265*(13), 7464–7471. https://doi.org/10.1016/s0021-9258(19)39136-7.

Batsios, P., Ishikawa-Ankerhold, H. C., Roth, H., Schleicher, M., Wong, C. C. L., & Müller-Taubenberger, A. (2019). Ate1-mediated posttranslational arginylation affects substrate adhesion and cell migration in Dictyostelium discoideum. *Molecular Biology of the Cell*, *30*(4), 453–466. https://doi.org/10.1091/mbc.E18-02-0132.

Carpio, M. A., Decca, M. B., Lopez Sambrooks, C., Durand, E. S., Montich, G. G., & Hallak, M. E. (2013). Calreticulin-dimerization induced by post-translational arginylation is critical for stress granules scaffolding. *The International Journal of Biochemistry & Cell Biology*, *45*(7), 1223–1235. https://doi.org/10.1016/j.biocel.2013.03.017.

Conibear, A. (2020). Deciphering protein post-translational modifications using chemical biology tools. *Nature Reviews Chemistry*, *4*, 674–695. https://doi.org/10.1038/s41570-020-00223-8.

Drazic, A., Myklebust, L. M., Ree, R., & Arnesen, T. (2016). The world of protein acetylation. *Biochimica et Biophysica Acta (BBA) - Proteins and Proteomics*, *1864*(10), 1372–1401. https://doi.org/10.1016/j.bbapap.2016.06.007.

Fina, M. E., Wang, J., Nikonov, S. S., Sterling, S., Vardi, N., Kashina, A., et al. (2021). Arginyltransferase (Ate1) regulates the RGS7 protein level and the sensitivity of light-evoked ON-bipolar responses. *Scientific Reports*, *11*(1). https://doi.org/10.1038/s41598-021-88628-3.

Humphrey, S. J., James, D. E., & Mann, M. (2015). Protein phosphorylation: A major switch mechanism for metabolic regulation. *Trends in Endocrinology and Metabolism*, *26*(12), 676–687. https://doi.org/10.1016/j.tem.2015.09.013.

Jiang, C., Moorthy, B. T., Patel, D. M., Kumar, A., Morgan, W. M., Alfonso, B., et al. (2020). Regulation of mitochondrial respiratory chain complex levels, organization, and function by arginyltransferase 1. *Frontiers in Cell and Development Biology*, *8*, 603688. https://doi.org/10.3389/fcell.2020.603688.

Kashina, A. (2014). Protein arginylation, a global biological regulator that targets actin cytoskeleton and the muscle. *The Anatomical Record*, *297*(9), 1630–1636. https://doi.org/10.1002/ar.22969.

Kishimoto, N., Okano, R., Akita, A., Miura, S., Irie, A., Takamune, N., et al. (2021). Arginyl-tRNA-protein transferase 1 contributes to governing optimal stability of the human immunodeficiency virus type 1 core. *Retrovirology*, *18*(1). https://doi.org/10.1186/s12977-021-00574-0.

Kumar, V., Thakur, J. K., & Prasad, M. (2021). Histone acetylation dynamics regulating plant development and stress responses. *Cellular and Molecular Life Sciences*, *78*(10), 4467–4486. https://doi.org/10.1007/s00018-021-03794-x.

Kwon, Y. T., Kashina, A. S., & Varshavsky, A. (1999). Alternative splicing results in differential expression, activity, and localization of the two forms of arginyl-tRNA-protein transferase, a component of the N-end rule pathway. *Molecular and Cellular Biology*, *19*(1), 182–193. https://doi.org/10.1128/mcb.19.1.182.

Saha, S., & Kashina, A. (2011). Posttranslational arginylation as a global biological regulator. *Developmental Biology*, *358*(1), 1–8. https://doi.org/10.1016/j.ydbio.2011.06.043.

Suto, K., Shimizu, Y., Watanabe, K., Ueda, T., Fukai, S., Nureki, O., et al. (2006). Crystal structures of leucyl/phenylalanyl-tRNA-protein transferase and its complex with an aminoacyl-tRNA analog. *The EMBO Journal*, *25*(24), 5942–5950. https://doi.org/10.1038/sj.emboj.7601433.

Tasaki, T., Sriram, S. M., Park, K. S., & Kwon, Y. T. (2012). The N-end rule pathway. *Annual Review of Biochemistry*, *81*(81), 261–289. https://doi.org/10.1146/annurev-biochem-051710-093308.

Timms, R. T., & Koren, I. (2020). Tying up loose ends: The N-degron and C-degron pathways of protein degradation. *Biochemical Society Transactions*, *48*(4), 1557–1567. https://doi.org/10.1042/bst20191094.

Van, V., Brown, J. B., Rosenbach, H., Mohamed, I., Ejimogu, N.-E., Bui, T. S., et al. (2021). Iron-sulfur clusters are involved in post-translational arginylation. *Biorxiv*. https://doi.org/10.1101/2021.04.13.439645.

Van, V., & Smith, A. T. (2020). ATE1-mediated post-translational arginylation is an essential regulator of eukaryotic cellular homeostasis. *ACS Chemical Biology*, *15*(12), 3073–3085. https://doi.org/10.1021/acschembio.0c00677.

Varshavsky, A. (2011). The N-end rule pathway and regulation by proteolysis. *Protein Science*, *20*(8), 1298–1345. https://doi.org/10.1002/pro.666.

Varshavsky, A. (2019). N-degron and C-degron pathways of protein degradation. *Proceedings of the National Academy of Sciences*, *116*(2), 358–366. https://doi.org/10.1073/pnas.1816596116.

Walsh, C. T., Garneau-Tsodikova, S., & Gatto, G. J. (2005). Protein posttranslational modifications: The chemistry of proteome diversifications. *Angewandte Chemie International Edition in English*, *44*(45), 7342–7372. https://doi.org/10.1002/anie.200501023.

Wang, J., Han, X., Saha, S., Xu, T., Rai, R., Zhang, F., et al. (2011). Arginyltransferase is an ATP-independent self-regulating enzyme that forms distinct functional complexes in vivo. *Chemistry & Biology*, *18*(1), 121–130. https://doi.org/10.1016/j.chembiol.2010.10.016.

Wang, J., Han, X., Catherine, W. C. L., Cheng, H., Aslanian, A., Xu, T., et al. (2014). Arginyltransferase ATE1 catalyzes midchain arginylation of proteins at side chain carboxylates in vivo. *Chemistry & Biology*, *21*(3), 331–337. https://doi.org/10.1016/j.chembiol.2013.12.017.

Watanabe, K., Toh, Y., Suto, K., Shimizu, Y., Oka, N., Wada, T., et al. (2007). Protein-based peptide-bond formation by aminoacyl-tRNA protein transferase. *Nature*, *449*(7164), 867–871. https://doi.org/10.1038/nature06167.

CHAPTER NINE

Testing anti-cancer drugs with holographic incoherent-light-source quantitative phase imaging

Daniel Zicha[a,b,*] and Radim Chmelik[a,b]

[a]CEITEC—Central European Institute of Technology, Brno University of Technology, Brno, Czech Republic

[b]Institute of Physical Engineering, Faculty of Mechanical Engineering, Brno University of Technology, Brno, Czech Republic

*Corresponding author: e-mail address: daniel.zicha@ceitec.cz

Contents

Abstract

Quantitative Phase Imaging is becoming an important tool in the objective evaluation of cellular responses to experimental treatment. The technique is based on interferometric measurements of the optical thickness of cells in tissue culture reporting on the distribution of dry mass inside the cells. As the measurement of the optical thickness is interferometric, it is not subjected to the Abbe resolution limit, and the use of an incoherent-light source further increases the accuracy practically achieving 0.93 nm in optical path difference corresponding to 4.6 femtograms/μm^2. Holographic mode reduces the exposure in comparison to phase-shifting or phase-stepping interference microscopy and allows observation of faster dynamics. An attractive application is in the development of novel anti-cancer drugs and there is an important potential for

Methods in Enzymology, Volume 679
ISSN 0076-6879
https://doi.org/10.1016/bs.mie.2022.08.017

255

pretesting chemotherapeutic drugs with biopsy material for personalized cancer treatment. The procedure involves the preparation of live cells in tissue culture, seeding them into suitable observation chambers, and time-lapse recording with an adjusted microscope. Subsequent image processing and statistical analysis are essential last steps producing the results, which include rapid measurements of cell growth in terms of dry-mass increase in individual cells, speed of cell motility and other dynamic morphometric parameters.

1. Introduction

Cancer is the leading cause of death worldwide. The global burden is expected to be 28.4 million cases in 2040, a 47% increase from 2020 (Sung et al., 2021). Cancer in Europe has also been a major burden with 3.45 million new cases and 1.75 million deaths in 2012 (Ferlay et al., 2013). Different types of cancers have not changed their incidence evenly across geographic regions. For example, the incidence of four common cancers in eastern and central European countries (prostate, postmenopausal breast, corpus uteri and colorectum) started to approach levels in northern and western Europe, where rates were already high (Arnold et al., 2015). Aggressive cancers increase their incidence generally more dramatically. For example, oropharyngeal cancer, significantly increased the incidence during 1983 to 2002 predominantly in economically developed countries (Chaturvedi et al., 2013). Growth of cancer cells is not the main problem in the aggressive cancers; the majority of deaths are due to invasion and metastasis. The reasons for this unfavorable situation lie in limited availability of treatments, inaccuracy of diagnosis and difficulties with the choice of the most effective treatment. Dynamic time-lapse analysis of live cancer cells in combination with fluorescence has been a very efficient research tool with at least 370 publications in the last 5 years. However, the currently used microscopy techniques lack speed and are difficult to quantify. Holographic microscopy (HM) is a promising candidate for improvement of the situation.

HM is a notable Quantitative Phase Imaging (QPI) technique such as wavefront sensing, ptychography, phase tomography or spatial light interference microscopy (Holden, Tarnok, & Popescu, 2017; Park, Depeursinge, & Popescu, 2018). Like the other QPI techniques, HM shows the shape and position of a live cell with high contrast and non-invasively without labeling and, primarily, measures the specimen-introduced phase shift of the illumination light at each image point. The phase shift is proportional to the area density of the mass of live cell's non-aqueous material (cell dry mass) according to the equation (adapted from Zangle & Teitell, 2014):

$$\rho = \frac{\phi\lambda}{2\pi\alpha}$$

where ρ is the area density distribution of cell dry mass in kg/m^2, ϕ the measured phase shift in radians, λ the wavelength of the illumination in meters, 2π is one wavelength in radians and α the specific refractive increment. This parameter has a meaningful average value of approximately $1.8 \times 10^{-4}\ m^3/kg$ for biomolecules that make up most of a typical cell's mass. Unlike the other QPI techniques, HM uses neither phase recovery from intensity image series nor integration from the differential-phase record. In HM, the measured phase is related directly to the form of recorded holographic pattern and is retrieved using Fourier transform methods. A complete full-frame phase image is reconstructed from just one holographic record. This provides frame rate limited by the camera and computer speeds only and allows observation of fast dynamics.

The original Gabor holographic setup is simplest to implement in a microscope but unsuitable for QPI because the reconstructed image overlaps with the conjugate and zero-order ones. This degrades the accuracy of the phase measurement. On the other hand, the off-axis arrangement created later by Leith and Upatnieks is optically more complicated but does not suffer from the multiple-image overlap. The main reason for the deterioration of accuracy then remains the coherent source of illumination, which is required by classic off-axis holographic systems. The coherent illumination causes strong coherence noise and unwanted interference and diffraction phenomena in the reconstructed image (Dohet-Eraly, Yourassowsky, El Mallahi, & Dubois, 2016). Low-coherent light eliminates these effects and, moreover, returns HM's resolving power to the wide-field microscopy standard (Chmelik et al., 2014). The most significant benefit of low-coherence illumination is the coherence-gate effect blocking the light scattered by the unwanted parts of a specimen. It occurs because the region in which the reference and object beams can interfere and produce a holographic pattern (the coherence volume) is strongly reduced. This effect is typically used for imaging through disordered media by least-scattered (ballistic) light. However, when the correlation of the object and reference beam is measured in more detail than with classical holography, new unique imaging capabilities of HM with low coherent illumination are unlocked. This way, we completed ballistic-light HM through a scattering medium with the non-ballistic light. Further, by combining ballistic and non-ballistic images, we synthesized images of quality superior to the ballistic-light-only approach (Ďuriš & Chmelík, 2021). Applying a similar procedure to the system of a specimen and diffraction grating in its proximity, we increased the lateral resolution of HM substantially beyond the Abbe diffraction limit while keeping a large field of view of low-NA objective lenses (Ďuriš, Bouchal, Rovenská, & Chmelík, 2022). The proposed approach affords more information about the observed object in comparison with the classical holographic recording, thus providing HM with the capability of improved resolution for both 2D and 3D objects.

As mentioned above, the main challenge is the implementation of the HM with incoherent illumination based on off-axis holography. Moreover, the reference arm must be separated from the object one to obtain the coherence-gate effect. A fully uncompromising solution is based on the proposal of Leith and Upatnieks (1967). The interference pattern of the image-plane hologram is formed as the image of a diffraction grating. The pattern is then invariant with respect to the source position and the illumination wavelength, so the interference fringes have a high contrast even for both a spatially and temporally incoherent source. We introduced this concept into HM as the Holographic Incoherent-light-source Quantitative Phase Imaging (hiQPI) technology. One of its possible implementations is a commercially available Telight Q-Phase microscope used in experimental methods described in this chapter.

The optical setup of this microscope (see Fig. 1, for the construction details) consists of the object and reference arms, which contain optically equivalent microscope systems (Slaby et al., 2013). The main component is the diffraction grating with groove frequency $150\,mm^{-1}$ placed in the reference arm, according to the Leith and Upatnieks principle, in the plane optically conjugated with the CCD chip surface. A halogen lamp or LED provides a broad source of illumination. The light beam is filtered by an interference filter with a central wavelength of 650 nm and 10 nm full width at half maximum to achieve a quasi-monochromatic spatially low-coherent illumination, which is most suitable for experiments in transmitted light. The source is imaged by a pair of achromatic doublets through the beam splitter to the front focal planes of the condensers (Nikon LWD condenser lenses NA 0.52 with adjustable aperture stop), forming the Köhler illumination. The culture and dummy chambers are imaged with objective lenses (e.g., Nikon CFI Plan Fluor, $10\times$ NA 0.3 WD 16 mm or $20\times$ NA 0.5 WD 2.1 mm) in couples with Nikon tube lenses to the intermediate image planes where the grating is

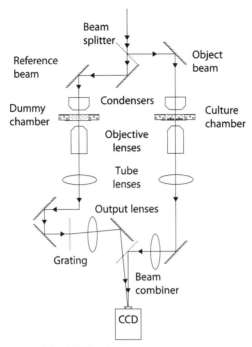

Fig. 1 Optical diagram of the hiQPI microscope.

placed in the reference arm. The intermediate image planes are finally imaged by the output lenses (objective lenses with focal lengths 35 mm and apertures f/2) on the CCD chip surface, where the object beam interferes with the reference one, thus producing the image-plane hologram captured by the CCD camera (e.g., MR4021MC-BH XIMEA).

The above-described setup is not the only possible implementation of the hiQPI concept. We have also developed a highly stable and easily adjustable single-path hiQPI setup in which the reference and object beams are separated by polarization encoding (Bouchal, Štrbková, Dostál, Chmelík, & Bouchal, 2019). This system works best for low-power applications.

2. Protocol

2.1 Before you begin

2.1.1 Cells

The choice of cells to be used in the testing depends on the purpose. Cell lines, such as the non-small-cell lung cancer cell line A549, are useful for pre-testing unknown responses to treatments such as putative anti-cancer drugs since they can be readily handled. An important aspect of testing anti-metastasis drugs is their ability to reduce cell motility significantly, which is usually coupled to a reduction of invasion and metastasis. Therefore, the cells from the chosen line need to exhibit a reasonable level of spontaneous motility under control conditions, otherwise, it is impossible to detect speed reduction. Primary cells from laboratory animals with experimental cancers are more complicated since the use of the animals requires special licenses. When the experimental tumors were induced by inoculation of cancer cells from cell lines, the culture of the primary cells is not challenging as the cells were already adjusted to tissue culture conditions. The most difficult primary cultures are from spontaneous tumors especially of human origin since they need to be adjusted to the tissue culture conditions.

2.1.2 Cell culture medium and procedure

An established cell line is adapted to a recommended medium, which is advisable to use. For example, the A549 cells grow and respond well when cultured in the commonly used Dulbecco's modified Eagle's medium (DMEM) requiring 5% CO_2. The CO_2-enriched atmosphere can be also provided during the microscopy time-lapse recording. However, this may lead to potential problems. The humidified CO_2-enriched atmosphere is often delivered to a small incubator chamber on the microscope stage. When multifield recording is used with a motorized stage, its tubing may snag and disrupt the recording unless the tubing is carefully attached to an elevated fixed point and delivered to stay sufficiently relaxed within the full range of the stage movement. We, therefore, prefer to avoid the CO_2 delivery during the time-lapse recording and use added buffer to stabilize pH. Our best experience is with non-volatile buffer TES and HEPES can also be used. The buffers may show some toxicity but many cells including A549 tolerate them well. Cells are regularly passaged, usually every 3 or 4 days, and not kept in culture for longer than 4 weeks, and seeded into observation chambers 1 day before the time-lapse recording. When cells are kept for longer than the 4 weeks, they tend to change their behavior, for example, reducing their ability to migrate, which is detrimental for the testing purpose. It is also important to keep the time between the seeding and recording constant. We have good experience with 1 day. The behavior of cells change in time after seeding making the reproducibility of the experiments difficult to achieve when cells are analyzed at different time points after seeding. The passage of the cells into the observation chamber requires proteolytic enzymes and the surface structures on the cells are affected and it takes about 1 day to recover. The passage also stimulates the cell activity, and we found that on the 2nd day, the cells are less active showing slower motility. The most demanding culture is with primary cells from patient biopsy and special medium, also supplemented with TES, can be used. Culture with the TES buffer is maintained at 37 °C in a humidified atmosphere and CO_2 can be reduced to 3.5%.

For cell behavior studies, it is important that the cells are in their best shape and therefore need to be treated with care. For example, it is beneficial to avoid unnecessary mechanical stress by avoiding centrifuge spin when the cells are defrosted and passaged. This can be achieved by placing the 1 mL of defrosted cell suspension into a small culture flask and adding 9 mL of culture medium. This reduces the DMSO concentration 10 ×, which is less detrimental than the spin. Cells become adherent overnight and the medium can be replaced next morning. Centrifuge spinning in a cell passage can be avoided by dry trypsinization, which is achieved by a PBS wash after the medium is removed and immediately after trypsin addition, most of it is removed and once the cells detach, a medium containing serum is added. The serum stops the activity of the residual amount of trypsin. The cells are then resuspended ready for seeding.

2.1.3 Observation chamber

Testing cell responses to a range of treatments, e.g., chemotherapeutic drugs or putative chemotherapeutic drugs, especially with primary cells, is best performed simultaneously since the cells may change in time. Multifield time-lapse is, therefore, desirable and the observation chamber should allow parallel multiple cultures. Quantitative phase imaging does not tolerate open surface of the culture medium, which would introduce far bigger changes to the phase than the cells. This is why standard multiwell plates should not be used. The best arrangement is with the medium around the observed cells held as a sandwich between two coverslips. We built a simple glass random walk chamber with 8 wells by drilling holes into a slide and securing coverslips to its faces (Fig. 2). The chamber has been successfully used. The overall dimensions are 7.5 cm × 5.0 cm. The large bottom cover slip is secured with Sylgard 184 (Dow Corning), which remains flexible and seals well even with the temperature change to 37 °C. The top coverslips with adherent cells need to be inverted over the wells and sealed down with a wax mixture made, for example, from beeswax (Fisher), soft yellow paraffin (Fisher), and paraffin wax (melting point 46 °C, Fisher) in the ratio 1:1:1.

Fig. 2 Random walk observation chamber with 8 wells.

Such a mixture remains flexible and does not leak. It is useful to keep the ready-made mixture heated with a temperature control system to prevent burning.

There is an alternative solution with Ibidi Channel Slides, which also feature top and bottom coverslips with the medium and cells in between. They can be easily purchased and handled but there is a range of setbacks. One problem is the evaporation of the medium in longer experiments when used with the provided port covers, which do not hermetically seal. It is possible to purchase sealed plugs but their introduction into the ports tends to force air bubbles into the chamber channel. A better solution is in use of mineral oil (designed for tissue culture, e.g., from Sigma-Aldrich) placed on top of the medium in the ports. This stops the evaporation and, when carefully handled without shaking, the oil does not mix with the medium. We found that, with the mineral oil, it was possible to use Ibidi μ-Slide with one 0.8 mm high channel and the μ-Slide VI 0.4 with six 0.4 mm high channels. Additional setbacks are specific for the highly critical hiQPI, which requires exquisite optical and mechanical properties for the observation chamber since the measurement accuracy is in the nanometer range. The top and bottom coverslips are plastic, which introduces light polarization effects and therefore reduces imaging quality to some extent, but the results can be acceptable. Another problem with the Ibidi Channel Slides is their intrinsic mechanical instability resulting from their production where melted plastic solidifies in a form leaving the product under tension. Different channels in the six-channel version often feature different tensions and, e.g., in one of the channels is almost impossible to achieve stable holographic imaging. Ibidi also provides glass bottom versions, which would reduce the adverse polarization effects but the glass coverslip is glued at room temperature and the transfer of the slide to 37 °C introduces additional tension, distortion, and instability. Therefore the all-plastic version is a better choice.

2.1.4 Microscopy and its environmental control

The commercially available hiQPI system, Q-Phase (Telight), allows multiposition/multiculture recording, as it features an accurate motorized stage. Multiple positions can be specified and the current system allows time-lapse recording of 16 positions (2 positions in each well of the previously described 8-well chamber) with a 5-min lapse interval. Our analysis of cell behavior with quantitative phase imaging revealed that a lapse-interval of 5 min was appropriate for measuring cell growth and motility with a range of adherent mammalian cells. Importantly, the system manages an auto-alignment procedure of the microscope for each position at each lapse interval essential for keeping a good signal during prolonged recordings. The entire Q-Phase microscope is encapsulated in an environmental 37 °C box with accurate temperature control. This arrangement not only keeps the cells at stable 37 °C but also keeps the microscope more stable since the room temperature is subjected to additional variations.

2.1.5 Image processing and data analysis

Our own software has been developed in Mathematica (Wolfram Research) and the procedures, which run particularly slow, have been re-implemented in C programming language and are called from Mathematica (Zicha, 2022; Zicha & Dunn, 1995). The software runs in batch processing mode and keeps identical parameters for all movies, which is important for comparative analysis. The software has been developed for multiple platforms running under Windows and Unix. Large numbers of movies are being processed under Unix since processing is faster even on the same hardware and dynamic stack allows algorithms with extensive recursion used in background detection. The image processing includes the following steps:

(1) Calculate the phase from the recorded holograms based on Fourier transform and phase unwrapping

(2) Subtract a specimen-free image to remove the wavefront distortion of the microscope alone keeping the full-size frames

(3) Reduce the number of pixels by averaging 4 neighbor pixels producing reduced size frames with half of the rows and half of the columns

(4) Flatten the residual dynamic distortions, introduced by microscope and specimen instability, of the backgrounds in all individual frames

    a. Flatten linear distortions of the background for each reduced-size frame by a bilinear transformation interpolated through image pixel values with an asymmetric penalty function

$$weight \times x^2 \text{ for } x < 0$$

and

$$\frac{x^2}{weight} \text{ for } x \geq 0$$

where x is the deviation of the measured phase values from the interpolation function. As *weight*, we used 8. This is ignoring the high phase shift values of cells as they protrude above the largely flat surrounding background. The standard least square method would produce a linear background interpolation shifted to higher levels.

    b. Produce estimates of backgrounds for all reduced-size frames in order to identify biquartic transformation functions to flatten the residual non-linear dynamic background distortions. The background area in the first reduced size frame is produced by finding a pixel with gray level nearest to the histogram maximum and adding recursively neighbor pixels with similar levels based on 4-connectivity. Identification of backgrounds in subsequent frames is based on processing frame differences and the background estimate for a frame is seeded by connectivity with its previous frame. The background mask is then adjusted along its edges by adding background pixels with similar levels recursively based on 4-connectivity

    c. Apply the bilinear functions from step 4a and the biquartic functions from step 4b to all full size frames

(5) Correct residual constant distortions introduced by the microscope and medium/chamber inhomogeneity. Reconstruct the stationary background by collecting all background levels for each background pixel and using their medians. Pixels, which have not been assigned, are tested whether they show similar levels to their neighbor pixels in time recursively. If similar levels ($\pm$ standard deviation) are found, their median is used in the additional areas. Since there is no sensible information about the rest, the missing areas are interpolated using the surrounding area. Holes in the background are filled from the edges using similar levels to the neighbor pixels and in the centers of the holes, the levels are anchored to the biquartic approximation using a rim of pixels around the holes. Unidentified areas at the edges start at similar levels to the neighbor pixels and simply continue flat since no better information is available. Background masks are used from the reduced size identification in 4b but full-size background is reconstructed

(6) Subtract the reconstructed stationary background from frames flattened for dynamic fluctuations in step 4. Large areas in the background show a standard deviation of 1.5 nm optical path difference, providing accuracy corresponding to 7.4 femtograms/μm^2

(7) Segment cells and calculate positions of their centroids, total mass, area, and other morphometric parameters

(8) Track cells through the time-lapse sequence. Identifying the same cell in the next frame on the basis of nearest distance and similar total dry mass. When the identical cell is not identified in the subsequent frame, the track is terminated. When a new cell appears in a frame without identified identical cell in the previous frame, a new track is started

Data derived from observations of live cells usually form a hierarchical structure. Typical levels in the hierarchy include treatments, cell cultures, observation fields, cells, and individual time-point measurements. One correct statistical approach to such data hierarchy is analysis of variance (ANOVA) with a nested model (Milliken & Johnson, 1992; Snedecor & Cochran, 1967). In this case, the nested model also has to be unbalanced because the numbers of data points at different levels are not equal since they depend on chance, e.g., the number of cells in the observation field. The ANOVA test calculates variability in the data introduced at individual levels and assigns statistical significance to the variability (at individual levels) taking into account interactions between the levels. The level of control/treatment is then usually expected to show a significant additional difference on top of the variability at the lower levels. When experimental conditions are well controlled then the level immediately below the culture level does not introduce significant variability. This means that individual cell cultures are reproduced with the same internal variability. The cell level usually gives a high significance since individual cells are highly variable even within individual cultures.

2.2 Materials and equipment

- Culture medium for the A549 cells: DMEM with 20 mM of non-volatile buffer TES and 10% fetal bovine serum
- Alternative medium for short-term cultures of human primary carcinoma cells: Hank's Minimum Essential Medium with non-essential amino acids supplemented with 0.5 µg/mL hydrocortisone, 5 µg/mL insulin, 5 ng/mL epidermal growth factor (EGF), 1 mM sodium pyruvate, 0.3 g/L L-glutamine, 10% bovine and 2% fetal bovine serum, 20 mM of non-volatile buffer TES, penicillin and streptomycin
- Incubator set to 37 °C with a humidified atmosphere enriched to 3.5% CO_2
- Cabinet for sterile work, usually biosafety cabinet class II
- Other tissue culture equipment and consumables: pipets, flasks, cuvettes, counting chamber
- Observation chamber and hot wax mixture with brush in case of the Random walk chamber
- Holographic microscope with environmental control

2.3 Step-by-step method

2.3.1 Preparation of A549 cells

- Detach cells by trypsin and re-suspend in medium
- Count cell density in suspension
- Seed cells on coverslips at low density to achieve subconfluent culture (e.g., seed 30,000 cells for a 35 mm Petri dish)
- Incubate overnight

2.3.2 Alternative preparation of human primary carcinoma cells

- Ethically obtained fragment from a tumor is handled with maximum sterility
- Once in the biosafety cabinet, cut the fragments with sterile scissors to small pieces
- Incubate with trypsin in tissue culture incubator for 2 h
- Remove the trypsin and add medium to the fragments
- Place the partially digested fragments on clean coverslips in Petri dishes and cover with small amount of medium so that the surface tension keeps the fragments in contact with the coverslip and incubate for several hours
- Add medium
- Incubate for several days until individual cells and cell clusters start crawling from the fragments and then remove the fragments

2.3.3 Observation chamber assembly

- Fill a well in the Random walk observation chamber with control medium or medium containing the test drug
- Invert a coverslip with adherent cells on top of the well, dry the surrounding with medical wipe and seal the coverslips with hot wax mixture using a small brush as shown in Fig. 2
- Repeat for all wells

2.3.4 Time-lapse microscopy imaging

- Place the assembled chamber onto the microscope stage with the raised condenser and secure in a suitable stage holder
- Place slides and coverslips with similar optical thickness to the observation chamber onto the stage in the reference arm of the microscope (e.g., one slide and 2 coverslips for the Random walk chamber) and lower the condenser
- Make sure that the main power switch under the microscope is set to "ON"
- Start the "Q-Phase" software which can be labeled "SophiQ"
- Choose the desired objective lens in the "Main Control" (e.g., 10 × NA 0.3 for larger field and better statistics with higher number of cells in the observation field)
- Press "QPI Active" in the "Main Control," which activates "Live" mode, switches on the "Lamp" and opens "Shutter" as indicated in the "Quantitative Phase Imaging" panel
- Make sure that "Averaging" in "Quantitative Phase Imaging" panel is set to 1 unless other value is required but the acquisition of each frame will be slower
- Set a suitable camera "Exposure" in the "Quantitative Phase Imaging" panel (e.g., 14 ms is a possibility with Andor Zyla 5.5 sensitive camera on the system)
- Set a suitable illuminator "Intensity" in the "Quantitative Phase Imaging" panel (e.g., 5%)
- Set a suitable condenser "Aperture" in the "Quantitative Phase Imaging" panel (e.g., 0.3, usually there is no need to go higher than the NA of the objective lens and a slight reduction improves the signal with only a small reduction of resolution)
- The Q-Phase software features a "Microscope Adjustment Wizard" in the "Main Toolbar" panel for setting-up the microscope, guiding the user through the individual steps and some of them are performed

automatically (manual steps include focusing the sample and checking the stacked coverslips on the stage of the reference arm of the microscope)

- In order to set up time-lapse recording, choose "Multidimensional Acquisition" labeled "REC" in the "Main Toolbar" panel
- Select "TimeLapse" in the "Multidimensional Acquisition" panel and "Position" for multifield recording
- Add a time-lapse specification with "+" button in the "TimeLapse" tab of the "Multidimensional Acquisition" panel and choose "Duration" and the "Main Interval," and the number of "Time Points" will be calculated
- Choose and store multifield positions with "Add current position and adjustment" labeled "+" in the "Position" tab of the "Multidimensional Acquisition" panel, the choice can be adjusted with "Use current position and adjustment for all selected positions"
- Align automatically all positions with "Automatically readjust all positions" in the "Position" tab of the "Multidimensional Acquisition" panel, this will take some time, e.g., 20 min with 16 positions
- Check focus and image quality for all positions and readjust if necessary
- Start recording by pressing "Run now" in the "Multidimensional Acquisition" panel

2.3.5 Quantitation and statistical analysis with the in-house developed software

- Start batch processing in Mathematica performing the previously described image processing steps, which produce the parameters for individual objects in individual frames including: cell number, frame number, centroid position x, centroid position y, area, total dry mass
- Load the parameters from the control movies and a set of treatments into a data processing Mathematica notebook, produce graphical representation in a form of scatter plots, line plots, box and whisker plots, and subject the data to ANOVA described above

2.3.6 Alternative analysis with Q-Phase software

The Q-Phase software provides interactive tools for image analysis, segmentation, and tracking of individual cells. It also calculates the growth and speed of cell motility. The data can be presented in graphical form within the Q-Phase software package.

2.4 Expected outcomes

The image processing part of the quantitative analysis produces images with segmented cells. Fig. 3 presents example processed images of A549 cells recorded in a presence of a PI3K inhibitor GDC0941 (GDC), dissolved in DMSO, and the solvent only as a control using lapse interval of 5 min. The thresholded background is in black and objects smaller than cells have been removed. In our current processing, we do not separate cells from clusters. This is because they tend to overlap partially and therefore a simple separating line does not produce the true outline. The separation can be achieved approximately and the Q-Phase software does exactly that.

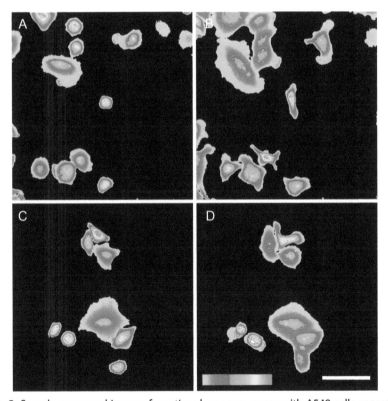

Fig. 3 Sample processed images from time-lapse sequences with A549 cells, recorded in the control medium with DMSO solvent only (A and B) and the treatment medium containing the GDC drug (C and D). Frames recorded at 0 h (A and C) and 24 h (B and D). The white scale bar is 100 μm. The pseudo-color bar represents the mapping of the range of mass densities between 0 and $2\,pg\,\mu m^{-2}$.

The data resulting from the subsequent data processing and statistical analysis report on fundamental aspects of live cells, their growth and motility, as well as other morphometric parameters. Fig. 4 shows growth curves of the A549 cells summarized from 4 movies with DMSO solvent-containing medium and 4 movies with GDC drug-containing medium, presenting the integrated dry mass as a function of time. The common dry mass of individual cells is around 700 pg and much heavier objects are cell clusters. The individual growth curves can be interpolated with exponential curves and dry-mass doubling time is estimated. The cell clusters do not complicate the doubling time parameter calculation since it is derived in principle from percentage increments. Statistical analysis showed that the presence of the GDC drug significantly reduced growth by increasing the doubling time from 42.1 ± 2.1 h for the DMSO control (number of movies 4, number of cells 55) to 55.6 ± 2.1 h for the GDC drug (number of movies 4, number of cells 20) with ANOVA P value <0.02. Box and whisker plot illustrating the difference in growth is in Fig. 6A.

The 3-h trajectories of the A549 cells are presented in Fig. 5. Longer tracks were partitioned into 3-h individual sequences and shorter tracks were discarded. It is otherwise difficult to compare visually the motile activity between the treatment and the control. The trajectories represent very accurate displacements of the dry-mass centroids.

The speed of cell motility can be calculated, in this case, as the 5-min displacement and multiplied by 12 to be expressed in μm/h. The GDC drug significantly reduced the speed from 14.7 ± 1.0 μm/h for the DMSO control (number of movies 4, number of cells 55, number of displacements 4951) to 3.5 ± 0.4 μm/h for the GDC drug (number of movies 4, number of cells 20, number of displacements 2105) with ANOVA P value <0.001. The mean was calculated for each full cell track first and means of all cells in the 4 movies for control and GDC treatment are presented. Box and whisker plot illustrating the difference in speed is in Fig. 6B.

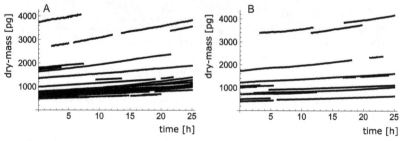

Fig. 4 Growth curves representing the dry-mass increase of cells and cell clusters in the movies with control DMSO medium (A) and GDC drug containing medium (B).

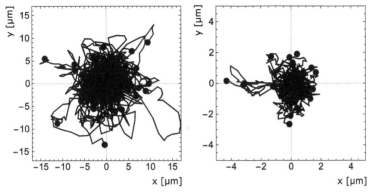

Fig. 5 Rose plots of cell trajectories shifted to the common origin.

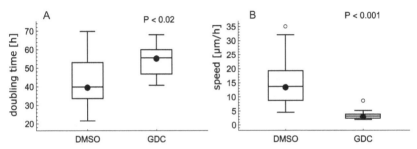

Fig. 6 Box-and-whisker plots illustrating significant changes in dry-mass doubling time (A) and speed of cell motility (B).

As an example of an additional parameter, we also calculated spreading, defined as cell area divided by its dry mass. In other words, it is reporting on the average spread area of 1 pg of dry mass in the cells. There was no significant change in this parameter resulting from the treatment. The spreading was $2.21 \pm 0.08\,\mu m^2/pg$ for the DMSO control (number of movies 4, number of cells 55, number of measurements 5006) and $1.93 \pm 0.14\,\mu m^2/pg$ for the GDC drug (number of movies 4, number of cells 20, number of displacements 2125) with ANOVA P value >0.85.

The example data, presented here, confirm the previously described effect of the GDC drug (Zicha, 2022).

2.5 Advantages

The unique advantage of the hiQPI is in rapid and ultimately accurate quantitation of cellular responses. The standard deviation in a background rectangular area (50×50 pixels) of the frame in Fig. 7 was 4.6 femtograms corresponding to a retardation of the light beam by 0.009 rad or 0.93 nm.

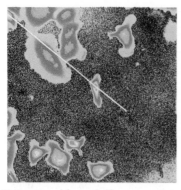 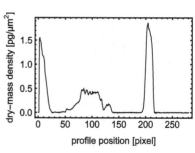

Fig. 7 Processed frame before segmentation and line profile along the yellow line. The frame corresponds to Fig. 3B. The pseudo-color look-up table mapping is identical to Fig. 3. The image contains negative values but they feature in black in order to demonstrate with high contrast that the small positive values in cyan fluctuations close to zero in the background.

The profile in Fig. 7 shows the accurate leveled background close to zero allowing low threshold of 0.04 rad, which includes even thin parts of the cell in the analysis and increases the accuracy of the measurements.

2.6 Limitations

The Q-Phase system works well for individual adherent cells in tissue culture. Thicker specimens, such as multilayer cultures and specimens with reduced transparency will be difficult. However, it will be possible to observe individual cells in scattering but fairly transparent environment, such as cells in collagen gel.

2.7 Optimization and troubleshooting

Problem: Initial focusing on the sample may be difficult when the microscope is out of alignment.

In this case, it is useful to switch to 4× objective lens. If this does not help, pre-align the microscope with the 4× lens. Keep focus at expected level and, if unknown, in the middle of the focusing range. Start closing the field iris and stop when the dimming is noticeable. Focus the field iris with the condenser control buttons. This might be easier when Reference branch is switched off. When in focus switch to the Reference branch and focus the second image of the iris with top jog on the control panel operating relative z position of the reference objective lens. Open the Objective branch and align the two branches of the microscope with relative x, y position of the reference objective lens and return back to the higher magnification. Focusing should now be easier.

Problem: The image signal and quality deteriorate during the time-lapse recording.

Increase the step-size for auto-signal procedure: parameter *contStep* in *[AutoAdjust]* section in C:\Telight\Q-PHASE\configurations\transmitted. ini file. The default value is 0.2 or 0.1 µm and can be changed, e.g., to 0.4 µm. The increased parameter value will allow bigger adjustments in the auto alignment procedure during time-lapse recording when Auto-signal is activated. As a result the image quality may change more but stay acceptable throughout the entire recording.

2.8 Safety considerations

One possible safety issue is biological. Use of genetically modified cells will require adequate GMO license and use of primary cells from patients usually requires biological safety level 2. The LED illumination for the holographic mode does not require extra care. Only, when the microscope is equipped with the fluorescence module, a possibility of UV light needs to be taken into account.

3. Summary and conclusions

Molecular biology and fluorescence imaging are very well defined and accurate. However, techniques for analyzing relevant cell behavior for cells taken as bulk are limited. For example, assessment of cell growth is usually based on counting cell divisions and measurement of cell motility relies on considerable translocation. Both of these approaches are slow and suffer by limited accuracy. Holographic quantitative phase imaging with incoherent-light source provides non-invasive characterization of the dynamic cell phenotype—as related to specific genetic manipulation or chemical treatment, which enables rigorous understanding of live cell activity resulting from the molecular mechanisms, mediated by essential and complementary information to fluorescence techniques. The most important measurements of the dynamic phenotype include rapid measurements of growth and motility in individual cells with ultimate accuracy. When used for testing the anti-cancer effects, it could be beneficial for finding novel drugs or in personalized cancer treatment with cells from biopsy.

Acknowledgments

This work was supported by the Ministry of Education, Youth and Sport of the Czech Republic (Large Research Infrastructures Project LM2018129 Czech-BioImaging).

References

Arnold, M., Karim-Kos, H. E., Coebergh, J. W., Byrnes, G., Antilla, A., Ferlay, J., et al. (2015). Recent trends in incidence of five common cancers in 26 European countries since 1988: Analysis of the European Cancer observatory. *European Journal of Cancer*, *51*, 1164–1187.

Bouchal, P., Štrbková, L., Dostál, Z., Chmelík, R., & Bouchal, Z. (2019). Geometric-phase microscopy for quantitative phase imaging of isotropic, birefringent and space-variant polarization samples. *Scientific Reports*, *9*, 3608.

Chaturvedi, A. K., Anderson, W. F., Lortet-Tieulent, J., Curado, M. P., Ferlay, J., Franceschi, S., et al. (2013). Worldwide trends in incidence rates for oral cavity and oropharyngeal cancers. *Journal of Clinical Oncology*, *31*, 4550–4559.

Chmelik, R., Slaba, M., Kollarova, V., Slaby, T., Lostak, M., Collakova, J., et al. (2014). The role of coherence in image formation in holographic microscopy. In E. Wolf (Ed.), *Vol. 59. Progress in optics* (pp. 267–335). Elsevier.

Dohet-Eraly, J., Yourassowsky, C., El Mallahi, A., & Dubois, F. (2016). Quantitative assessment of noise reduction with partial spatial coherence illumination in digital holographic microscopy. *Optics Letters*, *41*, 111–114.

Ďuriš, M., Bouchal, P., Rovenská, K., & Chmelík, R. (2022). Coherence-encoded synthetic aperture for super-resolution quantitative phase imaging. *APL Photonics*, *7*, 46105.

Ďuriš, M., & Chmelík, R. (2021). Coherence gate manipulation for enhanced imaging through scattering media by non-ballistic light in partially coherent interferometric systems. *Optics Letters*, *46*, 4486.

Ferlay, J., Steliarova-Foucher, E., Lortet-Tieulent, J., Rosso, S., Coebergh, J. W. W., Comber, H., et al. (2013). Cancer incidence and mortality patterns in Europe: Estimates for 40 countries in 2012. *European Journal of Cancer*, *49*, 1374–1403.

Holden, E., Tarnok, A., & Popescu, G. (2017). Quantitative phase imaging for label-free cytometry. *Cytometry Part A*, *91*, 407–411.

Leith, E. N., & Upatnieks, J. (1967). Holography with achromatic-fringe systems. *Journal of the Optical Society of America*, *57*, 975.

Milliken, G. A., & Johnson, D. E. (1992). Analysis of messy data volume I: Designed experiments. In *22. Journal of Marketing Research*. New York and London: Chapman and Hall.

Park, Y., Depeursinge, C., & Popescu, G. (2018). Quantitative phase imaging in biomedicine. *Nature Photonics*, *12*, 578–589.

Slaby, T., Kolman, P., Dostal, Z., Antos, M., Lostak, M., & Chmelik, R. (2013). Off-axis setup taking full advantage of incoherent illumination in coherence-controlled holographic microscope. *Optics Express*, *21*, 14747–14762.

Snedecor, G. W., & Cochran, W. G. (1967). *Statistical methods*. Iowa State University Press.

Sung, H., Ferlay, J., Siegel, R. L., Laversanne, M., Soerjomataram, I., Jemal, A., et al. (2021). Global cancer statistics 2020: GLOBOCAN estimates of incidence and mortality worldwide for 36 cancers in 185 countries. *CA: A Cancer Journal for Clinicians*, *71*, 209–249.

Zangle, T. A., & Teitell, M. A. (2014). Live-cell mass profiling: An emerging approach in quantitative biophysics. *Nature Methods*, *11*, 1221–1228.

Zicha, D. (2022). Addressing cancer invasion and cell motility with quantitative light microscopy. *Scientific Reports*, *12*, 1621.

Zicha, D., & Dunn, G. A. (1995). An image-processing system for cell behavior studies in subconfluent cultures. *Journal Of Microscopy Oxford*, *179*, 11–21.

CHAPTER TEN

A streamlined process for discovery and characterization of inhibitors against phenylalanyl-tRNA synthetase of *Mycobacterium tuberculosis*

Heng Wang and Shawn Chen*
Global Health Drug Discovery Institute, Haidian, Beijing, China
*Corresponding author: e-mail address: shuo.chen@ghddi.org

Contents

Methods in Enzymology, Volume 679
ISSN 0076-6879
https://doi.org/10.1016/bs.mie.2022.07.032

Abstract

Aminoacyl-tRNA synthetases (aaRSs) catalyze aminoacylation of tRNAs to produce aminoacyl-tRNAs for protein synthesis. Bacterial aaRSs have distinctive features, play an essential role in channeling amino acids into biomolecular assembly, and are vulnerable to inhibition by small molecules. The aaRSs continue to be targets for potential antibacterial drug development. The first step of aaRS reaction is the activation of amino acid by hydrolyzing ATP to form an acyladenylate intermediate with the concomitant release of pyrophosphate. None-radioactive assays usually measure the rate of ATP consumption or phosphate generation, offering advantages in high-throughput drug screening. These simple aaRS enzyme assays can be adapted to study the mode of inhibition of natural or synthetic aaRS inhibitors. Taking phenylalanyl-tRNA synthetase (PheRS) of *Mycobacterium tuberculosis* (Mtb) as an example, we describe a process for identification and characterization of Mtb PheRS inhibitor.

1. Introduction

Proteins are the biomolecules assembled in ribosome with amino acids carried by transfer RNAs (tRNA). Aminoacyl-tRNA synthetases (AaRSs) are a family of cellular enzymes responsible for specifically ligating an amino acid to its cognate tRNA (Gomez & Ibba, 2020). The aaRS-catalyzed aminoacylation of tRNA proceeds in two steps: an amino acid is first activated with the energy released from ATP hydrolysis to form an aminoacyl-adenylate intermediate and a pyrophosphate product; then a hydroxyl oxygen on the ribosyl group of $3'$-terminal adenosine (A76) of tRNA attacks the carbonyl carbon of aminoacyl-adenylate, resulting in two more products—adenosine monophosphate (AMP) and aminoacyl-tRNA in which the amino acid is linked to the $3'$-A76 of tRNA through an ester bond. Bacterial pathogens have at least 17 canonical aaRSs encoded in the genome. Gene essentiality studies performed on the genomic level have revealed aaRS genes are indispensable for the growth and survival of pathogen in vitro or in animals. Validated by natural antibiotics (e.g., mupirocin and albomycin) that inhibit the synthesis of aminoacyl-tRNA in bacteria (Yanagisawa & Kawakami, 2003; Zeng, Roy, Patil, Ibba, & Chen, 2009), aaRSs are therapeutic targets that have already delivered clinical benefits. Sequence divergence between bacterial aaRSs and the human counterparts, and subtle structural differences at the active sites make aaRSs high-ranking priorities in target-based antibacterial discovery (Francklyn & Mullen, 2019).

M. tuberculosis (Mtb), the etiological agent of tuberculosis (TB), is a bacterial pathogen of global health concern. One-quarter of the world's population is estimated to be latently infected with Mtb, and ~60 million deaths have been attributed to TB since 2000 (WHO, 2020). The United Nations have committed to end the TB epidemic globally by 2030, and new antibiotics are urgently needed to treat drug-resistant TB. Antibiotic resistance in Mtb is solely mediated by chromosomal mutations and rearrangement. The discovery and development of aaRS inhibitors against Mtb can set up a model for target-based approaches. Compared to environmental actinobacteria (Kulkarni et al., 2016), the genome of Mtb is much smaller. The aaRS genes have been unambiguously annotated with numerous genetic analyses, including experimental verifications (Dejesus et al., 2017). It is straightforward to identify an aaRS target of Mtb and use in vitro biochemical methods to screen for and characterize the inhibitors. Chemical synthesis of sulfamoyloxy-linked aminoacyl-AMP analogues to be used as tool compounds also facilitates the development of screening for novel aaRS inhibitors. Phenylalanyl-tRNA synthetase (PheRS) has been extensively studied for over 50 years (Moor, Klipcan, & Safro, 2011). Bacterial PheRS is a class II aaRS, and typically has a two-heterodimeric structure $(\alpha\beta)_2$. In the α subunit is the synthetic active site that includes L-Phe and ATP binding pockets. The β subunit has an editing domain that hydrolyzes improperly charged $tRNA^{Phe}$. Mtb PheRS specifically recognizes its cognate $tRNA^{Phe}$ through interaction with 5 nucleotide bases in the anticodon stem and loop, as well as contacts at the D- and T-loops. We recently identified the compound PF-3845 as a new inhibitor against Mtb PheRS (Wang et al., 2021). It is a Phe and ATP-competitive inhibitor, occupying two substrate-binding pockets when a sulfate ion is present. With the novel mode of inhibition, PF-3845 is a promising starting point for antituberculosis drug development.

In this chapter, using Mtb PheRS as an example, we present a set of generally applicable methods for the identification and characterization of aaRS inhibitors. A classical method of measuring tRNA aminoacylation is to incorporate a radiolabeled amino acid into tRNA, followed by capturing the radioactive tRNA on a filter to measure the counts (Francklyn, First, Perona, & Hou, 2008). To increase throughput, scintillation proximity technology involving beads containing scintillant is developed. These radiometric methods have limitations such as the high cost, generating hazardous waste, and growingly short supply of radioactive materials. More importantly, they cannot be used as a continuous assay. Our non-radioactive method builds on previous reports of continuous spectrophotometric assay for aaRS (Cestari & Stuart, 2013; Lloyd, Thomann, Ibba, & Söll, 1995; Upson, Haugland, Malekzadeh, & Haugland, 1996). An high-throughput screening is initially carried out with an end-point assay that uses malachite green reagent to detect the phosphate released from enzymatic degradation of pyrophosphate; alternatively, affordable commercial kit for ATP consumption can be used (Hewitt et al., 2017). The tRNA^{Phe} in our assay is enriched and largely purified from E. coli carrying an overexpression plasmid. Our method is suitable for screening and analysis of compounds that inhibit the overall synthesis activity of aaRS. It cannot be used to measure the editing activity of PheRS or others. Methods for analysis of potential inhibitors of the editing function are not covered here. Furthermore, the biochemical data obtained only represent the steady-state kinetics. They should be carefully interpreted in mechanistic studies of aaRS enzyme and its inhibitor. Other biophysical methods such as differential scanning fluorimetry, isothermal titration calorimetry, surface plasmon resonance and molecular docking or crystallization can be employed to measure the binding affinity of aaRS inhibitor and further illustrate the inhibition mechanism. Mycobacterial cell-based experiment involving genetic manipulation of the pathogen to verify the target engagement of aaRS inhibitor is not included in this chapter.

2. Expression and purification of PheRS protein

The sequences of Mtb phes (Rv1649) and pheT (Rv1649) identified in the genome are synthesized and cloned into the pET-30a vector as previously described (Wang et al., 2021). After comparing a few expression constructs, such as those having the subunits expressed separately or co-expressed with two promoters, we find the best choice is to place a ribosome-binding sequence ahead of the two consecutive genes, and a His6 tag fused only at the C-terminal of PheT for co-purification. E. coli BL21(DE3) cells transformed with the expression plasmid are grown to OD600 ~ 0.6, chilled and induced with isopropyl β-D-1-thiogalactopyranoside (IPTG) to produce the PheRS at 16 °C for 14 h before harvest and lysis of the cells. Cell lysates pass through three purification columns (HisTrap-FF, HiTrapQ-HP and Superdex-200); the fractions and protein purity are examined by an SDS-PAGE gel analysis (Fig. 1A), and the concentration of purified PheRS protein is determined by BCA protein assay (Cat. 23227, ThermoFisher).

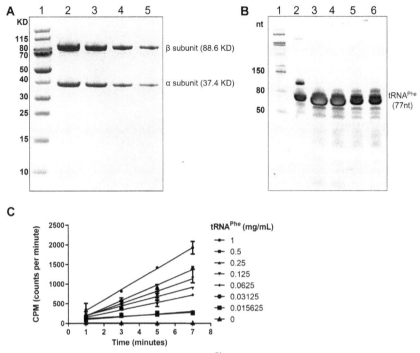

Fig. 1 Purified Mtb PheRS protein and tRNA^{Phe}. (A) Purified Mtb PheRS: Lane 1, protein ladder, molecular weight as indicated. Lanes 2–5, 4, 2, 1, or 0.5 µg Mtb PheRS protein in each column. The two subunits are marked. (B) Purified Mtb tRNA^{Phe}: 1, RNA ladder, low molecular weight ssRNA. 2, Mtb tRNA^{Phe} of in vitro transcription. 3–4, Eluted nucleic acid with buffer A. 5–6, Eluted nucleic acid with buffer B. Mtb tRNA^{Phe} (77 nt) is marked. (C) Aminoacylation of tRNA^{Phe} by Mtb PheRS is demonstrated with the radioactive SPA assay.

3. Expression and purification of Mtb tRNA^{Phe}

The purpose of this section is to overexpress Mtb tRNA^{Phe} in *E. coli* and obtain a huge amount of pure tRNA for high-throughput screening and other biochemical assays of PheRS. The sequences of Mtb tRNA^{Phe} are synthesized and cloned into the pTrc99a (Addgene) vector as previously described (Wang et al., 2021). The cell growth and IPTG induction are like in the protein overexpression part. Total RNA is extracted with acid-phenol and chloroform method. Transfer RNAs including the tRNA^{Phe} can be further enriched with an anion-exchange chromatography column such as Qiagen Q-tips.

3.1 Equipment

- Shaker.
- Cooling centrifuge (Beckman Coulter).
- Electrophoresis system (ThermoFisher).
- Nano Drop (ThermoFisher).

3.2 Materials

- Basic chemicals (Sigma).
- BL21 (DE3) competent cells (Transgen).
- TB (Terrific-Broth) medium (22711022, ThermoFisher).
- LB agar (Transgen).
- Ampicillin (Amersco).
- IPTG (R0392, ThermoFisher).
- Water-saturated phenol (Solarbio).
- Q-2500 column (Qiagen).
- Tris–borate–EDTA urea gel (EC6885BOX, ThermoFisher).
- Low M.W. ssRNA ladder (N0364S, NEB).
- Lysis buffer (100 mM Tris pH 7.0, 20 mM $MgCl_2$).
- Equilibration buffer (50 mM MOPS pH 7.0, 15% isopropanol, 1% Triton-X100).
- Washing buffer (50 mM MOPS pH 7.0, 200 mM NaCl).
- Elution buffer A (50 mM MOPS pH 7.0, 650 mM NaCl).
- Elution buffer B (50 mM MOPS pH 7.0, 700 mM NaCl).
- Regeneration buffer (50 mM MOPS pH 7.0, 1 M NaCl).
- UltraPure Water (10977015, ThermoFisher).

3.3 Procedure

1. The expression plasmid is transformed into BL21 (DE3) strain by heat shock method and selected on the LB agar plate with ampicillin incubating overnight in the incubator at 37 °C.
2. A single colony is picked up and inoculated into TB medium containing 100 µg/mL of ampicillin and incubated overnight at 37 °C in the shaker.
3. 10 mL fresh culture is inoculated into 1 L TB medium and grows at 37 °C until the OD600 reached 0.6–0.8. After 0.3 mM IPTG added, induction is kept for 16 h at 37 °C.
4. Cells are harvested by centrifugation (3800g, 15 min, 4 °C) and resuspended with lysis buffer.

5. An equal volume of water-saturated phenol is added and fully mixes at room temperature for 1 h. After centrifugation (14,000 g, 30 min, 25 °C), the upper aqueous layer is carefully transferred to be mixed with an equal volume of chloroform. The mixing is gently performed, and let the mixture stand for a while.

6. After a short centrifugation, carefully transfer the upper layer into multiple new tubes. Triple volumes of pre-cold ethanol are added in, and nucleic acid is precipitated at −20 °C for more than 2 h. The supernatant is discarded after centrifugation (14,000 g, 30 min, 4 °C).

7. The dissolved nucleic acid is mixed with 1 M MOPS pH 7.0 buffer to a final concentration of 0.1 mM. One Q-2500 column (for every 2 L starting culture) is pre-equilibrated with 50 mL equilibration buffer twice.

8. The nucleic acid is bound onto the column and washes four times with 50 mL washing buffer by gravity flow. Then each column is washed with 10 mL elution buffer A, followed by at least 2 × 10 mL of elution buffer A and B that elutes mainly tRNAs.

9. The elution fractions can be examined by Tris–borate–EDTA urea gel (Fig. 1B), and pure tRNA factions are precipitated by adding 1/10 volume of 3 M sodium acetate and an equal volume of isopropanol for 2 h at −20 °C.

10. The pellet is washed twice with 75% pre-cold ethanol by centrifugation, air-dried, and dissolved in ultrapure water. NanoDrop is used to determine the concentration, then tRNA is made to a final concentration of 10 mg/mL and examined by urea-PAGE. It is dispensed to 100 μL aliquots and stored at −20 °C.

11. The Q-2500 column can be washed with 50 mL regeneration buffer twice and 50 mL ddH$_2$O twice for reuse.

3.4 Notes

1. In step 7, to fully dissolve the nucleic acid, a mild vortex is needed. If there are insoluble matters after shaking, those can be removed by centrifugation.

2. In step 10, to accelerate the evaporation of ethanol, the pellets in the tubes can be put in an air oven.

3. We recommend confirming the aminoacylation activity of PheRS and testing the purified tRNA to be charged by radio-labeled L-Phe using classic scintillation proximity assay (SPA), where the charged and

uncharged tRNAs are adsorbed to positively charged beads, and the energy conversion of radioactive decay releases photons that can be detected by using devices such as the photomultiplier tubes of scintillation counter (Fig. 1C).

4. PPi production assay (kinetic assay)

The coupled pyrophosphate (PPi) production kinetic assay is developed using the commercial EnzChek™ Phosphate assay kit. Although the activities of other bacterial aaRSs can be detected with this assay in the absence of cognate tRNA, we found Mtb PheRS required tRNA to generate PPi continuously which leads to a signal increase in the absorbance reading at 360 nm. With this assay, we can determine the K_m's of the PheRS, the enzyme parameters needed in the screening. Later, this assay is used again to kinetically analyze the biochemical mechanism of an inhibitory compound.

4.1 Equipment

- EnVision plate reader (PerkinElmer).
- Xplorer plus electronic pipette (Eppendorf).
- Echo Liquid Handlers 550 (Beckman).

4.2 Materials

- Basic chemicals (Sigma).
- EnzChek Phosphate assay kit (E6646, ThermoFisher).
- ATP (A7699, Sigma).
- L-Phe (78019, Sigma).
- BSA (A1933, Sigma).
- Brij-35 (20150, ThermoFisher).
- TCEP (646547, Sigma).
- Pyrophosphatase, PPase (I1643, Sigma).
- 384-well microplate (3764, Corning).
- GDI05-001 (a tool compound customarily synthesized by WuXi AppTec).
- 2× reaction buffer (100 mM HEPES pH 7.4, 200 mM NaCl, 20 mM $MgCl_2$, 1 mM TCEP, 0.2 mg/mL BSA and 0.02% Brij-35).

4.3 Procedure

4.3.1 K_m determination

1. Fresh $2\times$ reaction buffer is prepared, and Mtb PheRS protein and $tRNA^{Phe}$ are thawed in the ice.

2. K_m of PheRS with respect to each substrate is determined by diluting one substrate at gradient concentrations with the other two substrates at saturating concentration (Saturating concentration: ATP, 1 mM; L-Phe, 200 µM; $tRNA^{Phe}$, 1 mg/mL. Gradient concentrations: ATP, $0 \sim 1$ mM; L-Phe, $0 \sim 200$ µM; $tRNA^{Phe}$, $0 \sim 1$ mg/mL; twofold dilution series).

3. Buffer I is prepared to contain the enzyme, saturated two of the three substrates, PPase, and the other two components, purine nucleoside phosphorylase (PNPase) and MESG, which are from the Enzchek Phosphate assay kit. Final concentrations are: 50 nM Mtb PheRS, 0.05 mM MESG, 0.5 unit/mL PPase, 0.1 unit/mL PNPase and saturating amounts of each substrate.

4. Buffer II is prepared to contain the gradient concentrations of the third substrate.

5. 10 µL buffer I is added to the 384-well microplate by electronic pipette, then 10 µL buffer II is added to the designated wells. Absorbance at 360 nm is read every minute for 30 min.

6. PPi standard curve is made at the same time with 0.05 mM MESG, 0.5 unit/mL PPase, 0.1 unit/mL PNPase, and 1.25–20 µM sodium PPi. The reactions are incubated for 15 min before being read at 360 nm.

7. The readout is converted to PPi production using the standard curve, and data analysis and plots are made by GraphPad Prism 9 (Fig. 2A).

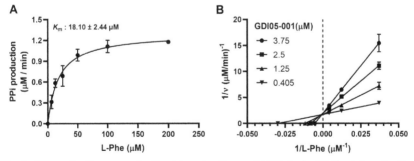

Fig. 2 PPi production assay for K_m determination and MOI study. (A) K_m determination of Mtb PheRS with respect to L-Phe. (B) Mode of inhibition analysis of tool compound GDI05-001 shows a competitive inhibition concerning L-Phe.

4.3.2 Mode of inhibition (MOI) analysis

Mtb PheRS with two saturating substrates and one variable substrate are used in the measurement of initial reaction velocities at a series of fixed concentrations of an inhibitor (Strelow et al., 2012). Here the analysis is demonstrated with a tool compound GDI-001, a phenylthiazolylurea sulfonamide compound previously published (Beyer et al., 2004). Generally, aminoacyl-AMP analogue mimicking aaRS reaction intermediate can also be used.

1. Compounds and DMSO are transferred into a 384-well microplate by using Echo liquid handlers.
2. Buffer I is prepared as described in K_m determination. 10 µL is transferred into the plate and mixed well with compounds or DMSO for 30 min of incubation.
3. Buffer II is prepared to contain variable substrate from $1/2 \times K_m$ to $5 \times K_m$, and 10 µL is added to the designated wells. Absorbance at 360 nm is read every minute for 30 min.
4. The readout is collected, and data are analyzed by GraphPad Prism. Lineweaver-Burk plots are fitted for modeling the most probable inhibition mode (competitive, non-competitive, mixed, or uncompetitive). Analysis of GDI-005 is a Phe-competitive inhibitor is shown (Fig. 2B).

4.4 Notes

1. In the reaction buffer, TCEP can be replaced by DTT, and 0.02% Brij-35 can be replaced by 0.008% Tween-20. BSA addition can increase the enzyme activity and reduce the amount of Mtb PheRS used.
2. Echo liquid handlers can be replaced by any other liquid handler or manual dilution with a pipette.
3. The linear reaction curves at each condition are selected for calculating initial reaction velocities.

5. ATP consumption assay (endpoint assay)

Numerous commercial kits are made for very sensitive quantification of ATP in biochemical reaction. To take the advantage of ATP consumption characteristic of aaRS reactions, we have developed and optimized a tRNA-dependent ATP consumption assay using the Kinase-Glo Luminescent Kit in a high-throughput screening against Mtb PheRS. This method can also be applied to other aaRSs.

5.1 Equipment

- Spectramax M5 plate reader (Molecular Devices).
- Multidrop dispensers (ThermoFisher).
- Xplorer plus electronic pipette (Eppendorf).
- Echo Liquid Handler 550 (Beckman).
- Incubator (ThermoFisher).

5.2 Materials

- Basic chemicals (Sigma).
- Kinase-Glo Max Luminescent Kit (V6073, Promega).
- ATP (A7699, Sigma).
- L-Phe (78019, Sigma).
- BSA (A1933, Sigma).
- Brij-35 (20150, ThermoFisher).
- TCEP (646547, Sigma).
- 384-well microplate (3570, Corning).
- Sealing Film (PCR-AS-200, Axygen).
- $2\times$ reaction buffer (100 mM HEPES pH 7.4, 200 mM NaCl, 20 mM $MgCl_2$, 1 mM TCEP, 0.2 mg/mL BSA and 0.02% Brij-35).
- Kinase-Glo Max dilution buffer (50 mM Tris pH 7.5, 5% glycerol).
- FDA-approved drug library (2148 compounds, Selleck).
- GDI05-001 (a tool compound customarily synthesized by WuXi AppTec).

5.3 Procedure

5.3.1 Enzyme titration

1. Buffer I is prepared to contain gradient concentrations of Mtb PheRS (0, 20, 40, 60, 80, and 100 nM).
2. Buffer II is prepared to contain three substrates (Final concentrations were: 1 μM ATP, 20 μM L-Phe, and 0.1 mg/mL tRNA^{Phe}).
3. Each 5 μL buffer I and II are added to a 384-well microplate and mixed well. Seal plate with aluminum film and incubate for 2 h at 37 °C in the incubator.
4. After Cooling down the reaction plate, the Kinase-Glo Max reagent is diluted 50-fold with buffer. Add 10 μL reagent to the reaction wells and incubate for 20 min at RT.
5. Record luminescence using SpectraMax M5. And data analysis and plots are made by GraphPad Prism 9.

5.3.2 Reaction linearity

1. Buffer I is prepared to contain 100 nM Mtb PheRS. Buffer II is prepared as above.
2. Mix 5 μL buffer I and II every 15 min within 135 min, no enzyme reactions are made as control.
3. Add 10 μL diluted Kinase-Glo reagent to each reaction well and incubate for 20 min.
4. Record luminescence using SpectraMax M5. And data analysis and plots are made by GraphPad Prism 9 (Fig. 3A).

5.3.3 IC$_{50}$ determination

1. The compound is prepared using Echo 550 in a 384-well microplate in a series of 10 concentrations from 0.01 to 202.5 μM.
2. 5 μL buffer I containing 100 nM Mtb PheRS is added and incubated for 30 min at RT.
3. 5 μL buffer II containing 1 μM ATP, 20 μM L-Phe, and 0.1 mg/mL tRNA^{Phe} is added and mixed well. Seal plate with aluminum film and incubated for 2 h at 37 °C in the incubator.

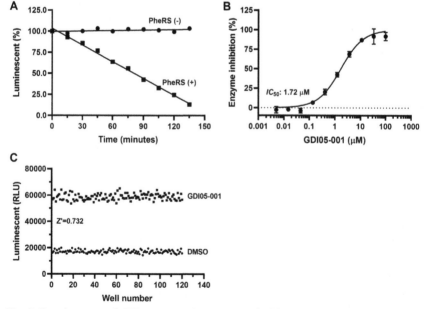

Fig. 3 Development of ATP consumption assay and HTS optimization. (A) Linearity of ATP consumption with or without PheRS. (B) IC$_{50}$ determination of GDI05-001. (C) Z prime factor determination of HTS.

4. After cooling down the reaction plate, add 10 µL diluted Kinase-Glo reagent and incubate for 20 min.
5. Plates are processed for luminescence (Lum) as aforementioned. And data analysis and plots are made by GraphPad Prism 9 (Fig. 3B).
6. The remaining activity is calculated as:

$$\text{activity } (\%) = \left(1 - \frac{\text{Lum}^{max} - \text{Lum}^{I}}{\text{Lum}^{max} - \text{Lum}^{min}} \right) \times 100$$

Lum^{max} means no enzyme control, Lum^{min} means no inhibitor well, and Lum^{I} means inhibitor well. The plot of IC_{50} is fit to the equation as below, and the H is the Hill slope factor.

$$y = y_{min} + \frac{y_{max} - y_{min}}{1 + \left(\frac{IC_{50}}{x} \right)^{H}}$$

5.3.4 Z' factor

1. 50 µM tool compound and DMSO are prepared using Echo 550 in a 384-well plate. Half wells of the plate are transferred tool compound, and the left wells fill with the same volume of DMSO.
2. 5 µL buffer I containing 100 nM Mtb PheRS is added and incubated for 30 min at RT.
3. 5 µL buffer II containing 1 µM ATP, 20 µM L-Phe, and 0.1 mg/mL tRNA^{Phe} is added and mixed well. Seal plate with aluminum film and incubate for 2 h at 37 °C in the incubator.
4. Plates are processed for luminescence as aforementioned. And data analysis and plots are made by GraphPad Prism 9 (Fig. 3C). Z' factor was calculated as:

$$Z' = 1 - 3 \times \frac{(SD^{+} + SD^{-})}{|Ave^{+} - Ave^{-}|}$$

SD^{+} means standard deviation of Lum values in tool compound wells, and SD^{-} means standard deviation of Lum values in DMSO wells; Ave^{+} means the average of Lum. values in tool compound wells, and Ave^{-} means the average of Lum values in DMSO wells.

5.3.5 High-throughput screening

1. Tool compound, DMSO and compounds from a screening library are arrayed in plates using Echo 550.

2. In each 384-well microplate, every side of two columns are transferred 50 μM tool compound or the same volume of DMSO, the remaining wells are filled with each compound from the library.

3. Add 5 μL buffer I containing 100 nM Mtb PheRS and incubate for 30 min at RT.

4. Add 5 μL buffer II containing 1 μM ATP, 20 μM L-Phe, and 0.1 mg/mL tRNA^{Phe} and mix well. Seal plate with aluminum film and incubates for 2 h at 37 °C in the incubator.

5. Luminescence is recorded using SpectraMax M5. The cut-off is set to 70% inhibition compared with 100% inhibition reaction wells (tool compound wells).

6. Hits from screening are cherry-picked for IC$_{50}$ determination (Fig. 4).

5.4 Notes

1. We use 1 μM ATP substrate in ATP consumption assay which is well below the $K_{m, ATP}$. But this low concentration offers an advantage for detecting ATP consumption with increasing sensitivity.

2. We confirm that the low ATP concentration is suitable for Mtb PheRS in the ATP consumption assay showing a high signal-to-noise ratio. However, some other bacterial aaRSs (e.g., Mtb TrpRS) require a high concentration of ATP to activate the catalytic center (Williams, Yin, & Carter, 2016), so the ATP consumption assay will not work well. Then an alternative assay presented in Section 6 can be chosen for an HTS with this type of aaRS.

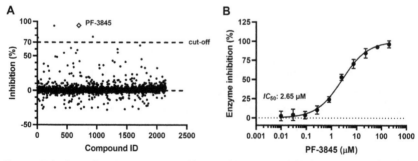

Fig. 4 High-throughput screening and hit confirmation. (A) Summary plot of the HTS result, the selected hit: PF-3845 is marked with diamond. (B) IC$_{50}$ of PF-3845 against Mtb PheRS by ATP consumption assay.

3. The stock concentration of most compounds in the Echo source plate is often 10 mM. To prepare a series of 10 concentrations from 0.01 to 202.5 μM, a middle concentration at 40 μM is made for lower concentrations.

6. Alternative assay

In an aaRS reaction, the rate of ATP consumption (PPi production) can be limited by the available amount of tRNA which breaks down the acyladenylate intermediate and accepts the aminoacyl group to form the product aminoacyl-tRNA. Due to the lack of tRNA recycling in traditional aaRS biochemical assay, the initial linear reaction time can be very short, rendering the assay not easy to operate. Hydroxylamine can be used as the substitutive acceptor of the aminoacyl group in studies of some adenylate-forming enzymes (Wilson & Aldrich, 2010). We also successfully implement a combination of hydroxylamine with the components of Enzcheck phosphate assay (kinetic) or malachite green (endpoint) in the investigation of indolmycin for inhibition of Mtb TrpRS (Yang et al., 2022). The assay provides an option to explore the aaRSs that have an ATP inducible binding pocket, or high binding affinity with the reaction intermediate (Fig. 5A).

6.1 Equipment
- EnVision plate reader (PerkinElmer).
- Xplorer plus electronic pipette (Eppendorf).
- Echo Liquid Handlers 550 (Beckman).

6.2 Materials
- Basic chemicals (Sigma).
- ATP (A7699, Sigma).
- L-Phe (78019, Sigma).
- BSA (A1933, Sigma).
- Brij-35 (20150, ThermoFisher).
- TCEP (646547, Sigma).
- Pyrophosphatase, PPase (I1643, Sigma).
- EnzChek Phosphate assay Kit (E6646, ThermoFisher).
- BIOMOL Green Kit (BML-AK111-1000, Enzo).
- 384-well microplate (3764, Corning).

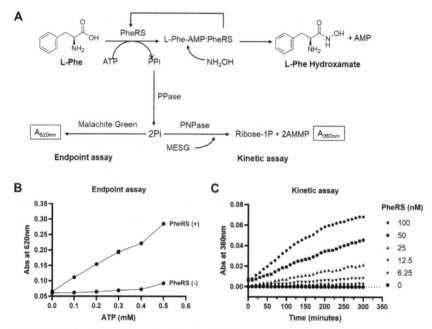

Fig. 5 Principles of the alternative biochemical assays and results. (A) Schematic presentation of assays detecting Pi and releasing PheRS with hydroxylamine to increase Pi production in the analysis of Phe-competitive inhibitor. PPi, pyrophosphate; Pi, phosphate; PPase, phosphatase; PNPase, purine nucleoside phosphorylase; MESG, 2-amino-6-mercapto-7-methyl-purine riboside; 2AMMP, 2-amino-6-mercapto-7-methylpurine; Ribose-1P, ribose 1-phosphate. (B) Endpoint PPi production assay with malachite green at indicated ATP concentrations. (C) Kinetic PPi production assay of gradient PheRS concentrations.

- Sealing Film (PCR–AS-200, Axygen).
- $2 \times$ reaction buffer (100 mM HEPES pH 7.4, 20 mM MgCl$_2$, 1 mM TCEP, 0.2 mg/mL BSA and 0.02% Brij-35).
- 7 M NaOH stock solution (Sigma).
- 4 M Hydroxylamine stock solution (Sigma).

6.3 Procedure

6.3.1 Endpoint assay

1. Prepare fresh 2 M hydroxylamine alkalescent with 7 M NaOH stock solution: every 1 mL buffer contains 500 μL 4 M hydroxylamine, 250 μL ddH$_2$O, and 250 μL 7 M NaOH. Fully mix and keep in the ice box.

2. Buffer I is prepared to contain 30 nM Mtb PheRS.
3. Buffer II is prepared to contain indicated concentration of ATP, 2 μM
   L-Phe, 250 mM hydroxylamine, and 0.5 unit/mL PPase.
4. Add each 10 μL of buffer I and II to a 384-well microplate, mix well, and
   incubated for 5 h at RT.
5. Add 40 μL BIOMOL green reagent to reaction wells and incubate for
   10 min, then record the absorbance at 620 nm using envision.
6. Collect and analyze the data by GraphPad Prism (Fig. 5B).

6.3.2 Kinetic assay

1. Prepare fresh 2 M hydroxylamine (alkalescent) with 7 M NaOH stock
   solution: every 1 mL buffer contains 500 μL 4 M hydroxylamine,
   250 μL ddH$_2$O, and 250 μL 7 M NaOH. Fully mix and keep in the
   ice box.
2. Buffer I is prepared to contain indicated concentration of Mtb PheRS.
3. Buffer II is prepared containing 0.1 mM ATP, 2 μM L-Phe, 250 mM
   hydroxylamine 0.05 mM MESG, 0.1 unit/mL PNPase and 0.5 unit/
   mL PPase.
4. Add each 10 μL of buffer I and II to a 384-well microplate, mix well, and
   record the absorbance at 360 nm every minute for at least 5 h at RT.
5. Collect readouts and analyze data by GraphPad Prism 9 (Fig. 5C).

6.4 Notes

1. Freshly prepared alkalescent hydroxylamine is important for this assay.
2. Do not replace Brij-35 with Tween-20 in the reaction buffer. Tween-20
   can slow down color development in the endpoint assay which leads to
   an increase in the non-enzymatic hydrolysis of ATP.
3. Not all aaRS enzymatic reactions progress at the approximately same rate
   as PheRS in this assay. Reaction temperature, the amounts of enzyme
   and substrates can change the time to reach reaction plateau.

7. Summary

This chapter presents a set of non-radioactive assays for Mtb PheRS,
which can be combined to build a streamlined process for the discovery
and characterization of small molecule inhibitors of PheRS. It starts
with the expression and purification of PheRS and tRNA^{Phe}. With the
Kinase-Glo Luminescent kit and a known PheRS inhibitor as tool compound, high-throughput screening for novel PheRS inhibitor is performed

in 384-well format. Then, a malachite green-based endpoint assay for phosphate detection can be used for hits confirmation. The following mode-of-inhibition analysis is performed with an MESG-based continuous assay. In the case of a Phe-competitive inhibitor, tRNA^{Phe} can be substituted for hydroxylamine in kinetic analysis because the phenylalanine hydroxamate produced does not inhibit the reaction. The same strategy can be used to study other aaRS reactions that have a bursting initial phase but subsequently become slow due to the tight-binding reaction intermediate. These methods can be generally applied to discover inhibitors against aaRSs of microbial pathogens for the discovery and development of target-based antimicrobial chemotherapy.

Acknowledgments

This work is supported by Bill & Melinda Gates Foundation and Global Health Drug Discovery Institute sponsors in China.

References

Beyer, D., Kroll, H. P., Endermann, R., Schiffer, G., Siegel, S., Bauser, M., et al. (2004). New class of bacterial phenylalanyl-tRNA synthetase inhibitors with high potency and broad-spectrum activity. *Antimicrobial Agents and Chemotherapy*, *48*(2), 525–532. https://doi.org/10.1128/AAC.48.2.525-532.2004.

Cestari, I., & Stuart, K. (2013). A spectrophotometric assay for quantitative measurement of aminoacyl-tRNA synthetase activity. *Journal of Biomolecular Screening*, *18*(4), 490–497. https://doi.org/10.1177/1087057112465980.

Dejesus, M. A., Gerrick, E. R., Xu, W., Park, S. W., Long, J. E., Boutte, C. C., et al. (2017). Comprehensive essentiality analysis of the Mycobacterium tuberculosis genome via saturating transposon mutagenesis. *MBio*, *8*(1), 1–17. https://doi.org/10.1128/mBio.02133-16.

Francklyn, C. S., First, E. A., Perona, J. J., & Hou, Y. M. (2008). Methods for kinetic and thermodynamic analysis of aminoacyl-tRNA synthetases. *Methods*, *44*(2), 100–118. https://doi.org/10.1016/j.ymeth.2007.09.007.

Francklyn, C. S., & Mullen, P. (2019). Progress and challenges in aminoacyl-tRNA synthetase-based therapeutics. *Journal of Biological Chemistry*, *294*(14), 5365–5385. https://doi.org/10.1074/jbc.REV118.002956.

Gomez, M. A. R., & Ibba, M. (2020). Aminoacyl-tRNA synthetases. *RNA*, *26*(8), 910–936. https://doi.org/10.1261/RNA.071720.119.

Hewitt, S. N., Dranow, D. M., Horst, B. G., Abendroth, J. A., Forte, B., Hallyburton, I., et al. (2017). Biochemical and structural characterization of selective allosteric inhibitors of the Plasmodium falciparum drug target, prolyl-tRNA-synthetase. *ACS Infectious Diseases*, *3*(1), 34–44. https://doi.org/10.1021/acsinfecdis.6b00078.

Kulkarni, A., Zeng, Y., Zhou, W., Van Lanen, S., Zhang, W., & Chen, S. (2016). A branch point of Streptomyces sulfur amino acid metabolism controls the production of albomycin. *Applied and Environmental Microbiology*, *82*(2), 467–477. https://doi.org/10.1128/AEM.02517-15.

Lloyd, A. J., Thomann, H.-U., Ibba, M., & Söll, D. (1995). A broadly applicable continuous spectrophotometric assay for measuring aminoacyl-tRNA synthetase activity. *Nucleic Acids Research*, *23*(15), 2886–2892. https://doi.org/10.1093/nar/23.15.2886.

Moor, N., Klipcan, L., & Safro, M. G. (2011). Bacterial and eukaryotic phenylalanyl-tRNA synthetases catalyze misaminoacylation of tRNA Phe with 3,4-dihydroxy-L-phenylalanine. *Chemistry and Biology*, *18*(10), 1221–1229. https://doi.org/10.1016/j.chembiol.2011.08.008.

Strelow, J., Dewe, W., Iversen, P. W., Brooks, H. B., Radding, J. A., Mcgee, J., et al. (2012). Mechanism of action assays for enzymes. In S. Markossian, A. Graddman, K. Brimacombe, & E. Al (Eds.), *Assay guidance manual* (pp. 1–20). Eli Lilly & Company and the National Center for Advancing Translational Sciences.

Upson, R. H., Haugland, R. P., Malekzadeh, M. N., & Haugland, R. P. (1996). A spectrophotometric method to measure enzymatic activity in reactions that generate inorganic pyrophosphate. *Analytical Biochemistry*, *243*(1), 41–45. https://doi.org/10.1006/abio.1996.0479.

Wang, H., Xu, M., Engelhart, C. A., Zhang, X., Yan, B., Pan, M., et al. (2021). Rediscovery of PF-3845 as a new chemical scaffold inhibiting phenylalanyl-tRNA synthetase in Mycobacterium tuberculosis. *Journal of Biological Chemistry*, *296*, 100257. https://doi.org/10.1016/j.jbc.2021.100257.

WHO. (2020). Global tuberculosis report 2019. In *Tuberculosis*. https://www.who.int/newsroom/fact-sheets/detail/tuberculosis.

Williams, T. L., Yin, Y. W., & Carter, C. W. (2016). Selective inhibition of bacterial tryptophanyl-tRNA synthetases by indolmycin is mechanism-based. *Journal of Biological Chemistry*, *291*(1), 255–265. https://doi.org/10.1074/jbc.M115.690321.

Wilson, D. J., & Aldrich, C. C. (2010). A continuous kinetic assay for adenylation enzyme activity and inhibition. *Analytical Biochemistry*, *404*(1), 56–63. https://doi.org/10.1016/j.ab.2010.04.033.

Yanagisawa, T., & Kawakami, M. (2003). How does Pseudomonas fluorescens avoid suicide from its antibiotic pseudomonic acid? Evidence for two evolutionarily distinct isoleucyl-tRNA synthetases conferring self-defense. *Journal of Biological Chemistry*, *278*(28), 25887–25894. https://doi.org/10.1074/jbc.M302633200.

Yang, Y., Xu, Y., Yue, Y., Wang, H., Cui, Y., Pan, M., et al. (2022). Investigate natural product indolmycin and the synthetically improved analogue toward antimycobacterial agents. *ACS Chemical Biology*, *17*(1), 39–53. https://doi.org/10.1021/acschembio.1c00394.

Zeng, Y., Roy, H., Patil, P. B., Ibba, M., & Chen, S. (2009). Characterization of two seryl-tRNA synthetases in albomycin-producing Streptomyces sp. strain ATCC 700974. *Antimicrobial Agents and Chemotherapy*, *53*(11), 4619–4627. https://doi.org/10.1128/AAC.00782-09.

CHAPTER ELEVEN

Site-specific quantitative cysteine profiling with data-independent acquisition-based mass spectrometry

Fan Yang[a] and Chu Wang[a,b,]*

[a]Synthetic and Functional Biomolecules Center, Beijing National Laboratory for Molecular Sciences, Key Laboratory of Bioorganic Chemistry and Molecular Engineering of Ministry of Education, College of Chemistry and Molecular Engineering, Peking University, Beijing, China
[b]Peking-Tsinghua Center for Life Sciences, Academy for Advanced Interdisciplinary Studies, Peking University, Beijing, China
*Corresponding author: e-mail address: chuwang@pku.edu.cn

Contents

Abstract

Chemical proteomics methods, such as activity-based protein profiling, have emerged as powerful and versatile tools to annotate the protein functions and targets of bioactive small molecules in complex biological systems. Incorporated with mass spectrometry (MS)-based quantitative proteomics method, changes of protein activities could be captured and investigated with site-specific precision. However, the semi-stochastic nature of data-dependent acquisition and high cost of the isotopic-labeled reagents make it challenging for chemical biology research to systematically and reproducibly analyze a large number of samples in multidimensional analysis and high-throughput screening. In this chapter, we describe an efficient quantitative chemical proteomic strategy, termed DIA-ABPP, with good reproducibility and high quantification accuracy. Cysteinome profiling was used as a proof-of-concept example with the detailed

Methods in Enzymology, Volume 679
ISSN 0076-6879
https://doi.org/10.1016/bs.mie.2022.07.037
295

protocol to demonstrate the workflow of the DIA-ABPP method, including dose-dependent analysis of cysteines that are sensitive to modification by a reactive metabolite, screening of a cysteine-reactive fragment library, and profiling of circadian cysteinome fluctuation. This quantitative chemoproteomic strategy would provide an opportunity for in-depth multi-dimensional chemical proteomic profiling and illuminate the function of bioactive small molecules and proteins in complex biological systems.

Abbreviations

ABP	activity-based probe
ABPP	activity-based protein profiling
DDA	data-dependent acquisition
DIA	data-independent acquisition
IA-alkyne	iodoacetamide-alkyne
LC-MS/MS	liquid chromatography tandem mass spectrometry
rdTOP-ABPP	reductive dimethylation-TOP-ABPP
TOP-ABPP	tandem orthogonal proteolysis-ABPP

1. Introduction

Activity-based protein profiling (ABPP) has emerged as a powerful and versatile chemoproteomic technology to characterize the reactivity of amino acids and report the activity of enzymes in complex biological system(Cravatt, Wright, & Kozarich, 2008; Wang & Chen, 2015). As the key element of the method, an activity–based probe (ABP) consists of three basic groups: a reactive group, a linker, and a reporter tag for visualization or enrichment such as biotin. By combining finely tuned ABPs, biotin enrichment and liquid chromatography tandem mass spectrometry (LC–MS/MS), functional sites and post-translational modifications in proteomes have been extensively profiled in various physiological systems and pathological processes (Fig. 1A) (Shannon & Weerapana, 2015; Yang & Wang, 2020).

As one of the most intrinsically nucleophilic amino acids in proteins, cysteine residues perform diverse biochemical functions in multiple enzyme families, such as nucleophilic catalysis, redox regulation and metal binding (Pace & Weerapana, 2013). An iodoacetamide probe (IA-alkyne) has been used as a broad-spectrum cysteine-reactive probe to perform the reactive cysteinome profiling and uncover the functional roles of the corresponding enzymes in biological processes (Weerapana, Simon, & Cravatt, 2008). A Tobacco Etch Virus (TEV) protease-based cleavable linker was further incorporated by click chemistry to generate a tandem orthogonal proteolysis

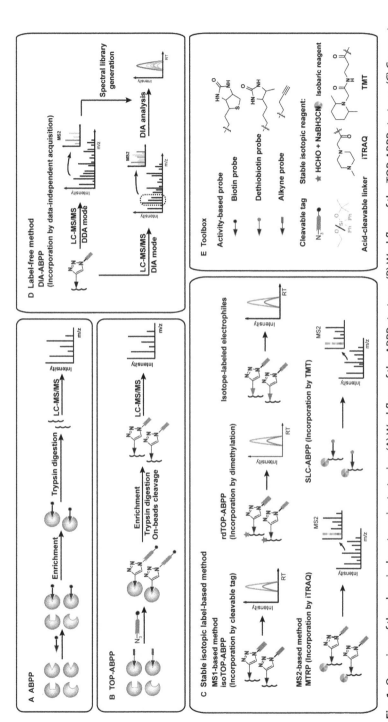

Fig. 1 Overview of the chemical proteomic strategies. (A) Workflow of the ABPP strategy. (B) Workflow of the TOP-ABPP strategy. (C) Concept of the label-based chemical proteomic strategies. (D) Concept of the label-free DIA-ABPP strategy. (E) Toolbox of ABPP. The structure of biotin, desthiobiotin, alkyne, acid-cleavable tag and isobaric reagent (iTRAQ, TMT) were shown as represented.

(TOP) strategy, named TOP-ABPP, to analyze the cysteinome with site-specific precision (Fig. 1B) (Speers, Adam, & Cravatt, 2003). An advanced version of TOP-ABPP platform termed isoTOP-ABPP has been successfully developed using a pair of isotopically labeled cleavable tags to quantitatively measure the intrinsic cysteine reactivity in native proteomes (Fig. 1C) (Weerapana et al., 2010). In addition to direct investigation of free cysteines, this method was further used to quantify the sensitivity towards modifications by endogenous or exogenous covalent bioactive small molecules in complexed biological systems using a competitive version of isoTOP-ABPP (Counihan, Ford, & Nomura, 2016; Qin, Yang, & Wang, 2020; Wang, Weerapana, Blewett, & Cravatt, 2014). For example, cysteinome reactivity with a series of lipid-derived electrophiles (LDE), which have been implicated in the pathogenesis of many diseases, were successfully investigated and an active-site proximal cysteine of the ZAK kinase could be modified by a represented LDE, 4-hydroxy-2-nonenal (HNE), to suppress the activation of the JNK pathway normally induced by oxidative stress. In addition to the original functionalized tag that can be cleaved by the TEV protease, the isotopically encoded valine residues have also been introduced to other widely used cleavable tags such as reductant, photolysis and acid-cleavable tags (Qin et al., 2018; Szychowski et al., 2010; Yang, Grammel, Raghavan, Charron, & Hang, 2010). Furthermore, isotopically labeled atoms were also directly incorporated into activity-based probes. By adding a pair of isotopically labeled phenyl groups to the nitrogen atom of the iodoacetamide probe (Fig. 1C), the probes were further applied for quantitative profiling of the extent of reversible cysteine oxidation (Abo, Li, & Weerapana, 2018). However, in all these cases described above, the customized synthesis of cleavable tags or probes with isotopic labeling are often not trivial and require significant efforts to achieve desired yields.

Certain quantitative proteomics methods, such as stable isotope dimethyl labeling, tandem mass tag (TMT) and isobaric tags for relative and absolute quantitation (iTRAQ) were also combined with ABPP methods to introduce the isotopic signatures into digested peptides to enable higher quantification multiplexity (Fig. 1C) (Tian et al., 2017; Vinogradova et al., 2020; Yang, Gao, Che, Jia, & Wang, 2018). While iTRAQ and TMT-based methods (SLC-ABPP, TMT-ABPP, MTRP) have higher multiplex capacity and less sample use, the costs associated with purchasing isobaric reagents correlate with multiplex capacity, which put limitations on their use in chemoproteomic profiling of large number of samples, such as high-throughput screening.

2. Quantitative cysteine profiling by DIA-ABPP

All the chemical proteomics studies reported to date have employed isotope-based quantification combined with data-dependent acquisition (DDA) methods, which is a traditional data collection mode of mass spectrometry for proteomic studies. However, the limited scanning speed of tandem mass spectrometry and the semi-stochastic sampling nature in the DDA mode compromised the reproducible and sensitive analysis of functional active-site residues across a large number of samples in the existing quantitative ABPP methods. The data loss would become even worse for residue-specific site identification in high-throughput applications, such as drug screening. Recently, data-independent acquisition (DIA) has emerged as a compelling alternative to DDA for proteomics analysis (Bruderer et al., 2015; Kelstrup et al., 2018). In contrast to DDA-based "shotgun" proteomic analysis, all co-eluting peptide ions within predefined mass-to-charge (m/z) windows are co-fragmented and the corresponding fragment ions are measured simultaneously in the DIA mode (Venable, Dong, Wohlschlegel, Dillin, & Yates, 2004). Therefore, all peptide ions could be comprehensively measured regardless of their intensity during the full mass scanning. Thus, DIA could in principle abolish the precursor selection in the DDA mode and provide a more reproducible identification and accurate quantitation. Due to the more complicated spectra from co-fragmented ions, DIA usually requires a pre-recorded DDA-based spectral library for spectral extraction and peptide quantification, in which the relationship between each peptide precursor with its corresponding fragment ions is constructed.

To date, DIA-based proteomics strategies have been successfully applied to O-glycoproteome, phosphoproteome and ubiquitinome for site-specific and large-scale analysis (Bekker-Jensen et al., 2020; Hansen et al., 2021; Ye, Mao, Clausen, & Vakhrushev, 2019). Given the superior performance for sensitive and reproducible analysis of site identification, we have combined DIA with TOP-ABPP to develop an efficient and powerful strategy, called "DIA-ABPP", for in-depth and multiplex quantitative chemical proteomics study with a site-specific resolution (Fig. 1D) (Yang, Jia, Guo, Liu, & Wang, 2022). In this chapter, we use the reactive cysteinome profiling as an example to describe the workflow of DIA-ABPP and demonstrate its applications in profiling functional cysteines that are sensitive towards dose-dependent HNE modifications, covalently modified by electrophilic ligand fragments from a synthetic library or subjected to redox fluctuation with the circadian clock.

To perform the in-depth reactive cysteine profiling by DIA-ABPP, we systematically investigated and optimized the key steps in the workflow, including sample preparation, chromatographic condition, spectral library construction and data processing (Fig. 2A). Notably, an open-source tool, termed "DIAcalc", was developed to integrate the precursors into distinct modified peptides and calculate the corresponding intensities. After optimization, more than one third of the probe-adducted peptides could be identified as compared to that by the original TOP-ABPP method, which suggested the necessity to optimized the working condition for each individual activity-based probe when it is used with DIA-ABPP.

The quantitative ability of DIA-ABPP using the label-free quantification was compared with the triplex rdTOP-ABPP quantification based on the isotopic labeling (Yang et al., 2018). High Pearson correlation coefficients were observed for ion intensities extracted from three replicated samples and the quantified ratios were also clustered around 1, which is similar to the results obtained in rdTOP-ABPP. In general, the quantification accuracy of DIA-ABPP outperformed that of rdTOP-ABPP, especially in the range of high quantification ratios. Moreover, narrower box distributions were found in DIA-ABPP, indicating its higher precision. The better quantitative ability of DIA-ABPP were probably due to the higher dynamic

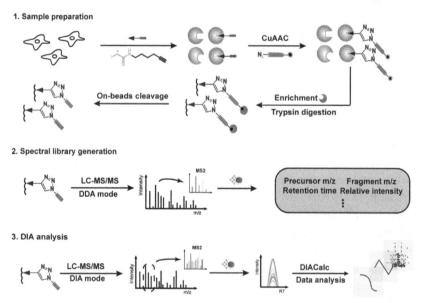

Fig. 2 The workflow of DIA-ABPP, including sample preparation, spectral library generation and DIA analysis.

range and lower noise interference in the DIA-based quantitation. Since it is a label-free approach with better quantification accuracy and sensitivity, the method would be highly suitable for quantitative ABPP profiling studies with a large number of samples such as high-throughput screening of fragments or clinical samples from a cohort study.

Three DIA-ABPP applications for reactive cysteinome profiling were demonstrated in this chapter. The first one is global analysis of HNE-sensitive cysteines in proteomes (Fig. 3A). As one of the common lipid-derived electrophiles generated upon oxidative stress, HNE can covalently modify proteins and alter their functions in signal transduction, cell proliferation, gene expression etc. (Poli, Schaur, Siems, & Leonarduzzi, 2008) Previously, quantitative profiling studies of HNE-reactive cysteines in proteomes were performed by the duplex isoTOP-ABPP and triplex rdTOP-ABPP (Wang et al., 2014; Yang et al., 2018), however, they suffered from severe data loss due to the semi-stochastic nature of DDA mode when multiple HNE doses were applied to compete the IA-alkyne probe labeling. In view of the high reproducibility and quantification ability of DIA-ABPP, we reason that this problem would be significantly mitigated. Here, a series of concentrations of HNE (0 µM, 5 µM, 10 µM, 50 µM and 100 µM) were chosen for the *in vitro* proteome treatment and three biological replicates for each HNE concentration were prepared for DIA analysis. Combined with a spectral library generated by three DDA-based samples (with no HNE competition), about 10,000 HNE-modified peptides were repeatedly quantified in 15 samples, which could be further used to fit the dose-response curves and get the EC_{50} value for each of the labeled cysteines. We found that most of the HNE-hyperreactive cysteines identified from the previous isoTOP-ABPP and rdTOP-ABPP studies were also accurately quantified by DIA-ABPP including C1101 of RTN4 (Fig. 3B and C). In addition to HNE as a model compound here, DIA-ABPP should be readily applicable to analyze the cysteine modifications by other electrophilic compounds including endogenous metabolites, natural products or synthetic ligands.

In addition to profiling targets of bioactive metabolites, DDA-based quantitative ABPP methods have also been extensively applied to proteome-wide ligand and drug discovery (Fig. 4A), aiming to mine new druggable proteins and illuminate protein function (Backus et al., 2016; Roberts, Ward, & Nomura, 2017; Vinogradova et al., 2020). To demonstrate the feasibility of applying DIA-ABPP in screening covalent fragment-based library at the proteome level, a total of 24 fragments (Fig. 4B), which

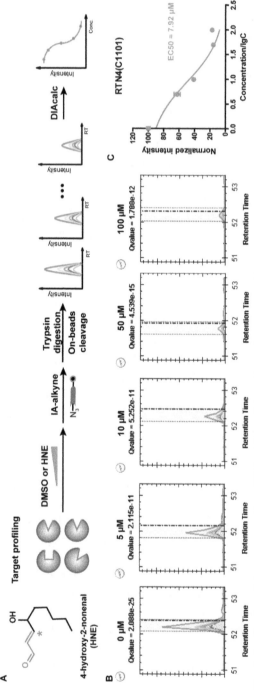

Fig. 3 Application of the target profiling. (A) Workflow of dose-dependent HNE profiling by DIA-ABPP. (B) Raw MS2 chromatographic peaks of C1101 with increasing concentration of HNE. (C) dose-response curve calculated EC50 value of C1101. Values represent means ± SD from three replicate experiments.

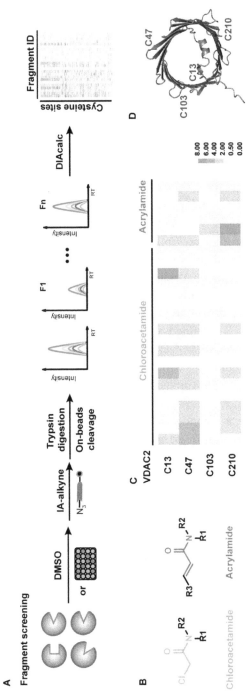

Fig. 4 Application of fragment screening. (A) Workflow of covalent fragment library screening by DIA-ABPP. (B) Functional groups of the covalent fragments. (C) Heat map analysis of the ligandability of 4 cysteines in VDAC2 against 24 fragments. (D) The localization of 4 cysteines based on the AlphaFold2-predicted structure.

have diverse molecular structures in the previous profiling study by isoTOP-ABPP (Backus et al., 2016), were chosen for a proof-of-concept study. The electrophilic fragment library, which contained ligands with the chloroacetamide or acrylamide "warheads" and an average molecular weight of 285 Da, were screened using the IA-alkyne probe by competitive DIA-ABPP in duplicates against Ramos cells. In total, it took about 7 days to analyze all the 54 DIA samples (2 samples per fragment plus 6 DMSO-treated control samples), which was only one fourth of the instrument time used by the isoTOP-ABPP screening. The DIA-ABPP screening quantified 8110 cysteine sites from 3734 proteins based on the initial DDA-based spectral library. When each of the 24 ligand fragments was used for competition, 6579 cysteines were quantified per sample on average with a competition ratio for that specific fragment, which is defined as the ratio of the peptide intensity from the DMSO-treated peptide versus that from the fragment-treated sample. In the previous isoTOP-ABPP screening (Backus et al., 2016), individual cysteines that were involved in at least three datasets were extracted, and only cysteines with more than two competition ratios that were larger than 4 and at least one competition ratio that was between 0.5 and 2 were defined as "ligandable" cysteines. Applying the same criteria to the DIA-ABPP data, a total of 563 ligandable cysteines from 458 proteins were identified, in which 85% were not found in the DrugBank database (Law et al., 2014). Taking advantages of the breakthrough in the field of protein structure prediction by AlphaFold2 (Jumper et al., 2021), our method shows great potential to systematically explore and deeply mine the new druggable proteins in proteomes.

As a representative example, four cysteines at different localization of VDAC2 (Fig. 4C and D) showed distinct ligandability towards fragments containing different warheads. C47 could be liganded by both chloroacetamide or acrylamide-based fragments but C103 could not be liganded by all fragments in the library. Notably, C210, one of the HNE-hypersensitive cysteines could be targeted only by the fragment containing the acrylamide, which is a similar reactive group as HNE. C13 of VDAC2 could only be targeted by the chloroacetamide fragment (Chen et al., 2018).

The high reproducibility and accuracy of DIA-ABPP technology should also enable temporal profiling of dynamic changes of functional cysteines along a time series (Fig. 5A). Inspired by the studies that cysteine redox states respond to the circadian rhythm (Bass & Takahashi, 2011; O'Neill & Feeney, 2014), we therefore apply DIA-ABPP in a third application to comprehensively profile the circadian cysteine sites in mouse liver proteomes.

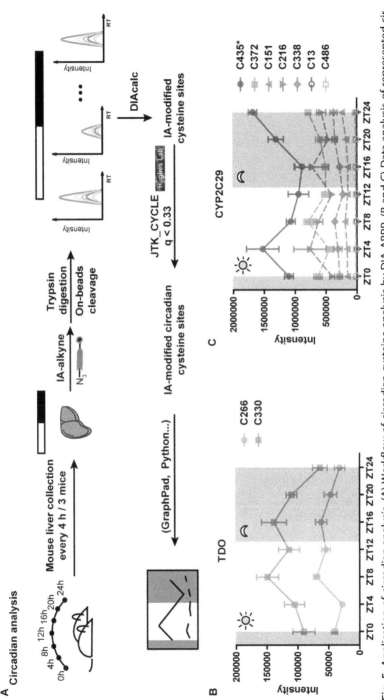

Fig. 5 Application of circadian analysis. (A) Workflow of circadian cysteine analysis by DIA-ABPP. (B and C) Data analysis of represented circadian cysteine. Circadian protein TDO (B) with two quantified rhythmic cysteines. Single circadian cysteine C435 found in CYP2C29 (C). In each figure, x-axis denotes the time points and y-axis denotes the peptide intensity. Light and dark cycles are shaded accordingly. Cysteines with rhythm (q < 0.33) are shown in solid lines and non-oscillating cysteines are shown in dotted lines. The protein name is shown on top and the cysteine sites are shown with colored symbols. Values represent means ± SEM from six replicate experiments.

After mice were kept in the constant darkness for 1 day, liver tissues from three mice were collected every 4 h to get 7 time points across 24 h. For each liver tissue sample at each time point, three aliquots were prepared with one sample for the DDA-based spectral library construction and the other two for the DIA-ABPP analysis. After filtering, we obtained 4214 cysteine sites mapping to 1934 proteins that were at least quantified in four time points. JTK_cycle, an efficient nonparametric algorithm for detecting cycling variables in large datasets (Hughes, Hogenesch, & Kornacker, 2010), was used to identify cysteines whose probe labeling are with a daily rhythmic fluctuation. Detailed analysis of these cysteines revealed that the rhythmic cysteine labeling could be attributed to either circadian protein expression, for example, protein tryptophan 2,3-dioxygenase (TDO) (Fig. 5B), or circadian cysteine reactivity like active site C435 of Cytochrome P450 2C29 (CYP2C29) (Fig. 5C). Overall, rhythmic cysteines were enriched in many functional important pathways such as the ROS pathway, ubiquitin-proteasome system, redox regulation, translation and metabolism. The data will provide a rich resource to study the mechanism of circadian clock for the community.

3. Protocol

3.1 Materials

3.1.1 Reagents and solvents

Dulbecco's modified Eagle's medium (Gibco, Life)

Penicillin-streptomycin (Thermo Fisher Scientific)

Fetal Bovine Serum (Premium, South America)

Triton X-100 (AMRESCO, 0694)

IA-alkyne (Made in house) (Weerapana et al., 2008)

4-hydroxy-2-nonenal (HNE) (Shyuanye, catalog number Y15980)

DADPS Biotin Alkyne (Click Chemistry Tools, catalog number 1330-25)

Tris-2-carboxyethyl phosphine (TCEP) (Sigma-Aldrich, catalog number C4706)

Tris[(1-benzyl-1H-1,2,3-triazol-4-yl)methyl]amine (TBTA) (Sigma-Aldrich, catalog number 678937)

Copper (II) sulfate (Sigma-Aldrich, catalog number C1297)

Methanol, Optima LC/MS grade (Fisher Scientific, catalog number A454SK-4)

Sodium dodecyl sulfate (SDS) (Sigma-Aldrich, catalog number L6026)

Streptavidin agarose beads (Thermo Scientific, catalog number 20353)

Urea (Sigma–Aldrich, catalog number U5378)

Dithiothreitol (DTT) (Shyuanye, catalog number S64054)

Iodoacetamide (Sigma–Aldrich, catalog number I1149)

Calcium chloride (Sigma–Aldrich, catalog number C1016)

Formic acid (Sigma–Aldrich, catalog number F0507)

Water, Optima LC/MS Grade (Fisher Scientific, catalog number W6-4)

Acetonitrile, Optima LC/MS Grade (Fisher Scientific, catalog number A955-4)

Dimethyl sulfoxide (DMSO) (Sigma–Aldrich, catalog number D5879)

t-Butanol (Sigma–Aldrich, catalog number 360538)

10 x phosphate buffer saline (PBS) buffer (Double-Helix)

Pierce BCA (Bicin-choninic Acid) protein assay kit (Thermo Fisher Scientific, catalog number 23225)

F2 (Santa Cruz Biotechnology, catalog number sc-345083)

F3 (Santa Cruz Biotechnology, catalog number sc-274729)

F8 (Santa Cruz Biotechnology, catalog number sc-273957)

F9 (Matrix scientific, catalog number 012688)

F10 (Santa Cruz Biotechnology, catalog number sc-345060)

F11 (Santa Cruz Biotechnology, catalog number sc-354895)

F12 (Enamine, catalog number R339568)

F13 (Santa Cruz Biotechnology, catalog number sc-274673)

F21 (Santa Cruz Biotechnology, catalog number sc-279681)

F27 (Santa Cruz Biotechnology, catalog number sc-342184)

F30 (Santa Cruz Biotechnology, catalog number sc-355362)

F32 (Santa Cruz Biotechnology, catalog number sc-354613)

F33 (Matrix scientific, catalog number 008532)

F52 (Santa Cruz Biotechnology, catalog number sc-279665)

F4, F5, F7, F14, F20, F23, F28, F31, F38, F56 (Made in house) (Yang et al., 2022)

3.1.2 Equipment and software

1.5 mL Eppendorf tubes (Axygen)

15 mL centrifuge tubes (Corning)

Screw-top microcentrifuge tubes (Sarstedt, catalog number 72.607)

Protein LoBind tubes (Eppendorf, catalog number 0030108116)

Fused silica capillary tubing (360 µm outer diameter, 75 µm inner diameter; Polymicro Technologies, catalog number TSP075375)

Fused silica capillary tubing (365 μm outer diameter, 100 μm inner diameter; Polymicro Technologies, catalog number TSP100375)

iRT Kit (Biognosys)

C18 reverse-phase resin (Phenomenex, catalog number 04A-4251)

Cell scraper (Biologix)

Glass dounce homogenizer (Kimble, 1 mL)

Ultrasonicator (Vibra-Cell™ ultrasonic processors)

Refrigerated centrifuge (Eppendorf Centrifuge 5810R)

Microcentrifuge (Eppendorf Centrifuge 5424, 5424R)

iMark microplate reader (Bio-Rad)

Thermomixer C (Eppendorf)

Heat block

Model P-2000 CO_2 laser puller (Sutter Instrument)

Q-Exactive plus Orbitrap mass spectrometer (Thermo Fisher Scientific)

Dionex Ultimate 3000 Nano LC System (Thermo Fisher Scientific)

Xcalibur (Thermo Fisher Scientific)

Spectronaut Pulsar X (Biognosys)

DIAcalc (Made in house, https://github.com/wangchulab/DIAcalc)

SciPy (https://github.com/scipy/scipy)

JTK_CYCLE (https://openwetware.org/wiki/HughesLab:JTK_Cycle)

AlphaFold2 (https://alphafold.ebi.ac.uk/)

Python 3.2 (https://www.python.org/)

RStudio (https://rstudio.com/products/rstudio/)

3.1.3 Animals, cells and enzymes

Male C57BL/6J mouse (Charles River Laboratories)

Ramos cells (American Type Culture Collection)

MDA-MB-231 cells (American Type Culture Collection)

Trypsin Gold (Promega, catalog number 161527)

3.1.4 Buffers and solutions

IA-alkyne: 20 mM solution in DMSO

DADPS Biotin Alkyne: 20 mM solution in DMSO

TBTA: 1.7 mM solution in t-butanol/DMSO (8:2 v/v)

TCEP: 50 mM solution in water

Copper sulfate: 50 mM solution in water

SDS: 1.2% and 0.2% solution in PBS buffer (w/v)

DTT: 200 mM solution in water

Iodoacetamide: 400 mM solution in water

Calcium chloride: 100 mM solution in water

Lysis buffer (0.1% Triton X-100/PBS)

Trypsin: 0.5 µg/µL solution in PBS buffer

iRT solution buffer: 20% acetonitrile in water

Eluting buffer A: 0.1% formic acid in water

Eluting buffer B: 80% acetonitrile, 0.1% formic acid in water

Loading buffer: 0.1% formic acid in water

3.2 Preparation of the proteomes

Proteomes derived from human cell lines as well as mouse tissue samples were applied for the method demonstration. Ramos cells and MDA-MB-231 cells were maintained in DMEM/High glucose media supplemented with 10% fetal bovine serum and 1% antibiotics (100 IU/mL penicillin and 100 µg/mL streptomycin) at 37 °C under 5% CO_2 atmosphere. For harvesting, cells were collected and washed twice using the PBS buffer. Pellets were then snap-frozen in liquid nitrogen and stored at −80 °C until lysis. C57BL/6J mice were housed in a temperature-controlled barrier facility with a 12-h light/dark cycle and were given free access to food and water. For circadian study, after 1 day in constant darkness, mice were sacrificed at defined intervals. Prior to liver excision, tissues were perfused with ice-cold PBS to remove blood contents and then quickly frozen in liquid nitrogen followed by storage at −80 °C until ready for the following process.

3.2.1 Preparation of human cell lysates (timing—2 h)

1. Thaw cell pellets in the Eppendorf tubes on ice
2. Resuspend cell pellet using a pre-chilled lysis buffer (0.1% Triton X-100/PBS) by gentle pipetting
3. Sonicate using a probe sonicator (3 × 10 pulses at 40% power) on ice and centrifuge at 20,000 g for 1 h at 4 °C
4. Transfer supernatants to clean 1.5 mL Eppendorf tubes
5. Measure protein concentration using the BCA protein assay kit and a microplate reader against a bovine serum albumin protein standard

3.2.2 Preparation of mouse tissue homogenates (timing—2 h)

1. Thaw liver tissue samples in Eppendorf tubes and rinse samples with 1 mL of ice-cold PBS Buffer to remove any remaining blood

2. Mince tissue samples with Iris scissors and transfer the samples into a pre-chilled glass dounce homogenizer using lysis buffer (0.1% Triton X-100/PBS).

Note: The ratio of lysis buffer to sample depends on the protein abundance of the tissue sample and should be optimized case by case. Less lysis buffer would cause insufficient lysis. Too much lysis buffer couldn't meet the required concentration for subsequent experiments (2 mg/mL).

3. Homogenize tissue samples to break the large tissue particles and transfer back to Eppendorf tubes

4. Sonicate using a probe sonicator (2 × 5 pulses at 40% power) on ice and centrifugate at 20,000 g for 1 h at 4 °C

5. Pass through the lipid layer and transfer supernatants to clean 1.5 mL Eppendorf tubes

6. Measure protein concentration using the BCA protein assay kit and a microplate reader against a bovine serum albumin standard

3.3 Protein labeling and cycloaddition (timing—4 h)

The DIA-ABPP workflow represents an efficient label-free and site-specific chemoproteomic strategy for in-depth and large-scale quantitative profiling of reactive residues in proteomes. Reactive cysteine profiling was used as a proof-of-concept and the method was applied to three biological systems, which were consider from different perspectives, including concentration, type and time. First, we applied DIA-ABPP to systematically analyze the sensitivity of cysteinome towards HNE and fitted the dose-response curves to obtain the EC_{50} value for each of the labeled cysteines. Second, a covalent fragment library was screened by DIA-ABPP method to comprehensively profile new ligandable cysteines and identify new druggable proteins with diverse functional annotation. Third, we applied DIA-ABPP to establish the global quantification of the circadian cysteinome in mammals to analyze the functional proteome changes at multiple time points.

3.3.1 Competitive protein labeling and cycloaddition for dose-dependent profiling of HNE-reactive cysteinome

1. Dilute the MDA-MB-231 cell lysates to 2 mg/mL solution using the lysis buffer and add 1 mL of the diluted proteome solution to each new Eppendorf tube for each condition

2. Incubate each sample with 0, 10, 50, 100, or 500 μM HNE for 1 h at 29 °C, respectively. Prepare another DMSO-treated sample (with 0 μM of HNE) for the spectral library construction

Note: Three biological replicates should be prepared for target analysis.

3. Add 100 μM IA-alkyne probe by the addition of 5 μL of 20 mM concentrated DMSO stock to each tube

4. Vortex the tubes briefly and incubate for 1 h at 29 °C in the dark.

5. For each sample, premix 60 μL of 1.7 mM TBTA solution, 20 μL of 50 mM copper sulfate and 20 μL of freshly prepared 50 mM TCEP in water and Vortex the tubes immediately

6. To each sample, add 5.5 μL of 20 mM DADPS (110 μM final concentration) and 100 μL premixed solution (final concentration: 1 mM TCEP, 100 μM TBTA and 1 mM copper sulfate). Vortex the tubes and incubate click chemistry reaction for 1 h at 29 °C in the dark

   Note: The use of alternative cleavable tags (i.e., photo-cleavable tag, TEV protease-cleavable tag) is not expected to impact the identification of labeled peptides. According to our experience, the acid-cleavable tag DADPS is the most efficient tag for IA-alkyne labeled peptide identification.

7. After click chemistry reaction, centrifuge for 5 min at 10,000 × g at 4 °C to pellet the proteins and remove the supernatant

   Note: use a suitable precipitation method to precipitate the proteins if there is no protein pellet after centrifugation.

8. Add 500 μL of pre-cooling methanol to the pellets and resuspend the proteins by gentle sonication (3 × 5 pulses at 30% power)

9. Centrifuge for 5 min at 10,000 × g at 4 °C to pellet the proteins and remove the supernatant

10. Repeat the washing Step 8–9 for a total of three methanol washes

3.3.2 Competitive protein labeling and cycloaddition for fragment-based ligand screening for functional cysteinome

1. Dilute the Ramos cell lysates to 2 mg/mL solution using lysis buffer and add 1 mL of diluted proteome solution to each new Eppendorf tubes for each fragment

2. Incubate the proteomes with 500 μM fragment for 1 h at 29 °C, respectively. Prepare at least three DMSO-treated samples for the spectral library construction

   Note: At least two biological replicates should be prepared for ligandable cysteine analysis.

3. Add 100 μM IA-alkyne probe by the addition of 5 μL of 20 mM concentrated DMSO stock to each tube.

4. Vortex the tubes briefly and incubate for 1 h at 29 °C in the dark
5. For each sample, premix 60 μL of 1.7 mM TBTA solution, 20 μL of 50 mM copper sulfate and 20 μL of freshly prepared 50 mM TCEP in water and Vortex the tubes immediately
6. To each sample, add 5.5 μL of 20 mM DADPS (110 μM final concentration) and 100 uL premixed solution (final concentration: 1 mM TCEP, 100 μM TBTA and 1 mM copper sulfate). Vortex the tubes and incubate click chemistry reaction for 1 h at 29 °C in the dark
7. After click chemistry reaction, centrifuge for 5 min at 10,000 g at 4 °C to pellet the proteins and remove the supernatant
   Note: use suitable precipitation method to precipitate the proteins if there is no protein pellet after centrifugation.
8. Add 500 μL of pre-cooling methanol to the pellets and resuspend the proteins by gently sonication (3 × 5 pulses at 30% power).
9. Centrifuge for 5 min at 10,000 × g at 4 °C to pellet the proteins and remove the supernatant
10. Repeat the washing Step 8–9 for a total of three methanol washes

3.3.3 Protein labeling and cycloaddition for profiling of circadian cysteinome fluctuation

1. Dilute each mouse liver lysates to 2 mg/mL solution using lysis buffer and add 1 mL of diluted proteome solution to each new Eppendorf tubes. For each liver tissue, prepare two replicates for DIA analysis and an additional sample for the spectral library construction
2. Add 100 μM IA-alkyne probe by addition of 5 μL of 20 mM concentrated DMSO stock to each tube
3. Vortex the tubes briefly and incubate for 1 h at 29 °C in the dark.
4. For each sample, premix 60 μL of 1.7 mM TBTA solution, 20 μL of 50 mM copper sulfate and 20 μL of freshly prepared 50 mM TCEP in water and Vortex the tubes immediately
5. To each sample, add 5.5 μL of 20 mM DADPS (110 μM final concentration) and 100 uL premixed solution (final concentration: 1 mM TCEP, 100 μM TBTA and 1 mM copper sulfate). Vortex the tubes and incubate click chemistry reaction for 1 h at 29 °C in the dark
6. After click chemistry reaction, centrifuge for 5 min at 10,000 × g at 4 °C to pellet the proteins and remove the supernatant
7. Add 500 μL of pre-cooling methanol to the pellets and resuspend the proteins by gently sonication (3 × 5 pulses at 30% power).

8. Centrifuge for 5 min at $10,000 \times g$ at $4\,^{\circ}\text{C}$ to pellet the proteins and remove the supernatant
9. Repeat the washing Step 7–8 for a total of three methanol washes

3.4 Affinity purification and tandem cleavage (timing—1.5 days)

1. For each sample in 3.3, add 1 mL of 1.2% SDS in PBS buffer to the pellets and resuspend the protein by sonication (3×10 pulses at 40% power)
2. Incubate the solution at $90\,^{\circ}\text{C}$ for 5 min to fully dissolve the proteins and cool samples on ice for 2 min
3. Centrifuge at $19,000 \times g$ for 5 min to pellet the excessive copper
4. Transfer the supernatant to clean 15 mL centrifuge tubes and add 5 mL PBS buffer
5. Use 100 µL of streptavidin agarose slurry for each sample. Add beads to 15 mL clean centrifuge tubes and wash the beads with 5 mL of PBS. Pellet the beads by centrifugation at $1400 \times g$ for 3 min and remove supernatant

   Note: 5 mL PBS could be used to wash as much as 1 mL beads slurry.
6. Repeat the washing Step 5 for a total of three PBS washes
7. Add the washed beads to the labeled proteome sample (from Step 4) and incubate for 4 h at $29\,^{\circ}\text{C}$

   Note: Beads should be resuspended using an equal amount of PBS by gentle pipetting and evenly divided into each sample. Repeat several times until all beads are moved from 15 mL centrifuge tube and the total amount of the PBS should be less than 300 µL.
8. Pellet the beads by centrifugation at $1700 \times g$ for 10 min and remove the supernatant from beads
9. Wash the beads with 5 mL of PBS buffer three times and 5 mL of water three times three times by centrifugation at $1400 \times g$ for 3 min
10. Transfer the beads to screw-top Eppendorf tubes using 1 mL water and centrifuge at $2000 \times g$ for 10 min to pellet the beads completely

    Note: If beads are not settled, centrifuge for a longer time to avoid the sample loss.
11. Pipette off the supernatant and add 500 µL of 6 M urea in PBS to the beads
12. Add 25 µL of freshly prepared 200 mM dithiothreitol and incubate the sample for 30 min at $37\,^{\circ}\text{C}$ using thermomixer at 1000 rpm

Note: The beads should be agitated for complete reduction of the disulfide bonds.

13. Add 25 μL of freshly prepared 400 mM iodoacetamide and incubate samples using thermomixer at 35 °C for 30 min in the dark to block the reduced cysteines

14. Add 950 μL of PBS carefully and pellet the samples by centrifugation at $1400 \times g$ for 3 min

15. Remove the supernatant and resuspend the beads using 200 μL of 2 M Urea in PBS

16. Add 4 μL of 0.5 μg/μL of trypsin in PBS and 2 μL of 100 mM calcium chloride in water

    Note: trypsin buffer containing acetic acid should be avoided in case of the DADPS cleavage.

17. Incubate the samples at 37 °C for 16–17 h using thermomixer with agitation

18. Centrifuge the beads at $1400 \times g$ for 3 min and pipette off the supernatant to remove the unlabeled tryptic peptides

19. Wash the beads with 1 mL of PBS buffer three times and 1 mL of water three times by centrifugation at $1400 \times g$ for 3 min. Remove the supernatant

20. Centrifuge at $1400 \times g$ for 1 min and pipet off the supernatant thoroughly

    Note: Pipet carefully to avoid the sample loss.

21. Release the samples from beads using 200 μL of 2% formic acid in water at 29 °C for 1 h with rotation. Centrifuge the samples at $1400 \times g$ for 3 min and transfer the eluents to clean protein Lo-bind tubes

22. Repeat the cleavage step 20 and combine the eluents with the previous fractions

23. Wash the beads twice using 200 μL of 50% acetonitrile in water containing 1% formic acid. Centrifuge the samples at $1400 \times g$ for 3 min and collect all the supernatants to the tubes

24. Concentrate the cleavage fraction using a vacuum centrifuge and store at −20 °C until mass spectrometry analysis

3.5 Liquid chromatography-tandem mass spectrometry (LC-MS/MS) analysis

Samples were analyzed on a Q-Exactive plus Orbitrap mass spectrometer coupled with Ultimate 3000 LC system using library-based DIA approach, which has been widely used in DIA analysis. At least three DDA samples should be applied to spectral library construction. "Library-free" methods like directDIA could be used as alternatives to enable analysis directly using DIA data, but it might result in lower identification ability. Libraries with sample pre-fractionations could loss of the labeled peptides with lower intensity, resulting in inefficient peptide identification. Notably, column oven for chromatography and newer high-performance mass spectrometer will likely improve the performance of peptide identification and following quantification.

3.5.1 Packing column (timing—1 day)

1. Cut 60 cm of 75 μm fused-silica capillary tubing. Use flame to burn off 2 cm of polyimide coating in the center of the capillary tubing and wipe off polyimide coating using an ethanol-soaked Kimwipe
2. Use a laser tip puller to pull capillary tubing at the exposed silica to make two smooth tips
3. Pack the tips in Step 2 with 28 cm of 3 μ C18 reverse-phase resin using a pressure injection cell to construct the analytical column
4. Pack the desalting column with 4 cm of 5 μ C18 reverse-phase resin using 100 μm fused-silica capillary tubing
5. Equilibrate the analytical column and the desalting column using an LC system with oscillating gradient

Note: All samples are recommended to be subjected to the mass spectrometry analysis at one continuous round to ensure the best data reproducibility. Otherwise, all samples from one project are highly recommended to be analyzed using the same set of trap column and analytical column.

3.5.2 Spectral library construction (timing—16 h)

1. Combine the analytical column with the desalting column and attach to the LC-MS/MS system
2. Set up the gradient indicated in Table 1 with 0.3 μL/min for sample eluting. Mobile phase A was 0.1% formic acid in water, and mobile phase B

Table 1 The solvent gradient for both DDA and DIA-based LC-MS/MS system.

Time (min)	Phase A (%)	Phase B (%)
0	93	7
40	80	20
120	70	30
140	20	80

was 0.1% formic acid 80% acetonitrile in water. Sample was loaded at 3 μL/min using autosampler

3. For mass spectrometer, full-scan mass spectra were acquired over the m/z range from 350 to 1800 using the Orbitrap mass analyzer with a mass resolution of 70,000 under the positive-ion mode. MS/MS fragmentation is performed in a data-dependent mode, of which 20 most intense ions are selected for MS/MS analysis a resolution of 17,500 using collision mode of HCD. Other important mass parameters: isolation window, 2.0 m/z units; default charge, 2 +; normalized collision energy, 28%; maximum IT, 50 ms; dynamic exclusion, 20.0 s

4. Resuspend the sample in 17 μL of 0.1% formic acid in water supplemented with iRT peptides (45/1 v/v) and centrifuge at 20,000 × g for 1 h at 4 °C

5. Load 13 μL sample solution for LC-MS/MS measurement using the chromatographic method and mass method described in Step 3–4

    Note: equal volume of peptide samples should be loaded for further DIA analysis.

6. Open Spectronaut and insert the FASTA file of protein sequence database from uniport (e.g., human and mouse proteomes) and the cysteine modification (280.18993 Da for IA-alkyne conjugated acid-cleavable tag as "IA_acid").

7. Open the Library module and click "enerate Library from Pulsar/Search Archives" to insert the DDA raw data from Step 6

    Note: Multiple search engines and formats are supported by Spectronaut for library generation (e.g., MaxQuant, Proteome Discoverer, ProteinPilot, Mascot, or mzIdentML). Spectral library could be obtained using the corresponding output files.

8. Assign the fasta file in the database and set the variable modifications of applied modification in BGS factory settings as "57.0215 Da (carbamidomethylation) and 280.18993 Da (IA_acid)". Other important parameters: PTM localization in BGS factory settings was activated and site confidence score cutoff set to 0.75. Protein Q value cut off was set to 1. Other parameters were used default setting

    Note: Project-specific spectral library is highly recommended for DIA data analysis and publicly available spectral libraries would result in poorer DIA quantification. A spectral library acquired before in the search archive could be added to further enrich the depth of the library. Gene annotation could be assigned by loading the mgi file in the databases perspective to gain the biological insight.

3.5.3 DIA sample acquisition and data analysis (timing—several days)

1. Export the spectral library and filter the precursors containing IA-alkyne modifications
2. Divide the precursor equally into 70 windows and calculate the mean and width of each window
3. For the DIA method, insert the mean values into the inclusion list and set the isolation window as the width of window +1 Th to assure the optimal quantification of precursors with m/z values at the edges of isolation windows. For full MS analysis, resolution was set to 70,000 and full MS AGC target was set as 3E6 with an IT of 20 ms. For each DIA window, resolution was set to 17,500. AGC target value for fragment spectra was set at 1E6 with an auto IT. Normalized CE was set at 27%. Default charge was 3 and the fixed first mass was set to 200 Th. Apply the same chromatographic method as in the spectral library construction
4. Resuspend the samples in 17 μL of 0.1% formic acid in water supplemented with iRT peptides (45/1 v/v) and centrifuge at 20,000 g for 1 h at 4 °C
5. Load each sample using 13 μL supernatant for the DIA analysis with the chromatographic method and mass method described in Step 3
6. Open Spectronaut and Go the "Analysis" module
7. Click "Setup a DIA analysis from File" and insert the DIA raw data from Step 3
8. Choose the spectral library generated from Step 7 in 3.4.2. Set the minor grouping of quantification in the BGS factory setting as "by modified sequence". Other important parameters: Protein Q value cutoff was set to 1. Differential abundance grouping was set as Minor Group (Quantification setting). Other parameters were set as default

    Note: Gene annotation could be applied to gain further biological insights. Replicate and condition could be set to perform the statistical tests during post analysis.
9. Go to the "Post analysis" module and open the "Peptide Quant" in the "Run Pivot Report". Activate the "EG.PrecusorID" and "PG.ProteinAssesions" part and export the data as the "xls" format
10. Rename the column header of each sample by separation of the condition name and replicate name using "_" and save as a new txt file
11. Load the file to "DIAcalc" and choose the fasta file used in the DDA and DIA analysis

12. Input the "IA_acid" as the target modification and group the modified peptides to get the normalized peptide intensity of each sample

   Note: The peptide intensities among samples were retrieved from the pair-wise intensity ratio by least-squares analysis and normalized to the same intensity level as that of the total original precursors. Precursors from the same peptide with identical modification structures and modification sites would be integrated. The redundancy of charge and the miss cleavage would be also grouped.

3.5.3.1 Data analysis of the EC_{50} value of the cysteines toward HNE

1. Open the DIAcalc output file and rename the column header of each replicate by only using the value of the concentration and save as a new txt file
2. Use the peptide intensity values of all three biological replicates as a function of HNE concentrations and fit to a sigmoidal curve using the curve_fit function from SciPy
3. Filter the fitting curves with $R^2 \geq 0.8$
4. The EC_{50} value for each cysteine was obtained according to the equation $f(EC_{50}) = 50\%$
5. Set the EC_{50} value which is larger than $100\,\mu M$ as "$EC_{50} > 100\,\mu M$"
6. Export the cysteines with the corresponding EC_{50} value
7. Apply the HNE concentration and peptide intensity to a statistic software (e.g., GraphPad Prism) and generate the dose-response curves

3.5.3.2 Data analysis of the ligandable cysteines

1. Open the DIAcalc output file and calculate the ratio of each cysteines toward the fragments (peptide intensity of the DMSO-treated sample over peptide intensity of the fragment-treated sample)
2. For each compound, set the lowest ratio as the final ratio when at least one R value of cysteine is less than 4 and set the mean ratio as the final ratio when all R values of cysteine are larger than 4
3. Filter the cysteines which have quantification ratios in 3 fragments
4. Categorize cysteines as liganded cysteines if they had at least two ratios of $R \geq 4$ (hit fragments) and one ratio between 0.5 and 2 (control fragments).
5. Analyze the localization of the liganded cysteines using the experimental protein structure or the predicted protein structure from AlphaFold2

3.5.3.3 Data analysis of the circadian cysteines

1. Open the DIAcalc output file and filter out the peptide without intensities in less than 4 intervals
2. Rename the column header of each sample by separation of the time point and replicate name using "_" and save as a new csv file
3. Perform the JTK_CYCLE algorithm with a fixed period range of 24 h, a fixed interval and replicate number with amplitude and phase as free parameters
4. Open the output txt files and filter the cysteines as circadian cysteine with the q value less than 0.33
5. Apply the time point and peptide intensity to the statistic software (e.g., GraphPad Prism) and generate the circadian curves

4. Summary and discussion

As a representative chemical proteomic method, ABPP was used to characterize the intrinsic reactivity of amino acid side chains and elucidate the function of bioactive small molecules in native biological systems. By converting the standard DDA mode into the DIA mode with comprehensive optimization of the workflow, DIA-ABPP could solve the problem of stochastic data loss and save from the trouble of stable isotope labeling in current quantitative chemical proteomic methods, with high coverage, reproducibility and accuracy.

Here we have demonstrated the DIA-ABPP method to characterize reactive cysteine sites by three specific applications, including dose-dependent, structure-dependent and time-dependent variations, respectively. In this chapter, we mainly showcased DIA-ABPP in cysteines profiling, owing to the well-designed reactive probe and essential roles of this special amino acid. Notably, as ABPP is a modularized platform which is applicable with different kinds of ABPs and the diverse biological systems, DIA-ABPP also has high compatibility and can be customized for specific applications. For example, other residue-specific probes could be further developed and directly implemented with DIA-ABPP to explore novel function roles of these amino acids and expand the scope of druggable proteomes for drug discovery (Gehringer & Laufer, 2019; Hacker et al., 2017; Hahm et al., 2020; Lin et al., 2017; Long & Aye, 2017; Singh, Petter, Baillie, & Whitty, 2011; Tower, Hetcher, Myers, Kuehl, & Taylor, 2020). Additionally, some chemical probe for disease-associated

PTMs, like phenylglyoxal for protein citrullination (Thygesen, Boll, Finsen, Modzel, & Larsen, 2018; Tilvawala et al., 2018), could be combined with DIA-ABPP to investigate the level of related PTMs in large-scale patient samples and develop diagnostic biomarkers.

With recent advances in ABP design, quantification method incorporation and mass spectrometry instrumentation, ABPP has been successfully applied for systematically characterizing protein function and post-translational modifications. As an efficient method for large-scale chemoproteomic sample quantification, DIA-ABPP mainly focuses on optimizing the data acquisition of mass spectrometer. However, sample preparation, which is another key component of ABPP sample analysis, has lagged far behind. Current ABPP workflow mainly relies on 3-day preparation with manual operation. Therefore, multistep sample transfer and operation can easily introduce sample loss and variation, especially with the growing needs for analyzing a large number of biological samples with limited starting materials. Integration and automation of the ABPP sample preparation are highly demanded for significantly reducing variation of manual operation and sample loss in large-scale sample processing. In the future, ABPP method could be integrated with high-throughput analysis platforms like chips or autosamplers to provide an automated analysis workflow for chemical proteomic applications with high demand for throughput, reproducibility and sensitivity.

Acknowledgments

This work is supported by the National Natural Science Foundation of China (No. 21925701, No. 91953109 and No. 92153301) to C.W.

References

Abo, M., Li, C., & Weerapana, E. (2018). Isotopically-labeled iodoacetamide-alkyne probes for quantitative cysteine-reactivity profiling. *Molecular Pharmaceutics*, *15*, 743–749.

Backus, K. M., Correia, B. E., Lum, K. M., Forli, S., Horning, B. D., Gonzalez-Paez, G. E., et al. (2016). Proteome-wide covalent ligand discovery in native biological systems. *Nature*, *534*, 570–574.

Bass, J., & Takahashi, J. S. (2011). Circadian rhythms: Redox redux. *Nature*, *469*, 476–478.

Bekker-Jensen, D. B., Bernhardt, O. M., Hogrebe, A., Martinez-Val, A., Verbeke, L., Gandhi, T., et al. (2020). Rapid and site-specific deep phosphoproteome profiling by data-independent acquisition without the need for spectral libraries. *Nature Communications*, *11*, 787.

Bruderer, R., Bernhardt, O. M., Gandhi, T., Miladinović, S. M., Cheng, L. Y., Messner, S., et al. (2015). Extending the limits of quantitative proteome profiling with data-independent acquisition and application to acetaminophen-treated three-dimensional liver microtissues. *Molecular & Cellular Proteomics*, *14*, 1400–1410.

Chen, Y., Liu, Y., Lan, T., Qin, W., Zhu, Y., Qin, K., et al. (2018). Quantitative profiling of protein carbonylations in ferroptosis by an aniline-derived probe. *Journal of the American Chemical Society*, *140*, 4712–4720.

Counihan, J. L., Ford, B., & Nomura, D. K. (2016). Mapping proteome-wide interactions of reactive chemicals using chemoproteomic platforms. *Current Opinion in Chemical Biology*, *30*, 68–76.

Cravatt, B. F., Wright, A. T., & Kozarich, J. W. (2008). Activity-based protein profiling: From enzyme chemistry to proteomic chemistry. *Annual Review of Biochemistry*, *77*, 383–414.

Gehringer, M., & Laufer, S. A. (2019). Emerging and re-emerging warheads for targeted covalent inhibitors: Applications in medicinal chemistry and chemical biology. *Journal of Medicinal Chemistry*, *62*, 5673–5724.

Hacker, S. M., Backus, K. M., Lazear, M. R., Forli, S., Correia, B. E., & Cravatt, B. F. (2017). Global profiling of lysine reactivity and ligandability in the human proteome. *Nature Chemistry*, *9*, 1181–1190.

Hahm, H. S., Toroitich, E. K., Borne, A. L., Brulet, J. W., Libby, A. H., Yuan, K., et al. (2020). Global targeting of functional tyrosines using sulfur-triazole exchange chemistry. *Nature Chemical Biology*, *16*, 150–159.

Hansen, F. M., Tanzer, M. C., Brüning, F., Bludau, I., Stafford, C., Schulman, B. A., et al. (2021). Data-independent acquisition method for ubiquitinome analysis reveals regulation of circadian biology. *Nature Communications*, *12*, 254.

Hughes, M. E., Hogenesch, J. B., & Kornacker, K. (2010). JTK_CYCLE: An efficient nonparametric algorithm for detecting rhythmic components in genome-scale data sets. *Journal of Biological Rhythms*, *25*, 372–380.

Jumper, J., Evans, R., Pritzel, A., Green, T., Figurnov, M., Ronneberger, O., et al. (2021). Highly accurate protein structure prediction with AlphaFold. *Nature*, *596*, 583–589.

Kelstrup, C. D., Bekker-Jensen, D. B., Arrey, T. N., Hogrebe, A., Harder, A., & Olsen, J. V. (2018). Performance evaluation of the Q Exactive HF-X for shotgun proteomics. *Journal of Proteome Research*, *17*, 727–738.

Law, V., Knox, C., Djoumbou, Y., Jewison, T., Guo, A. C., Liu, Y., et al. (2014). DrugBank 4.0: Shedding new light on drug metabolism. *Nucleic Acids Research*, *42*, D1091–D1097.

Lin, S., Yang, X., Jia, S., Weeks, A. M., Hornsby, M., Lee, P. S., et al. (2017). Redox-based reagents for chemoselective methionine bioconjugation. *Science*, *355*, 597–602.

Long, M. J. C., & Aye, Y. (2017). Privileged electrophile sensors: A resource for covalent drug development. *Cell Chemical Biology*, *24*, 787–800.

O'Neill, J. S., & Feeney, K. A. (2014). Circadian redox and metabolic oscillations in mammalian systems. *Antioxidants & Redox Signaling*, *20*, 2966–2981.

Pace, N. J., & Weerapana, E. (2013). Diverse functional roles of reactive cysteines. *ACS Chemical Biology*, *8*, 283–296.

Poli, G., Schaur, R. J., Siems, W. G., & Leonarduzzi, G. (2008). 4-Hydroxynonenal: A membrane lipid oxidation product of medicinal interest. *Medicinal Research Reviews*, *28*, 569–631.

Qin, W., Yang, F., & Wang, C. (2020). Chemoproteomic profiling of protein-metabolite interactions. *Current Opinion in Chemical Biology*, *54*, 28–36.

Qin, K., Zhu, Y., Qin, W., Gao, J., Shao, X., Wang, Y. L., et al. (2018). Quantitative profiling of protein O-GlcNAcylation sites by an isotope-tagged cleavable linker. *ACS Chemical Biology*, *13*, 1983–1989.

Roberts, A. M., Ward, C. C., & Nomura, D. K. (2017). Activity-based protein profiling for mapping and pharmacologically interrogating proteome-wide ligandable hotspots. *Current Opinion in Biotechnology*, *43*, 25–33.

Shannon, D. A., & Weerapana, E. (2015). Covalent protein modification: The current landscape of residue-specific electrophiles. *Current Opinion in Chemical Biology*, *24*, 18–26.

Singh, J., Petter, R. C., Baillie, T. A., & Whitty, A. (2011). The resurgence of covalent drugs. *Nature Reviews. Drug Discovery, 10*, 307–317.

Speers, A. E., Adam, G. C., & Cravatt, B. F. (2003). Activity-based protein profiling in vivo using a copper(i)-catalyzed azide-alkyne [3 + 2] cycloaddition. *Journal of the American Chemical Society, 125*, 4686–4687.

Szychowski, J., Mahdavi, A., Hodas, J. J., Bagert, J. D., Ngo, J. T., Landgraf, P., et al. (2010). Cleavable biotin probes for labeling of biomolecules via azide-alkyne cycloaddition. *Journal of the American Chemical Society, 132*, 18351–18360.

Thygesen, C., Boll, I., Finsen, B., Modzel, M., & Larsen, M. R. (2018). Characterizing disease-associated changes in post-translational modifications by mass spectrometry. *Expert Review of Proteomics, 15*, 245–258.

Tian, C., Sun, R., Liu, K., Fu, L., Liu, X., Zhou, W., et al. (2017). Multiplexed thiol reactivity profiling for target discovery of electrophilic natural products. *Cell Chemical Biology, 24*, 1416–1427.e1415.

Tilvawala, R., Nguyen, S. H., Maurais, A. J., Nemmara, V. V., Nagar, M., Salinger, A. J., et al. (2018). The rheumatoid arthritis-associated citrullinome. *Cell Chemical Biology, 25*, 691–704.e696.

Tower, S. J., Hetcher, W. J., Myers, T. E., Kuehl, N. J., & Taylor, M. T. (2020). Selective modification of tryptophan residues in peptides and proteins using a biomimetic electron transfer process. *Journal of the American Chemical Society, 142*, 9112–9118.

Venable, J. D., Dong, M. Q., Wohlschlegel, J., Dillin, A., & Yates, J. R. (2004). Automated approach for quantitative analysis of complex peptide mixtures from tandem mass spectra. *Nature Methods, 1*, 39–45.

Vinogradova, E. V., Zhang, X., Remillard, D., Lazar, D. C., Suciu, R. M., Wang, Y., et al. (2020). An activity-guided map of electrophile-cysteine interactions in primary human T cells. *Cell, 182*, 1009–1026.e1029.

Wang, C., & Chen, N. (2015). Activity-based protein profiling. *Acta Chimica Sinica, 73*, 657.

Wang, C., Weerapana, E., Blewett, M. M., & Cravatt, B. F. (2014). A chemoproteomic platform to quantitatively map targets of lipid-derived electrophiles. *Nature Methods, 11*, 79–85.

Weerapana, E., Simon, G. M., & Cravatt, B. F. (2008). Disparate proteome reactivity profiles of carbon electrophiles. *Nature Chemical Biology, 4*, 405–407.

Weerapana, E., Wang, C., Simon, G. M., Richter, F., Khare, S., Dillon, M. B., et al. (2010). Quantitative reactivity profiling predicts functional cysteines in proteomes. *Nature, 468*, 790–795.

Yang, F., Gao, J., Che, J., Jia, G., & Wang, C. (2018). A dimethyl-labeling-based strategy for site-specifically quantitative chemical proteomics. *Analytical Chemistry, 90*, 9576–9582.

Yang, Y. Y., Grammel, M., Raghavan, A. S., Charron, G., & Hang, H. C. (2010). Comparative analysis of cleavable azobenzene-based affinity tags for bioorthogonal chemical proteomics. *Chemistry & Biology, 17*, 1212–1222.

Yang, F., Jia, G., Guo, J., Liu, Y., & Wang, C. (2022). Quantitative chemoproteomic profiling with data-independent acquisition-based mass spectrometry. *Journal of the American Chemical Society, 144*, 901–911.

Yang, F., & Wang, C. (2020). Profiling of post-translational modifications by chemical and computational proteomics. *Chemical Communications (Cambridge, England), 56*, 13506–13519.

Ye, Z., Mao, Y., Clausen, H., & Vakhrushev, S. Y. (2019). Glyco-DIA: A method for quantitative O-glycoproteomics with in silico-boosted glycopeptide libraries. *Nature Methods, 16*, 902–910.

CHAPTER TWELVE

Generation of nucleotide-linked resins for identification of novel binding proteins

Shikha S. Chauhan[a] and Emily E. Weinert[a,b,]*
[a]Department of Biochemistry and Molecular Biology, Pennsylvania State University, University Park, PA, United States
[b]Department of Chemistry, Pennsylvania State University, University Park, PA, United States
*Corresponding author: e-mail address: emily.weinert@psu.edu

Contents

Abstract

Organisms use numerous nucleotide-containing compounds as intracellular signals to control behavior. Identifying the biomolecules responsible to sensing and responding to changes in signaling molecule concentration is an important area of research. However, identifying the binding proteins can be challenging when there is no prior information available about binding motifs. In this chapter, we describe a straightforward method to generate nucleotide-linked resins for use in pull-down experiments to identify binding proteins. The protocol outlined in this chapter also can be adapted to generate custom resins linked to other molecules of interest.

Methods in Enzymology, Volume 679
ISSN 0076-6879
https://doi.org/10.1016/bs.mie.2022.08.052

323

1. Introduction

Organisms throughout the domains of life utilize nucleotide-containing signaling molecules to modulate intracellular pathways and behaviors (Botsford & Harman, 1992; Denninger & Marletta, 1999; Kalia et al., 2013; Pesavento & Hengge, 2009; Romling, Galperin, & Gomelsky, 2013; Sutherland, 1972). For example, 3′,5′-cyclic adenosine monophosphate (3′,5′-cAMP) has been demonstrated to control steroidogenesis in mammals and carbon utilization in prokaryotes, while 3′,5′-cyclic guanosine monophosphate (3′,5′-cGMP) regulates visual transduction and vasodilation in mammals (Botsford & Harman, 1992; Lucas et al., 2000; Sutherland, 1972). A plethora of additional nucleotide-based signaling molecules have been discovered in prokaryotes, including guanosine tetra/pentaphosphate ((p)ppGpp; regulates stringent response), 3′,5′-cyclic dimeric guanosine monophosphate (c-di-GMP; controls biofilm formation, and 3′,5′-cyclic dimeric adenosine monophosphate (c-di-AMP; regulates sporulation) (Fahmi, Port, & Cho, 2017; Kalia et al., 2013; Pesavento & Hengge, 2009; Romling et al., 2013). The recent discoveries of additional nucleotide-based signaling molecules, including 2′,3′-cyclic adenosine-guanosine monophosphate (Zaver & Woodward, 2020), diadenosine tetraphosphate (Ap4A) (Ferguson, McLennan, Urbaniak, Jones, & Copeland, 2020), and 2′,3′-cyclic nucleotide monophosphates (2′,3′-cNMPs) (Duggal et al., 2022; Fontaine et al., 2018), have highlighted the possibility of numerous other undiscovered cellular signals.

Each time a novel signaling molecule is discovered, it is then necessary to identify binding proteins and downstream pathways, which can be quite challenging if there is no prior information on binding pocket requirements. Generation of deletion/overexpression strains of putative binders can provide insight, but phenotypic changes can be due to other secondary effects within the signaling pathway. Differential radial chromatography (DRaCALA) (Roelofs, Wang, Sintim, & Lee, 2011), which directly probes migration of the signaling molecule following incubation with overexpression strain lysates, is very powerful for identifying interactions, but typically relies on radiolabeled small molecules, which are not always readily available. In contrast, using signaling molecule-linked resins to isolate binding proteins from cell lysate is generally quite rapid. However, many of the newly identified signaling molecules are not available as commercial resins, necessitating the generation of custom resins in-house.

The recent finding that 2',3'-cNMPs alter *Escherichia coli* motility and biofilm formation (Pesavento & Hengge, 2009), as well as modulating transcription in both *E. coli* and *Salmonella enterica* Typhimurium (Duggal et al., 2022), suggested that cellular proteins may serve as 2',3'-cNMP sensors. However, 2',3'-cNMP-binding proteins had not been previously identified, which precluded using bioinformatic approaches. The dearth of knowledge about how 2',3'-cNMPs were bound within sensing proteins posed a challenge; linking 2',3'-cNMPs to a resin could potentially interfere with binding depending on the site of attachment. Coupling 2',3'-cNMPs to an epoxide resin results in potential reaction at multiple attachment sites, as well as a facile coupling protocol (Chauhan, Marotta, Karls, & Weinert, 2022). In the case of 2',3'-cNMPs, the most likely sites of attachment are the nucleophilic nitrogen atoms of the nucleobases and the free 5'-hydroxyl group of the ribose (Fig. 1; highlighted in red). Following the generation of the resin, binding proteins (or other biomolecules) can be identified by passing cellular lysate over the resin, washing, and then eluting with either salt or the molecule of interest (Fig. 2). The elution fractions can then be analyzed by mass spectrometry-based proteomics techniques (readily available as core facilities at many universities) or by Western blot to identify the binding proteins.

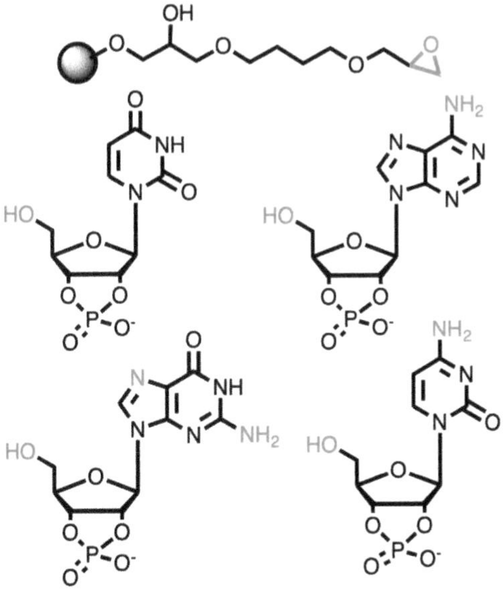

Fig. 1 Structures of the epoxide-linked resin and the four 2',3'-cNMPs. Potential sites of reaction on the 2',3'-cNMPs are highlighted in red.

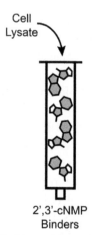

2',3'-cNMP
Binders

Fig. 2 Binding proteins are identified by passing cell lysate (or fractions thereof) over the nucleotide-linked resin, and then collecting the elution.

2. Materials required

1. Epoxy-activated Sepharose™ 6B purchased from Sigma-Aldrich (or equivalent) 0.1 g freeze-dried powder gives about 3.5 mL final volume of medium
2. All four 2',3'-cyclic nucleotide monophosphates (2',3'-cNMPs): 2',3'-cAMP, 2',3'-cCMP, 2',3'-cGMP and 2',3'-cUMP were purchased from Carbosynth (Note 1)
3. 50 mM phosphate buffer pH 7.5 and pH 9.5 (coupling buffer).
4. Acetate buffer (0.1 M, pH 4) and coupling buffer (pH 8.3; Note 3) each containing 0.5 M NaCl
5. *E. coli* BW25113 Δ*cpdB* (strain with known 2',3'-cyclic phosphodiesterase gene deleted) or other bacterial strain of interest (Note 2)
6. M9 media
7. TRIS buffer (50 mM, pH 7.5)
8. Wash buffer: 50 mM TRIS buffer pH 7.5 (with 20 mM NaCl)
9. Elution buffer: 50 mM TRIS buffer pH 7.5 (with 200 mM, 500 mM, and 1 M NaCl)

3. Equipment

1. UV spectrophotometer
2. End-on-end rotator

3. Fritted column
4. Incubator shaker
5. Sonicator (or other instrument for cell lysis)
6. Microcentrifuge

4. Resin coupling procedure

1. 1 mg/mL solution of the 2′,3′-cNMP was prepared in the coupling buffer (pH 9.5).
2. 1 g of epoxy activated Sepharose 6B was weighed out and suspended in distilled water (swells immediately). The resin was then transferred in a fritted column and washed with 200 mL water
3. For coupling, the media prepared in step 2 was dissolved in 3 mL of buffer prepared at step 1. The suspension was rotated end to end at room temperature for 16 h (overnight).
4. Next day, the resin coupled with 2′,3′-cNMP was loaded in a fritted column and allowed to drain. The resin was then washed with 10 mL of coupling buffer
5. The resin was then suspended in 60% glycerol and the absorbance of a 50% slurry measured at 265 nm and compared to the absorbance of the uncoupled control resin (step 8). The increased absorbance of the nucleotide-coupled resin at 265 nm confirmed the completion of reaction
6. Thin layer chromatography (ethyl acetate:acetonitrile:water:methanol: ammonium hydroxide 6:1.5:1.5:2:0.25 using silica plate) of the flow through was performed to determine if the 2′,3′-cNMP were intact and had not been hydrolyzed to the linear monophosphates
7. The 2′,3′-cNMP-bound resin was suspended in 3 mL of 10 mM DTT in the coupling buffer (pH 7.5) and the reaction was allowed to rotate end-to-end in the incubator at 45 °C overnight to block any remaining epoxy groups
8. For the control experiment the resin was prepared by reacting the epoxy groups with DTT by following step 7, as described above
9. Finally, both the 2′,3′-cNMP coupled and the control resin were washed with alternating buffers: acetate buffer (0.1 M, pH 4) and coupling buffer (pH 8.3), each containing 0.5 M NaCl, four times each to remove any excess uncoupled 2′,3′-cNMPs
10. The prepared resins were stored in coupling buffer at 4 °C

5. Pull-down procedure

1. *E. coli* BW25113 Δ*cpdB* (Keio Collection) was streaked onto LB agar containing kanamycin (50 μg/mL) and incubated overnight at 37 °C

2. Isolated colonies were picked and used to inoculate 10 mL of M9 media + kanamycin (50 mg/mL; M9 + Kan) and then incubated at 37 °C with 220 rpm shaking overnight

3. The resulting culture was inoculated in 1:100 ratio into 100 mL of M9 + Kan in a sterile 250 mL Erlenmeyer flask and incubated with shaking at 220 rpm, 37 °C until the OD_{600} reached 0.5–0.6

4. Cells were harvested by centrifugation at 4000 rpm, 4 °C for 20 min in an Allegra X-30 R centrifuge (Beckman) and the resultant cell pellets were frozen at −80 degree until further use

5. Lysates were prepared fresh on the same day that the pull-down was performed. Lysate was prepared by thawing the frozen cell pellets on ice, resuspending in 10 mL TRIS buffer (50 mM, pH 7.5) and then lysed by sonication (QSonica). The sonication was conducted at 50% amplitude for 100 s at 5 s on and 3 s off condition

6. Cellular debris from the lysed cells was removed by centrifugation at $10,000 \times g$ at 4 °C for 20 min on a tabletop microcentrifuge and the supernatant was collected for pull-down assays

7. The pull-downs can be performed on a small scale very efficiently. To do so, 200 μL of the 2′,3′-cNMP-linked Sepharose resin (stored in phosphate coupling buffer) was washed with at least 1 mL of 50 mM TRIS buffer pH 7.5

8. The resin was then suspended in 500 μL of *E. coli* cell BW25113 Δ*cpdB* lysate and rotated on an end-to-end rotator at 4 °C for 4 h

9. The protein-bound resin was loaded in a small gravity column for the subsequent chromatography steps and the lysate was allowed to drain from the column by gravity

10. The column was washed with 1 mL of wash buffer (50 mM TRIS buffer, pH 7.5, 20 mM NaCl; Note 4) to remove unbound cellular molecules

11. This was followed by elution of the column with 400 μL elution buffer at different concentrations of salt (200 mM, 500 mM, and 1 M NaCl) in 50 mM TRIS buffer, pH 7.5. Fractions were collected from each elution condition

12. Finally, the beads were heated at 56 °C for 10 min in 50 mM TRIS buffer, pH 7.5, 1 M NaCl and the supernatant was collected to ensure that all the bound protein was eluted

13. All of the samples were stored at $-80\,^{\circ}\text{C}$ for later submission for proteomic analysis by quantitative mass spectrometry or for analysis by Western blot, etc.

6. Notes

1. The standard procedure can be used for other nucleotide-containing compounds. For compounds that do not contain a nucleotide but do contain a nucleophilic substituent (hydroxyl group, amine, etc.), the general procedure can be followed but the coupling buffer pH may need to be adjusted for optimal coupling efficiency
2. The same pull-down procedure has been conducted for *Salmonella* and *Vibrio* cells and could be adopted for any other cell type
3. The pH of the coupling buffer should be adjusted so that it is above the pK_a of the nucleophilic substituents on the molecule of interest to ensure adequate coupling. However, increased pH will result in decreased reactivity of the epoxide (acidic conditions yield higher epoxide reactivity). Testing the coupling reaction on the small scale at a number of pHs is recommended to identify the optimal coupling parameters
4. If the compound of interest is available in sufficient quantity, the elution can be performed using wash buffer with increasing concentrations of the compound. Using the compound for the elution will result in specific elution of binding proteins, rather than general elution due to increased salt concentration

Acknowledgments
This work was supported by NIH NIGMS grant # R01GM125842 (E.E.W.).

Author contributions
S.S.C. and E.E.W. conceived and designed the protocol. S.S.C. performed the experiments. S.S.C. and E.E.W. wrote the paper.

References
Botsford, J. L., & Harman, J. G. (1992). Cyclic AMP in prokaryotes. *Microbiological Reviews*, *56*(1), 100–122.

Chauhan, S. S., Marotta, N. J., Karls, A. C., & Weinert, E. E. (2022). Binding of 2′,3′-cyclic nucleotide monophosphates to bacterial ribosomes inhibits translation. *ACS Central Science*, *8*, 1518–1526. https://doi.org/10.1021/acscentsci.2c00681.

Denninger, J. W., & Marletta, M. A. (1999). Guanylate cyclase and the •NO/cGMP signaling pathway. *Biochimica et Biophysica Acta*, *1411*(2–3), 334–350.

Duggal, Y., Kurasz, J. E., Fontaine, B. M., Marotta, N. J., Chauhan, S. S., Karls, A. C., et al. (2022). Cellular effects of 2′,3′-cyclic nucleotide monophosphates in Gram-negative bacteria. *Journal of Bacteriology, 204*(1), e0020821. https://doi.org/10.1128/JB.00208-21.

Fahmi, T., Port, G. C., & Cho, K. H. (2017). c-di-AMP: An essential molecule in the signaling pathways that regulate the viability and virulence of Gram-positive bacteria. *Genes (Basel), 8*(8). https://doi.org/10.3390/genes8080197.

Ferguson, F., McLennan, A. G., Urbaniak, M. D., Jones, N. J., & Copeland, N. A. (2020). Re-evaluation of diadenosine tetraphosphate (Ap4A) from a stress metabolite to bona fide secondary messenger. *Frontiers in Molecular Biosciences, 7*, 606807. https://doi.org/10.3389/fmolb.2020.606807.

Fontaine, B. M., Martin, K. S., Garcia-Rodriguez, J. M., Jung, C., Southwell, J. E., Jia, X., et al. (2018). RNase I regulates *Escherichia coli* 2′,3′-cyclic nucleotide monophosphate levels and biofilm formation. *The Biochemical Journal, 478*(8), 1491–1506.

Kalia, D., Merey, G., Nakayama, S., Zheng, Y., Zhou, J., Luo, Y. L., et al. (2013). Nucleotide, c-di-GMP, c-di-AMP, cGMP, cAMP, (p)ppGpp signaling in bacteria and implications in pathogenesis. *Chemical Society Reviews, 42*(1), 305–341. https://doi.org/10.1039/c2cs35206k.

Lucas, K. A., Pitari, G. M., Kazerounian, S., Ruiz-Stewart, I., Park, J., Schulz, S., et al. (2000). Guanylyl cyclases and signaling by cyclic GMP. *Pharmacological Reviews, 52*(3), 375–414.

Pesavento, C., & Hengge, R. (2009). Bacterial nucleotide-based second messengers. *Current Opinion in Microbiology, 12*(2), 170–176. https://doi.org/10.1016/j.mib.2009.01.007. S1369-5274(09)00006-X [pii].

Roelofs, K. G., Wang, J., Sintim, H. O., & Lee, V. T. (2011). Differential radial capillary action of ligand assay for high-throughput detection of protein-metabolite interactions. *Proceedings of the National Academy of Sciences of the United States of America, 108*(37), 15528–15533. https://doi.org/10.1073/pnas.1018949108. Epub 2011 Aug 29.

Romling, U., Galperin, M. Y., & Gomelsky, M. (2013). Cyclic di-GMP: The first 25 years of a universal bacterial second messenger. *Microbiology and Molecular Biology Reviews, 77*(1), 1–52. https://doi.org/10.1128/MMBR.00043-12.

Sutherland, E. W. (1972). Studies on the mechanism of hormone action. *Science, 177*(4047), 401–408.

Zaver, S. A., & Woodward, J. J. (2020). Cyclic dinucleotides at the forefront of innate immunity. *Current Opinion in Cell Biology, 63*, 49–56. https://doi.org/10.1016/j.ceb.2019.12.004.

CHAPTER THIRTEEN

Functional assay of light-induced ion-transport by rhodopsins

Shoko Hososhima[a], Rei Abe-Yoshizumi[a], and Hideki Kandori[a,b],*

[a]Department of Life Science and Applied Chemistry, Nagoya Institute of Technology, Nagoya, Japan
[b]OptoBioTechnology Research Center, Nagoya Institute of Technology, Nagoya, Japan
*Corresponding author: e-mail address: kandori@nitech.ac.jp

Contents

Abstract

Microbial rhodopsins are photoreceptive membrane proteins found from diverse microorganisms such as archaea, eubacteria, eukaryotes and viruses. Many microbial rhodopsins possess ion-transport activity by light, such as channels and pumps, and ion-transporting rhodopsins are important tools in optogenetics that control animal behavior by light. Historically, molecular mechanism of rhodopsins has been studied by spectroscopic methods for purified proteins. On the other hand, ion-transport function has to be studied by different methods. This chapter introduces two methods of functional assay of ion-transporting rhodopsins by light. One is a patch clamp method using mammalian cells, and another is an ion-transport assay using pH electrode and microbial cells. These functional assay provides fundamental data of ion-transporting rhodopsins, and thus contributes to evaluation for optogenetic tools.

Methods in Enzymology, Volume 679
ISSN 0076-6879
https://doi.org/10.1016/bs.mie.2022.08.018

331

1. Introduction

Rhodopsins are photoactive membrane proteins containing a retinal chromophore in animals and microbes (Ernst et al., 2014). Rhodopsins, composed of seven transmembrane helices, are now found in all domains of life and are classified into two groups, animal and microbial rhodopsins. While animal rhodopsins are exclusively photosensory receptors as a specialized subset of G-protein coupled receptors, microbial rhodopsins have various functions, including as photosensory receptors, a light-switch for gene expression, photoactivatable enzymes, light-driven ion pumps and light-gated ion channels.

The first found microbial rhodopsin was a light-driven proton-pump bacteriorhodopsin (BR) from *Halobacterium salinarum* in 1971 (Fig. 1A) (Oesterhelt & Stoeckenius, 1971). Then, a light-driven inward chloride pump halorhodopsin (HR) was found in 1977 (Fig. 1A) (Matsuno-Yagi & Mukohata, 1977). As pumps, BR and HR contribute to the formation of

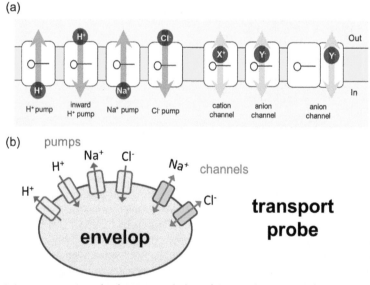

Fig. 1 Ion-transporting rhodopsins and their functional assay. (A) Ion-transporting microbial rhodopsins. Pump and channel rhodopsins transport specific ions uni- and bi-directionally, respectively. Atomic structures of the seven rhodopsins are similar, but different functions emerge. Black circle and straight line inside each rhodopsin schematically illustrate the all-trans retinal chromophore. Ion-transport pathway is inside rhodopsin, except for bestrhodopsin (right). (B) To analyze ion-transport function of rhodopsins, rhodopsin environment such as envelop and transport probe are needed.

a membrane potential and thus function in light-energy conversion. Although microbial rhodopsins were epitomized by haloarchaeal proteins for the first 30 years, so many related photoactive proteins were identified in archaea, eubacteria, eukaryotes and viruses for the last 20 years (Ernst et al., 2014; Govorunova, Sineshchekov, Li, & Spudich, 2017; Grote, Engelhard, & Hegemann, 2014; Rozenberg, Inoue, Kandori, & Béjà, 2021), including a light-driven outward sodium pump (Inoue et al., 2013; Kandori, Inoue, & Tsunoda, 2018) and inward proton pump (Inoue et al., 2016; Shevchenko et al., 2017) (Fig. 1A). Channelrhodopsin (ChR), a microbial rhodopsin found in green algae, functions as a light-gated cation channel. Discovery of ChRs in 2002 and 2003 (Fig. 1A) (Nagel et al., 2002, 2003) quickly led to an emergence of optogenetics (Boyden, Zhang, Bamberg, Nagel, & Deisseroth, 2005; Deisseroth & Hegemann, 2017; Zhang et al., 2007), in which light-gated ion channels and light-driven ion pumps are used to depolarize and hyperpolarize selected cells of neuronal networks, respectively. Anion channelrhodopsin (ACR) was discovered in 2015 (Govorunova, Sineshchekov, Janz, Liu, & Spudich, 2015), which is used as a hyperpolarization tool in optogenetics (Fig. 1A). Although all these ion-transporting rhodopsins possess ion pathways inside seven transmembrane helices, recently discovered bestrhodopsin shows that the rhodopsin does not transport ions inside the domain, but the rhodopsin domain activates and opens the anion channel domains by light (Fig. 1B) (Rozenberg et al., 2022).

When rhodopsins are activated by light, several intermediates appear during their functional processes (Ernst et al., 2014; Govorunova et al., 2017; Grote et al., 2014; Rozenberg et al., 2021). Therefore, molecular mechanism during each function has been studied by spectroscopic methods for purified proteins in the history of rhodopsins research. On the other hand, ion-transport function has to be studied by different methods, where one side and another should be spatially isolated. Two sides can be divided by use of planar lipid bilayer, which is also called black membrane (Darszon, 1986). On the other hand, envelop such as living and artificial cells is commonly used (Fig. 1B). Liposomes are typical artificial cells, to which isolated membrane proteins are inserted, where control of membrane topology is not easy. On the other hand, living cells are suitable systems of envelope, where a rhodopsin gene is introduced, and heterologously expressed. Mammalian cells such as ND7/23 cells, Xenopus oocytes, *Pichia pastoris* cells, and *E. coli* cells are typically used for functional assay of ion-transporting rhodopsins.

Ion-transport functions of rhodopsins have to be measured by some probes. General electrophysiological method so called "patch clamp" is to record light-induced ion currents by use of two electrodes, inside and outside cells, in glass micropipettes (Fig. 2A). Among several patch clamp configurations, the most commonly used is the whole-cell mode, in which the membrane patch is disrupted by briefly applying strong suction to establish electrical and molecular access to the intracellular space. In the voltage clamp method, the voltage is held constant allowing the study of ionic currents, while the current is controlled enabling the study of changes in membrane potential in the current-clamp mode.

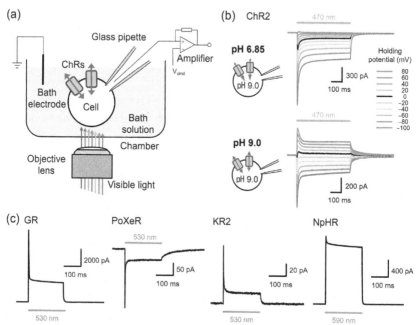

Fig. 2 Ion transport assay of microbial rhodopsins expressed in Mammalian cells. (A) Schematic diagram for whole-cell patch clamp. (B) Electrophysiological measurements of a light-gated cation channel (ChR2) driven photocurrent in ND7/23 cells. The cells were illuminated with blue light ($\lambda = 470$ nm, 6.8 mW/mm^2) during the time region shown by the blue bars. The membrane voltage was clamped from -100 to $+80$ mV for every 20-mV step. The pipette solution was 110 mM NMG-Cl, pH 9.0, the bath solution was 140 mM NMG-Cl, pH 6.85 (top) or 9.0 (bottom). (C) Representative photocurrent traces of light-driven outward proton pump (GR), inward proton pump (PoXeR), outward sodium pump (KR2) and inward chloride pump (NpHR) recorded in ND7/23 cells. The cells were illuminated with green or orange light ($\lambda = 530$ or 590 nm, 25 mW/mm^2 or 10 mW/mm^2) during the time region shown by the green or orange bars. The membrane voltage was clamped 0 mV. The pipette solution was 110 mM NaCl, pH 7.4, the bath solution was 140 mM NaCl, pH 7.4.

Advantage of the whole-cell patch clamp recordings is that intracellular solution can be exchanged in addition to extracellular solution, by which we can control the concentration of ions and pH both inside and outside the envelop (Figs. 1B and 2B). In case of Xenopus oocytes, cell size (diameter $\sim$ 1 mm) is larger than mammalian cells ($\sim$ 20 μm) such as ND7/23 and HEK293 cells. While large light-induced current is advantageous to characterize the rhodopsins with small signals, it is not easy to exchange intracellular solution. Thus, we standardly use whole-cell patch clamp recording to microbial rhodopsins expressed in ND7/23 cells, whose protocol is first described in this chapter. Fig. 2B and C show typical patch clamp data for light-gated cation channel (channelrhodopsin 2 from *Chlamydomonas reinhardtii*; ChR2), and light-driven ion pumps (outward proton pump GR, inward proton pump PoXeR, outward sodium pump KR2, and inward chloride pump NpHR), respectively.

Sizes of *Pichia pastoris* cells (2–5 μm) and procaryote cells (<2 μm) are smaller than mammalian cells, which are not suitable for patch clamp recordings. In this case, information inside cells cannot be gained, but we are able to record concentration of solute in extracellular solution using ion-specific electrodes. pH electrode is most commonly used in the functional assay of light-induced ion-transport of microbial rhodopsins (Fig. 3A). It should be noted that this method enables ion-transport of not only protons, but also any ions. The reason is that ion-transport of non-proton cations or anions normally alters chemical potential (so-called Nernst potential) across the membrane, which causes driving force of the movement of protons in the system. Proton leak across the membrane is generally very small, but addition of protonophore, reagents that increases membrane permeability to protons, enlarges pH changes. Carbonyl cyanide 3-chlorophenylhydrazone (CCCP) is a typical protonophore in the study, and CCCP diminishes signals of proton pumps/channels, but increases signals of non-proton pumps/channels. *E. coli* cells are convenient host, as they do not respond to visible light. When rhodopsins are heterologously expressed and retinal is added, illumination of the *E. coli* cells changes pH. This method can be applied to even native living cells, as we reported different expression period for proton pump and sodium pump in *Krokinobacter eikastus* (Fig. 3B) (Inoue et al., 2013). The protocol of ion transport assay of microbial rhodopsins expressed in microbial cells is next introduced in this chapter. Fig. 3C shows typical data for light-driven ion pumps; outward proton pump GR, inward proton pump PoXeR, outward sodium pump KR2, and inward chloride pump NpHR.

(a)   (b)

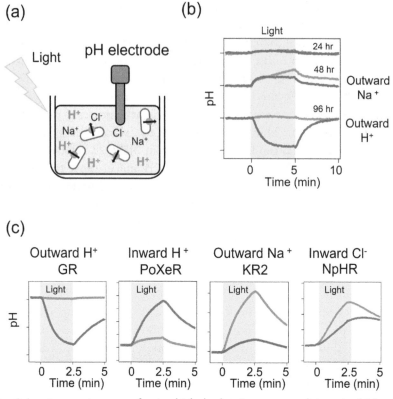

Fig. 3 Ion transport assay of microbial rhodopsins expressed in microbial cells. (A) Schematic diagram for microbial cells. (B) Light-induced pump activity in native *K. eikastus* cell suspensions without (blue) and with (green) CCCP. *K. eikastus* possesses two types of rhodopsin (outward proton pump and outward sodium pump), and their ion transport activity varies with growth time. (C) Various light-induced pump activities heterologously expressed in *E. coli* without (blue) and with (green) CCCP.

2. Whole-cell voltage clamp for recording of microbial rhodopsins expressed in mammalian cultured cells

The electrophysiological assays of microbial rhodopsins were performed on ND7/23 cells (DS Pharma Biomedical, Osaka, Japan), hybrid cell lines derived from neonatal rat dorsal root ganglia neurons fused with mouse neuroblastoma (Wood et al., 1990). ND7/23 cells were grown on a collagen-coated coverslip (12 mm dia., Matsunami Glass, Osaka, Japan) in Dulbecco's modified Eagle's medium (FUJIFILM Wako Pure Chemical Corporation, Osaka, Japan) supplemented with 2.5 μM all-*trans* retinal,

5% fetal bovine serum under a 5% CO_2 atmosphere at 37 °C. The expression plasmids were transiently transfected by using Lipofectamine 2000 (Invitrogen, Carlsbad, CA, USA), Lipofectamine 3000 (Invitrogen) or FuGENE HD (Promega, Madison, WI, USA) according to the manufacturer's instructions. Electrophysiological recordings were then conducted 16–36 h after the transfection. Successfully transfected cells were identified by fluorescence tag under a microscope prior to the measurements.

2.1 Expression plasmids

The CMV or CAG promoter-based expression plasmids were used for expression of microbial rhodopsins. A fluorescence protein (e.g., eGFP, Venus) was tagged at the C-terminus of a microbial rhodopsin for identifying the cells expressed microbial rhodopsins (Ishizuka, Kakuda, Araki, & Yawo, 2006; Tsunoda et al., 2017). However, HeRs were converted membrane topology, a fluorescence protein was tagged at the N-terminus of HeR. For improving the photocurrents of HeRs, P2A peptide sequences (GSGATNFSLLKQAGDVEENPGP) was inserted in between a fluorescence protein and HeRs. For improving membrane expression, microbial rhosopsins fused with a membrane trafficking signal (KSRITSEGEYIPLDQIDINV), an enhanced yellow fluorescent protein (eYFP), and an ER export signal (FCYENEV) (Gradinaru et al., 2010; Hoque et al., 2016; Hososhima, Kandori, & Tsunoda, 2021). Both signals were derived from a Kir2.1 potassium channel.

2.2 Seeding of ND7/23 cells

(1) Before passaging, warmed the cultured medium to 37 °C in water bath
(2) Collect ND7/23 cells by pipetting to a centrifuge tube. ND7/23 cells were easy to peel off the cultured dish. It is recommended that cultures be split after 70–80% confluence
(3) Centrifuge at 1000 rpm for 5 min
(4) Remove the supernatant by aspiration
(5) Gently resuspend the pellet in the cultured medium
(6) Before seeding, put on coverslip in 4 well-dish (176740, Thermo Fisher Scientific, Waltham, MA, USA or SPL-30004, SPL Life Sciences, Gyeonggi-do, Korea) and coated collagen or Poly-L-Lysin. Seed 0.5×10^5 cells in 0.5 mL cultured medium in each well
(7) Incubated the dish overnight under a 5% CO_2 atmosphere at 37 °C

2.3 Transfection

(1) Remove the cultured medium from each well
(2) Add a new cultured medium supplemented all-*trans* retinal at a final concentration of 1–5 µM (Usually 2.5 µM).
(3) Transfected by reagents (e.g., Lipofectamine 2000, Lipofectamine 3000 or FuGENE HD) according to the manufacturer's instructions

2.4 Electrophysiology

All electrophysiological experiments were carried out at room temperature ($23 \pm 2\,°C$). Photocurrents were recorded using an Axopatch 200B amplifier (Molecular Devices, Sunnyvale, CA, USA) or IPA amplifier (Sutter instrument, Novato, CA, USA) under a whole-cell patch clamp configuration. Data were filtered at 5 kHz and sampled at 10 kHz and stored in a computer (Digdata1550 and pClamp10.6, Molecular Devices or SutterPatch, Sutter instrument). Irradiation at visible light was carried out using Colibri7 or collimated LED (Carl Zeiss, Oberkochen, Germany or Mightex, Toronto, Canada) controlled by computer software. The light power was directly measured at an objective lens of microscopy by a visible light-sensing thermopile (MIR-100Q, SSC Inc., Mie, Japan).

The standard internal pipette solution for whole-cell voltage-clamp contained (in mM) 110 N-methyl-D-glucamine or 110 mM NaCl, 2 $MgCl_2$, 1 $CaCl_2$, 10 HEPES, 10 EGTA, 3 glucose, adjusted to pH 7.4 with HCl. The standard extracellular solution for whole-cell voltage-clamp contained (in mM) 140 NaCl, 2 $MgCl_2$, 2 $CaCl_2$, 10 HEPES, 11 glucose, adjusted to pH 7.4 with N-methyl-D-glucamine. For investigation of cation selectivity, composition of pipette and extracellular buffers were described in the following (Shigemura, Hososhima, Kandori, & Tsunoda, 2019; Tashiro et al., 2021). The ion selectivity internal pipette solution contained (in mM) 110 N-methyl-D-glucamine, 1 NaCl, 1 KCl, 2 $CaCl_2$, 2 $MgCl_2$, 10 CHES, 10 EGTA, adjusted to pH 9.0. The ion selectivity extracellular solution contained (in mM) 140 N-methyl-D-glucamine, 1 NaCl, 1 KCl, 2 $CaCl_2$, 2 $MgCl_2$, 10 CHES or 10 MES, adjusted to pH 9.0 or 6.85. 140 NaCl, 1 KCl, 2 $CaCl_2$, 2 $MgCl_2$, 10 CHES, adjusted to pH 9.0. 140 KCl, 1 NaCl, 2 $CaCl_2$, 2 $MgCl_2$, 10 CHES, adjusted to pH 9.0. 140 CsCl, 1 NaCl, 1 KCl, 2 $CaCl_2$, 2 $MgCl_2$, 10 CHES, adjusted to pH 9.0. 70 $CaCl_2$, 1 NaCl, 1 KCl, 2 $MgCl_2$, 10 CHES, adjusted to pH 9.0. 70 $MgCl_2$, 1 NaCl, 1 KCl, 2 $CaCl_2$, 10 CHES, adjusted to pH 9.0.

These solutions of pH were adjusted with *N*-methyl-ᴅ-glucamine or HCl. The liquid junction potential was calculated and compensated by pClamp 10.6 software.

3. Ion transport assay of microbial rhodopsins expressed in microbial cells

The ion transport assays were performed on microbial cells expressed microbial rhodopsins, *Escherichia coli* (Abe-Yoshizumi, Inoue, Kato, Nureki, & Kandori, 2016; Inoue et al., 2016; Konno, Inoue, & Kandori, 2021), *Pichia pastris* (Kikuchi et al., 2021) and *Krokinobacter eikastus* (Inoue et al., 2013). Microorganisms were grown in their optimal liquid medium and temperature. In the case of the heterologous expression system, all-*trans* retinal was added when inducing expression.

3.1 Expression plasmids for recombinant rhodopsins in *E. coli* and *P. pastris*

For the heterologous expression of microbial rhodopsin, the T7 promoter was used for *E. coli* and the methanol-inducible promoter, *AOX1*, for *P. pastris*. Addition of the tag was not mandatory because cells expressing rhodopsin have been found to be chromogenic.

3.2 Growth of microbial cells

1. In the case of *E. coli* cells
    (1) Grow 3 mL of *E. coli* strain harboring expression plasmids in $2 \times$ YT in 15 mL glass tube at 37 °C overnight
    (2) Inoculate 100 mL of fresh medium in a 300 mL baffled flask with 1 mL of the overnight culture. Incubate the *E. coli* cells with shaking at 37 °C until OD_{660} reaches 0.4–1 (0.6 recommended)
    (3) Add all-*trans* retinal and IPTG from a 1 M stock to a final concentration of 1 mM and continue the incubation for 3–4 h
2. In the case of *P. pastris* cells
    (1) Grow 3 mL of *P. pastris* strain possessing the rhodopsin gene in BMGY in a 15 mL tube at 30 °C overnight
    (2) Inoculate 100 mL of BMMY medium in a 300 mL baffled flask with the overnight culture. Expression was induced by the addition of 0.5% methanol every 24 h in the presence of 30 µM all-*trans*-retinal. Incubate the *P. pastris* cells with shaking 250 rpm at 30 °C for 48 h

3. In the case of *K. eikastus* cells

    (1) Grow 3 mL of *K. eikastus* cells in Marine Broth 2216 at 20 °C overnight

    (2) Inoculate 100 mL of MB2216 medium in a 300 mL baffled flask with 2 mL of the overnight culture

    (3) Incubate the *K. eikastus* cells with shaking 180 rpm at 20 °C for 72 h

3.3 Ion transport assays

Ion transport assays were carried out at 20 °C or 4 °C. The measurement temperature was changed depending on the expression level of rhodopsin, and when the expression level of rhodopsins was high, for example, *E. coli* was measured at 20 °C, while when the expression was low, *K. eikastus* was measured at 4 °C. The number of microbial cells expressing rhodopsins was estimated by the apparent optical density at 660 nm (OD_{660}), and 7.5 mL cell culture ($OD_{660} = 2$) were used for the experiment. The cells were washed with an unbuffered solution three times by centrifugation (4800 g, 2 min) and resuspended in the same solution. As the standard solution 100 mM NaCl was used, and KCl, Na_2SO_4, etc. at the same concentration were used to identify the ionic species to be transported. The cell suspension was placed in the dark and then illuminated at $\lambda > 500$ nm, by the output of a 1-kW tungsten–halogen projector lamp (Rikagaku) through a glass filter (Y-52, AGC Techno Glass). The light-induced pH changes were measured with a pH electrode (HORIBA, Ltd., Japan). Measurements were repeated under the same conditions with the addition of 30 μM CCCP, a protonophore molecule.

Ion transport activity was estimated by the initial slope after light illumination for the time-dependent pH changes. The initial slopes were calculated using Igor software (HULINKS Inc. Japan).

Acknowledgment
We thank S. Shigemura for his contribution to the data of Fig. 2B.

References
Abe-Yoshizumi, R., Inoue, K., Kato, H. E., Nureki, O., & Kandori, H. (2016). Role of Asn112 in a light-driven sodium ion-pumping rhodopsin. *Biochemistry*, *55*(41), 5790–5797.

Boyden, E. S., Zhang, F., Bamberg, E., Nagel, G., & Deisseroth, K. (2005). Millisecond-timescale, genetically targeted optical control of neural activity. *Nature Neuroscience*, 8(9), 1263–1268.

Darszon, A. (1986). Planar bilayers: A powerful tool to study membrane proteins involved in ion transport. In *Vol. 36. Methods in enzymology* (pp. 486–502). Academic Press.

Deisseroth, K., & Hegemann, P. (2017). The form and function of channelrhodopsin. *Science, 357*(6356), eaan5544. https://doi.org/10.1126/science.aan5544.

Ernst, O. P., Lodowski, D. T., Elstner, M., Hegemann, P., Brown, L. S., & Kandori, H. (2014). Microbial and animal rhodopsins: Structures, functions, and molecular mechanisms. *Chemical Reviews, 114*(1), 126–163.

Govorunova, E. G., Sineshchekov, O. A., Janz, R., Liu, X., & Spudich, J. L. (2015). Natural light-gated anion channels: A family of microbial rhodopsins for advanced optogenetics. *Science, 349*(6248), 647–650.

Govorunova, E. G., Sineshchekov, O. A., Li, H., & Spudich, J. L. (2017). Microbial rhodopsins: Diversity, mechanisms, and optogenetic applications. *Annual Review of Biochemistry, 86*, 845–872.

Gradinaru, V., Zhang, F., Ramakrishnan, C., Mattis, J., Prakash, R., Diester, I., et al. (2010). Molecular and cellular approaches for diversifying and extending optogenetics. *Cell, 141*(1), 154–165.

Grote, M., Engelhard, M., & Hegemann, P. (2014). Of ion pumps, sensors and channels - perspectives on microbial rhodopsins between science and history. *Biochimica et Biophysica Acta, 1837*(5), 533–545.

Hoque, M. R., Ishizuka, T., Inoue, K., Abe-Yoshizumi, R., Igarashi, H., Mishima, T., et al. (2016). A chimera Na+-pump rhodopsin as an effective optogenetic silencer. *PLoS One, 11*(11), e0166820.

Hososhima, S., Kandori, H., & Tsunoda, S. P. (2021). Ion transport activity and optogenetics capability of light-driven Na+-pump KR2. *PLoS One, 16*(9), e0256728.

Inoue, K., Ito, S., Kato, Y., Nomura, Y., Shibata, M., Uchihashi, T., et al. (2016). A natural light-driven inward proton pump. *Nature Communications, 7*, 13415.

Inoue, K., Ono, H., Abe-Yoshizumi, R., Yoshizawa, S., Ito, H., Kogure, K., et al. (2013). A light-driven sodium ion pump in marine bacteria. *Nature Communications, 4*, 1678.

Ishizuka, T., Kakuda, M., Araki, R., & Yawo, H. (2006). Kinetic evaluation of photosensitivity in genetically engineered neurons expressing green algae light-gated channels. *Neuroscience Research, 54*, 85–94.

Kandori, H., Inoue, K., & Tsunoda, S. P. (2018). Light-driven sodium-pumping rhodopsin: A new concept of active transport. *Chemical Reviews, 118*(21), 10646–10658.

Kikuchi, C., Kurane, H., Watanabe, T., Demura, M., Kikukawa, T., & Tsukamoto, T. (2021). Preference of Proteomonas sulcata anion channelrhodopsin for NO3- revealed using a pH electrode method. *Scientific Reports, 11*(1), 1–13.

Konno, M., Inoue, K., & Kandori, H. (2021). Ion transport activity assay for microbial rhodopsin expressed in *Escherichia coli* cells. *Bio-Protocol, 11*(15), e4115-e4115 https://doi.org/10.21769/bioprotoc.4115.

Matsuno-Yagi, A., & Mukohata, Y. (1977). Two possible roles of bacteriorhodopsin; a comparative study of strains of Halobacterium halobium differing in pigmentation. *Biochemical and Biophysical Research Communications, 78*, 237–243.

Nagel, G., Ollig, D., Fuhrmann, M., Kateriya, S., Musti, A. M., Bamberg, E., et al. (2002). Channelrhodopsin-1: A light-gated proton channel in green algae. *Science, 296*(5577), 2395–2398.

Nagel, G., Szellas, T., Huhn, W., Kateriya, S., Adeishvili, N., Berthold, P., et al. (2003). Channelrhodopsin-2, a directly light-gated cation-selective membrane channel. *Proceedings of the National Academy of Sciences of the United States of America, 100*(24), 13940–13945.

Oesterhelt, D., & Stoeckenius, W. (1971). Rhodopsin-like protein from the purple membrane of Halobacterium halobium. *Nature: New Biology, 233*, 149–152.

Rozenberg, A., Inoue, K., Kandori, H., & Béjà, O. (2021). Microbial rhodopsins: The last two decades. *Annual Review of Microbiology*, *75*, 427–447. https://doi.org/10.1146/annurev-micro-031721-020452.

Rozenberg, A., Kaczmarczyk, I., Matzov, D., Vierock, J., Nagata, T., Sugiura, M., et al. (2022). Rhodopsin-bestrophin fusion proteins from unicellular algae form gigantic pentameric ion channels. *Nature Structural & Molecular Biology*, *29*, 592–603.

Shevchenko, V., Mager, T., Kovalev, K., Polovinkin, V., Alekseev, A., Juettner, J., et al. (2017). Inward H+ pump xenorhodopsin: Mechanism and alternative optogenetic approach. *Science Advances*, *3*(9), e1603187.

Shigemura, S., Hososhima, S., Kandori, H., & Tsunoda, S. P. (2019). Ion channel properties of a cation channelrhodopsin, Gt_CCR4. *Applied Sciences*, *9*.

Tashiro, R., Sushmita, K., Hososhima, S., Sharma, S., Kateriya, S., Kandori, H., et al. (2021). Specific residues in the cytoplasmic domain modulate photocurrent kinetics of channelrhodopsin from Klebsormidium nitens. *Communications Biology*, *4*(1), 235.

Tsunoda, S. P., Prigge, M., Abe-Yoshizumi, R., Inoue, K., Kozaki, Y., Ishizuka, T., et al. (2017). Functional characterization of sodium-pumping rhodopsins with different pumping properties. *PLoS One*, *12*(7), e0179232.

Wood, J. N., Bevan, S. J., Coote, P. R., Dunn, P. M., Harmar, A., Hogan, P., et al. (1990). Novel cell lines display properties of nociceptive sensory neurons. *Proceedings of the Biological Sciences*, 187–194.

Zhang, F., Wang, L.-P., Brauner, M., Liewald, J. F., Kay, K., Watzke, N., et al. (2007). Multimodal fast optical interrogation of neural circuitry. *Nature*, *446*(7136), 633–639.

CHAPTER FOURTEEN

Assay design for analysis of human uracil DNA glycosylase

Rashmi S. Kulkarni, Sharon N. Greenwood, and Brian P. Weiser*
Department of Molecular Biology, Rowan University School of Osteopathic Medicine, Stratford, NJ, United States
*Corresponding author: e-mail address: weiser@rowan.edu

Contents

Abstract

Human uracil DNA glycosylase (UNG2) is an enzyme whose primary function is to remove uracil bases from genomic DNA. UNG2 activity is critical when uracil bases are elevated in DNA during class switch recombination and somatic hypermutation, and additionally, UNG2 affects the efficacy of thymidylate synthase inhibitors that increase genomic uracil levels. Here, we summarize the enzymatic properties of UNG2 and its mitochondrial analog UNG1. To facilitate studies on the activity of these highly conserved proteins, we discuss three fluorescence-based enzyme assays that have informed much of our understanding on UNG2 function. The assays use synthetic DNA oligonucleotide substrates with uracil bases incorporated in the DNA, and the substrates can be single-stranded, double-stranded, or form other structures such as DNA hairpins or junctions. The fluorescence signal reporting uracil base excision by UNG2 is detected in different ways: (1) Excision of uracil from end-labeled oligonucleotides is measured by visualizing UNG2 reaction products with denaturing PAGE; (2) Uracil excision from dsDNA substrates is detected in solution by base pairing uracil with 2-aminopurine, whose intrinsic fluorescence is enhanced upon uracil excision; or (3) UNG2 excision of uracil from a hairpin molecular beacon substrate changes the

Methods in Enzymology, Volume 679
ISSN 0076-6879
https://doi.org/10.1016/bs.mie.2022.07.033

structure of the substrate and turns on fluorescence by relieving a fluorescence quench. In addition to their utility in characterizing UNG2 properties, these assays are being adapted to discover inhibitors of the enzyme and to determine how protein–protein interactions affect UNG2 function.

1. Introduction: Biomedical relevance of uracil DNA glycosylase activity

The human uracil DNA glycosylase (UNG) gene encodes the primary proteins responsible for removing uracil bases from DNA in cells. Two protein isoforms (UNG1 and UNG2) arise from the single UNG gene through the use of separate promoters and alternative splicing (Nilsen et al., 1997). The two isoforms have an identical catalytic domain of 20 kDa whose globular structure and catalytic mechanism are highly conserved across all of biology (Fig. 1B) (Olsen, Aasland, Wittwer, Krokan, & Helland, 1989; Parikh et al., 1998; Schormann, Ricciardi, & Chattopadhyay, 2014; Xiao et al., 1999). UNG enzymes cleave the N-glycosidic bond on DNA between deoxyribose and a uracil base to produce free uracil and an abasic site that is subsequently repaired by base excision repair machinery. Besides their conserved catalytic domain, the human UNG isoforms have different N-terminal domains that encode targeting sequences to the mitochondria (UNG1) or the nucleus (UNG2) (Fig. 1A) (Nilsen et al., 1997). UNG1 is processed after mitochondrial import to produce an enzyme that largely resembles a truncated UNG2 protein that is missing part of its N-terminal domain (Fig. 1A) (Bharati, Krokan, Kristiansen, Otterlei, & Slupphaug, 1998). The 92 residue N-terminal domain of UNG2 is largely disordered (Buchinger et al., 2017; Rodriguez, Orris, Majumdar, Bhat, & Stivers, 2020).

The presence of uracil in DNA has primarily been examined in the context of nuclear genomic processes. The uracil base substrates of UNG2 can mistakenly be incorporated by polymerases into genomic DNA when dUTP is used instead of dTTP, resulting in the formation of non-mutagenic U/A base pairs (Fig. 1C). Clinically used chemotherapeutics that inhibit the enzyme thymidylate synthase increase nuclear levels of dUTP compared to dTTP, and this elevates the level of uracil incorporated into newly synthesized DNA (Bulgar et al., 2012; Grogan, Parker, Guminski, & Stivers, 2011; Seiple, Jaruga, Dizdaroglu, & Stivers, 2006; Weeks, Zentner, Scacheri, & Gerson, 2014). Metabolites of 5-fluorouracil and 5-fluorodeoxyuridine not only inhibit thymidylate synthase, but can also

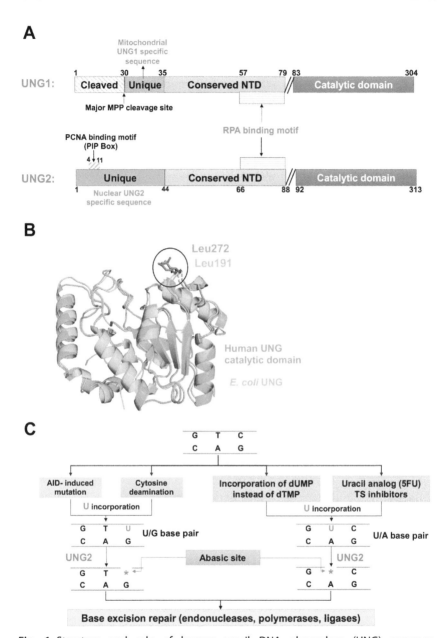

Fig. 1 Structure and role of human uracil DNA glycosylase (UNG) enzymes. (A) Scheme for human UNG isoforms 1 and 2. The mitochondrial UNG1 isoform is cleaved at a mitochondrial processing peptidase (MPP) site after organellar import. Mature UNG1 has a short unique sequence of six residues on its N-terminal domain (NTD), and shares the remainder of its N-terminal domain with nuclear UNG2. UNG2

(Continued)

be inserted as 5-fluorouracil bases in DNA which also serve as a substrates of UNG2 (Grogan et al., 2011). Uracil and 5-fluorouracil bases cause genomic instability especially during replication, and thus the cellular activity of UNG2 is related to the efficacy of thymidylate synthase inhibitors (Bulgar et al., 2012; Christenson et al., 2021; Huehls et al., 2016; Weeks et al., 2014; Yan, Qing, Pink, & Gerson, 2018). Additionally, uracil is found in U/G base pairs that occur from the deamination of cytosine, which is an enzymatically accelerated process during class switch recombination and somatic hypermutation (Fig. 1C) (Peled et al., 2008; Saha, Sundaravinayagam, & Di Virgilio, 2021). UNG2 is the primary isoform involved in removing uracil bases during these antibody diversification processes that occur in the nuclei of B cells (Imai et al., 2003; Zan et al., 2012). On the other hand, the activity of UNG1 helps maintain the integrity of the mitochondrial genome, but its importance in health and disease is less characterized (Akbari, Otterlei, Peña-Diaz, & Krokan, 2007; Kozhukhar, Spadafora, Fayzulin, Shokolenko, & Alexeyev, 2016; Liu et al., 2016; Wollen Steen et al., 2012).

Because of its established importance in several biomedical fields, UNG2 and its isolated catalytic domain have been examined in most biochemical studies on human UNG proteins. Therefore, we refer extensively to the characteristics of UNG2 in this chapter and summarize methods specifically related to this isoform. However, we also note generalizable features of UNG1 and UNG2 that arise from their conserved catalytic domain, and distinguish certain properties of enzyme function that may be dependent on their unique N-terminal domains. Many properties of the human UNG catalytic domain are also conserved across UNG proteins from diverse biological taxa, and therefore similar experimental approaches can be used to

Fig. 1—Cont'd   has a longer N-terminal domain that remains unprocessed and also contains the binding sites for PCNA and RPA. The isoforms share a conserved catalytic domain. (B) Structural alignment of the human UNG catalytic domain (PDB code 1AKZ) and the *E. coli* UNG protein (PDB code 1EUG). The root-mean-square deviation of the Cα backbone atoms from the alignment was 0.743 Å (determined using PyMOL). The conserved leucine (circled) is in the active site, and its role in DNA intercalation during uracil excision from dsDNA is discussed in Section 2.1. (C) Incorporation of uracil in DNA occurs through cytosine deamination or through the misincorporation of uracil bases into DNA instead of thymidine. Cytosine deamination results in potentially mutagenic U/G base pairs and is enzymatically accelerated by activation-induced cytidine deaminase (AID). The misincorporation of uracil by polymerases is enhanced during treatment with thymidylate synthase (TS) inhibitors including 5-fluorouracil (5FU). Both U/G and U/A base pairs can create genomic instability and are recognized as substrates by UNG proteins.

examine all of these different enzymes. Specific emphasis in this chapter is placed on the design of assays and their limitations in their ability to elucidate specific functions of UNG enzymes by themselves and in the presence of regulators such as protein binding partners and small molecules.

2. UNG2 characteristics and enzyme activity assays

2.1 Features of human UNG2 that influence assay design

Like its isolated catalytic domain, UNG2 will remove uracil bases from both ssDNA and dsDNA (Grogan et al., 2011; Hagen et al., 2008). In dsDNA, UNG2 has only a slight kinetic preference for uracil in U/G base pairs compared to U/A base pairs ($\leq$two-fold selectivity) (Grogan et al., 2011; Pettersen et al., 2011; Weiser, 2020). UNG2 will also remove the artificial nucleobase 5-fluorouracil from ssDNA and dsDNA, but with much lower efficiency than uracil bases (Grogan et al., 2011; Pettersen et al., 2011).

The catalytic domain of UNG2 interacts with DNA even in the absence of uracil bases through electrostatic interactions that occur between the protein and the phosphate backbone, and additionally, the catalytic domain makes weak carbon–carbon interactions with the deoxyribose moiety of DNA (Cravens, Hobson, & Stivers, 2014; Parker et al., 2007). For this reason, UNG2 will bind non-specifically to non-uracilated DNA, and the presence of excess DNA hinders UNG2's ability to identify uracil bases and reduces its overall activity. The electrostatic interactions that occur between the UNG2 catalytic domain and the DNA backbone are heavily influenced by the presence of ions, and thus the salt concentration in solution will affect UNG2 steady-state kinetic parameters. K_m especially increases at higher salt concentrations which reflects a lower binding affinity of UNG2 for DNA, driven primarily by a slower association rate (k_{on}) (Cravens et al., 2014). The kinetics of UNG2 activity are also influenced by weak electrostatic interactions that occur between its N-terminal domain and DNA (Rodriguez et al., 2020; Weiser, Rodriguez, Cole, & Stivers, 2018). A cluster of basic residues on the N-terminal domain of UNG2 that interact with DNA are also found on UNG1 (Rodriguez et al., 2020). The N-terminal domain of UNG2 also interacts with ssDNA sections next to ssDNA-dsDNA junctions, and this interaction spatially targets the activity of the catalytic domain near the junction (Weiser et al., 2018).

High resolution structures of the UNG2 catalytic domain bound to DNA indicate that the protein does not make specific contacts with non-uracil DNA bases (Cravens et al., 2014; Parker et al., 2007). When a uracil base is encountered, however, it is flipped from the DNA helix into the enzyme active site, and a conserved leucine residue from the active site intercalates into the DNA base stack to replace the uracil (Fig. 1B) (Cravens et al., 2014; Parikh et al., 2000). Interestingly, the nucleotide sequence surrounding the uracil base will affect the activity of UNG2 whether the DNA substrate is single- or double-stranded (Bellamy, Krusong, & Baldwin, 2007; Eftedal, Guddal, Slupphaug, Volden, & Krokan, 1993; Nilsen, Yazdankhah, Eftedal, & Krokan, 1995). This is important to consider when designing experiments that measure the ability of the enzyme to remove uracil bases from different sites in DNA. In these cases, at least 2 nucleotides upstream and downstream of different uracil sites should be identical to mitigate changes in enzyme activity at any site that could arise from flanking sequence preferences. Experiments where UNG2 activity is compared at different sites can be used to identify factors that influence uracil site selectivity such as local DNA structure or interactions with protein binding partners (Weiser et al., 2018). DNA substrates containing more than one uracil base can also be used to measure the ability of UNG2 to transfer between the different uracil sites (Porecha & Stivers, 2008). UNG2's mechanism of transfer between different uracil sites in DNA can be described as taking either a sliding or hopping pathway (Porecha & Stivers, 2008; Schonhoft & Stivers, 2012). The sliding pathway involves localized associative movement along the DNA backbone, whereas the hopping pathway involves complete dissociation of the enzyme from the DNA strand before reassociating elsewhere. The N-terminal domain of UNG2 enhances the ability of its catalytic domain to slide or transfer along DNA in an associative manner (Rodriguez et al., 2017). The enzyme's ability to slide along DNA is also promoted by macromolecular crowding (Cravens et al., 2015; Rodriguez et al., 2017).

2.2 Design of synthetic substrates for UNG2 assays

Early enzyme assays examining UNG activity from different species often used *in vitro* generated DNA substrates with [^{3}H]deoxyuracil bases incorporated randomly in the sequence, for example, by using nick translation or primer extension (Krokan & Wittwer, 1981; Talpaert-Borlé, Clerici, & Campagnari, 1979). [^{3}H]Uracil released by the enzyme after catalysis can be extracted from quenched reactions and quantified following a thin layer chromatography step. Besides the constant use of radioactivity, a key disadvantage of this method is the heterogeneous nature of the uracilated DNA substrate. As summarized in Section 2.1, UNG2 activity can be affected by a variety of factors including substrate sequence, local DNA structure, and the presence of excess DNA. To control for these factors, enzyme activity assays are more simply performed with defined synthetic DNA substrates where uracil bases are site-specifically incorporated into the sequence. Such synthetic substrates are commercially available as oligonucleotides that can also contain site specific modifications such as fluorescent tags or artificial nucleobases. The versatility of synthetic oligonucleotide substrates is apparent from their applications discussed below.

A straightforward time point assay that measures UNG2 activity can be performed using a single end-labeled oligonucleotide substrate that contains an internal uracil base (Fig. 2A). Fluorescein end-labeling is commonly used on the uracilated substrate and does not typically interfere with assays, often yielding identical results as DNA substrate end-labeled with $[^{32}P]$ in the form of phosphate (Weiser et al., 2018). In a simple experiment, UNG2's excision of uracil from the ssDNA substrate shown in Fig. 2A results in the formation of an abasic site in the DNA (also referred to as an apurinic/apyrimidinic (AP) site). The enzyme reaction is quenched under alkali conditions with moderate heating, and this chemical quench serves an additional purpose as an effective way to chemically cleave the DNA backbone at the abasic site (Haldar et al., 2022). Thus, two end-labeled DNA strands are found in the final mixture: (1) the remaining unprocessed, uracil-containing oligonucleotide that served as the substrate and (2) a smaller fragment that resulted from UNG2's excision of uracil from the substrate and subsequent cleavage of the abasic site. The small DNA oligonucleotides are separated by denaturing (TBE-Urea) PAGE prior to imaging, and the rate of UNG2 activity is determined by quantifying the relative ratio of the processed and unprocessed substrate. As shown in Fig. 2B, end-labeled oligonucleotides containing uracil can also be annealed to complementary strands to produce dsDNA substrates that may also contain other structural features such as a ssDNA-dsDNA junction. Additionally, the relative selectivity of UNG2 for uracil bases located at two specific sites on a DNA strand can be determined if that DNA strand contains an end-label on both the 5' and 3' ends (Fig. 2B) (Schonhoft & Stivers, 2012; Weiser et al., 2018). The probability that UNG2 excises both uracils from the same strand during a single encounter with the DNA molecule measures the enzyme's translocation between the two sites, and can be used to explore the mechanism of transfer between the uracil bases (e.g., hopping or sliding) (Cravens & Stivers, 2016; Schonhoft & Stivers, 2012, 2013).

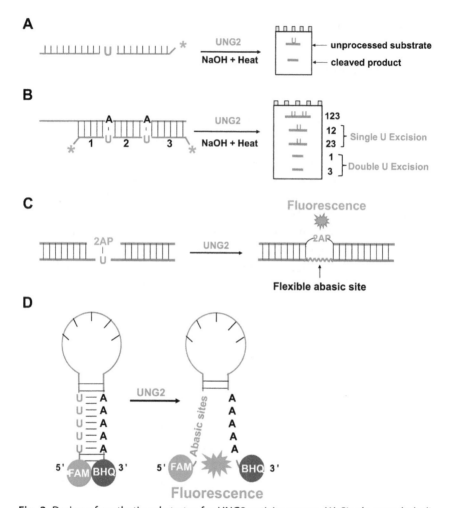

Fig. 2 Design of synthetic substrates for UNG2 activity assays. (A) Single-stranded oligonucleotide substrate with a single internal uracil is fluorescently end-labeled. Uracil excision by UNG2 followed by alkaline treatment with moderate heat results in a smaller product oligonucleotide that can be separated from the unprocessed substrate by denaturing PAGE. The fluorescent bands are imaged in the gel then quantified. (B) End-labeled oligonucleotide substrates as in panel A can be annealed to complementary stands, including to position uracil bases near DNA structures such as ssDNA-dsDNA junctions. By including two uracil bases in the substrate with two end-labels, one can determine the selectivity of UNG2 for either uracil site if the DNA fragments that result from uracil excision are different lengths. DNA fragments that are generated from uracil excision at a single site or at both sites are represented on the hypothetical gel. Note that such duplex substrates could also contain U/G base pairs instead of U/A base pairs. (C) The fluorescent, artificial nucleobase 2-aminopurine is incorporated in a dsDNA substrate directly opposite the uracil site. Excision of uracil creates a highly flexible abasic site opposite of 2-aminopurine, which enhances its fluorescence. Continuous enzyme activity measurements can be made with 2-aminopurine substrates in a cuvette. (D) A uracilated hairpin molecular beacon substrate with the terminal ends of the hairpin labeled with a fluorophore and a quencher (FAM, fluorescein; BHQ, black hole quencher). The U/A base pairs stabilize the hairpin duplex, and the abasic sites that result from uracil excision destabilize the duplex leading to a separation of the fluorophore from its quencher. This turn-on fluorescence substrate can also be used to make continuous measurements of UNG2 activity.

Assays allowing continuous measurement of UNG2 activity have been developed using fluorescence that "turns on" or enhances as a result of uracil base excision. A classic example involves the incorporation of the fluorescent, artificial nucleobase 2-aminopurine in a dsDNA substrate at the position that is directly opposite a uracil site (Fig. 2C). The fluorescence of 2-aminopurine is low when base paired with uracil and stacked in the helix of a duplex (Stivers, 1998). Excision of uracil creates a highly flexible abasic site opposite of 2-aminopurine which unstacks the fluorescent base from the helix and enhances its fluorescence by reducing excitation energy transfer to adjacent bases (Bellamy et al., 2007; Stivers, 1998). An interesting aspect of the fluorescence signal from 2-aminopurine in this assay is its dependence on divalent cations including Mg^{2+} (Stivers, 1998). Divalent cations further destabilize the local stacking structure of the DNA surrounding the abasic site and 2-aminopurine leading to a further enhancement of fluorescence intensity (Stivers, 1998). For perspective, the fluorescence of a duplex substrate containing a uracil/2-aminopurine base pair increases $\sim$2-fold when the uracil is excised and 2-aminopurine transitions from stacked to unstacked in the absence of divalent cations, but the fluorescence increases $\sim$8-fold if $10\,mM\ MgCl_2$ is included in the assay. Interestingly, the UNG analog from E. coli which resembles the human catalytic domain binds abasic sites in dsDNA ($K_d = 0.4\,\mu M$), and it will restack a 2-aminopurine base positioned opposite the abasic site (Stivers, 1998). In this way, the affinity of the protein for abasic sites can be measured as the high fluorescence of unstacked 2-aminopurine becomes reduced upon UNG binding. However the affinity of the human catalytic domain for abasic sites ($K_d = 50\,\mu M$) is much lower than the E. coli enzyme (Esadze, Rodriguez, Cravens, & Stivers, 2017). Considering that typical activity assays with UNG2 are performed with low nM enzyme concentrations, the occupancy of abasic sites by UNG2 following uracil excision is negligible and does not significantly affect functional measurements.

An alternative design for a fluorescent UNG2 substrate in a continuous assay is a uracilated DNA hairpin or molecular beacon. Here, a $\sim$24 nucleotide oligo forms a hairpin structure through self-complementation, and included in the annealed section are U/A base pairs that promote the double-stranded structure (Fig. 2D) (Jiang, Krosky, Seiple, & Stivers, 2005; Maksimenko et al., 2004; Mehta, Raj, Sundriyal, Gopal, & Varshney, 2021; Weil et al., 2013). The hairpin oligonucleotide is also labeled on both the 5' and 3' ends with a fluorescent group such as fluorescein and a quencher such as dabcyl (Fig. 2D). With this design, the fluorescence is quenched until

UNG2 excises the uracil bases, which destabilizes the hairpin duplex and ultimately separates the fluorophore from the quencher to generate strong fluorescence. Uracilated DNA hairpin substrates have some advantages compared to duplex 2-aminopurine substrates including the use of higher excitation/emission wavelengths that reduce background fluorescence, and the ability to produce strong fluorescence without divalent cations. This latter point is important because the activity of UNG2 is reportedly enhanced by Mg^{2+} ions in a manner dependent on its N-terminal domain (Doseth, Ekre, Slupphaug, Krokan, & Kavli, 2012; Kavli et al., 2002; Ko & Bennett, 2005). On the other hand, the hairpin design of molecular beacon substrates limits the variety of DNA structures that can be examined and can easily be disrupted by other proteins such as RPA that may be present during UNG2 assays (discussed in Section 2.3).

As a final note, the activity of UNG2 in cell lysates or nuclear extracts can also be measured with uracilated molecular beacon/DNA hairpin substrates as well as end-labeled DNA substrates, further easing the analysis of UNG2 without radiolabeling (Esadze, Rodriguez, Weiser, Cole, & Stivers, 2017; Kaiser & Emerman, 2006; Skjeldam et al., 2010). In some cases, the stability of the synthetic substrates to cellular nucleases might be increased by including phosphorothiolate linkages in the DNA backbone away from the uracil sites, especially on the terminal ends of the DNA (Eckstein, 2014; Weil et al., 2013). In addition to UNG1 and UNG2, several other mammalian enzymes also have the ability to excise uracil bases from DNA which could be a confounding factor in assays that examine cell lysates. These enzymes lack the broad activity of UNG2 and are selective for uracil bases in specific contexts such as U/G mismatches (thymidine DNA glycosylase (TDG) and methyl-CpG-binding domain protein 4 (MBD4)), or alternatively, only have uracil excision activity on ssDNA substrates (single-stranded selective monofunctional uracil DNA glycosylase (SMUG1)) (Schormann et al., 2014). UNG1 and UNG2 are orders of magnitude faster than these alternative glycosylases and account for the vast majority of uracil base excision activity in cells (Esadze, Rodriguez, Weiser, et al., 2017; Grogan et al., 2011; Kavli et al., 2002). Nonetheless, the simplest control to confirm that uracil excision activity in lysates derives from UNG1 or UNG2 is to treat the reaction with uracil DNA glycosylase inhibitor protein UGI, which derives from bacteriophage and is highly specific for binding the UNG active site (Mol et al., 1995; Sanderson & Mosbaugh, 1996; Wang & Mosbaugh, 1989). The inhibitor protein is stable when expressed in mammalian cells or *E. coli* and is commonly commercially available (Christenson et al., 2021; Grogan et al., 2011).

2.3 Effects of protein-protein interactions on UNG2 activity

The N-terminal domain of UNG2 facilitates its interactions with its well-known binding partners proliferating cell nuclear antigen (PCNA) and replication protein A (RPA). The binding site for PCNA on UNG2 is at its distal N-terminus (Fig. 1A) (Otterlei et al., 1999). PCNA is a processivity clamp that encircles dsDNA and slides along it, and UNG2 has higher activity when PCNA is bound to uracilated dsDNA (Ko & Bennett, 2005). Conversely, RPA is a high affinity ssDNA binding protein that interacts with UNG2's N-terminal domain through a short helix that forms immediately adjacent to its catalytic domain (Fig. 1A) (Buchinger et al., 2017; Otterlei et al., 1999). The interaction site for RPA on UNG2's N-terminal domain coincidentally overlaps with the DNA binding region of the N-terminal domain and is also found on UNG1 despite the absence of RPA in mitochondria (Rodriguez et al., 2020). RPA can simultaneously interact with ssDNA and UNG2, and RPA targets UNG2 activity toward uracil bases located in nearby DNA regions (Kavli et al., 2021; Weiser, 2020; Weiser et al., 2018). Several other human proteins have been proposed to bind UNG2, but their interaction sites and influence on UNG2 function are less characterized. These include proteins that reportedly promote UNG2 activity (Rev1 and XRCC1) or reduce UNG2 activity (FAM72A) in different contexts (Akbari et al., 2010; Feng et al., 2021; Guo et al., 2008; Rogier et al., 2021; Zan et al., 2012).

The structure of the uracilated DNA substrate and the origin of the activity measurement should be considered in assays that analyze how other proteins influence UNG2. For example, UNG2's binding partner RPA interacts with ssDNA, but if ssDNA sites are not available, RPA will destabilize helical dsDNA (including hairpins) causing it to melt or bubble, thereby creating a new ssDNA section that it can bind (Chen, Le, Basu, Chazin, & Yan, 2015; De Vlaminck et al., 2010; Eckerich, Fackelmayer, & Knippers, 2001; Lao, Lee, & Wold, 1999; Pestryakov, Khlimankov, Bochkareva, Bochkarev, & Lavrik, 2004). RPA binding also results in significant changes to the local structure and dynamics of nucleotides. For these reasons, end-labeled uracilated substrates containing ssDNA sections of defined length have most successfully been employed to understand RPA's effects on UNG2 (Kavli et al., 2021; Weiser, 2020; Weiser et al., 2018). Assays examining the effects of PCNA on UNG2 activity have different requirements. First, PCNA requires the clamp loader replication factor C (RFC) in order to efficiently encircle dsDNA. In the absence of RFC, PCNA cannot encircle dsDNA unless it slips onto the end of a helix, which is an extremely inefficient process (Burgers & Yoder, 1993). Moreover, RFC requires a primer/template junction for efficient PCNA loading onto dsDNA, and PCNA will slide off the ends of the dsDNA it encircles unless a physical barrier impedes its movement (Burgers & Yoder, 1993; Hedglin, Aitha, & Benkovic, 2017). Thus far, early experiments examining PCNA's stimulatory effects on UNG2 activity using recombinant proteins have been limited to assays using long dsDNA substrates (~6000 base pairs) with randomly incorporated [^{3}H]deoxyuracil bases (Ko & Bennett, 2005).

2.4 Poor substrates of UNG2

UNG2 does not efficiently remove uracil bases from certain types of DNA that deviate from conventional single-stranded and double-stranded structures. For example, UNG2 activity is reduced on uracilated dsDNA that is wrapped around histones and assembled into nucleosomes (Ye et al., 2012). Excision of uracil bases from nucleosomal DNA is affected by the surface accessibility of the uracil site, changes to the local structure of DNA grooves, steric hindrance from histones tails near the uracil site, and the local dynamics and flexibility of the DNA (Cole, Tabor-Godwin, & Hayes, 2010; Hinz, Rodriguez, & Smerdon, 2010; Ye et al., 2012). Uracil bases positioned adjacent to G-quadruplex structures are also less favorable for UNG2, as are uracil bases in DNA loops regions (Holton & Larson, 2016; Kumar & Varshney, 1994). Together, the strong preference of UNG2 for uracil bases in exposed DNA supports its protective function on unwound DNA at the replication fork in cells, and also supports its pro-mutagenic action at transcription bubbles and R-loops during somatic hypermutation and class-switch recombination (Otterlei et al., 1999; Sohail, Klapacz, Samaranayake, Ullah, & Bhagwat, 2003). The roving activity of UNG2 during replicative senescence is low when chromatin is compact, and indeed, UNG2 protein levels are significantly reduced in the cell after S-phase (Fischer, Muller-Weeks, & Caradonna, 2004; Hagen et al., 2008).

2.5 Production of recombinant human UNG2 for enzyme assays

The enzymatic properties of UNG2 on different DNA substrates are ideally determined in assays performed with purified protein. Full-length UNG2 and its catalytic domain can be produced for enzyme assays using standard procedures for recombinant protein expression and purification (Bharati et al., 1998; Cravens et al., 2014; Weiser, Stivers, & Cole, 2017). The catalytic domain is especially stable and robustly produced in bacterial culture where it resembles its *E. coli* analog UNG (Fig. 1B). However, the expression of full-length UNG2 in *E. coli* is greatly enhanced by coexpressing a SUMO domain fused to the protein upstream of its N-terminal domain (Weiser et al., 2017). The SUMO domain "protects" the disordered N-terminus of UNG2, and without it present, the bacterial cells degrade the vast majority of the N-terminal domain to yield a mixture of truncated proteins. The SUMO domain can be removed from the N-terminus of UNG2 during chromatographic purification of the full-length enzyme using the SUMO protease ULP1 from *S. cerevisiae* (Weiser et al., 2017). A variety of truncated UNG2 proteins lacking sections of the disordered N-terminal domain have also been produced using this method, and these proteins behave as stable as the full-length protein and catalytic domain (Weiser, 2020; Weiser et al., 2017).

3. Future utility for assays in discovering UNG2 inhibitors and assessing novel protein-protein interactions

The important role of UNG2 in both normal physiology and disease has become clear through our understanding of its enzymatic properties. UNG2 and base excision repair ordinarily serve protective functions to maintain genomic integrity. However, one important area for future research is whether inhibition of UNG2 has therapeutic potential to enhance the effectiveness of thymidylate synthase inhibitors used to treat cancer. The mechanism of toxicity for thymidylate synthase inhibitors involves the elevation of nuclear dUTP levels and the prolific incorporation of uracil bases into the replicating genome, which creates instability through replication fork stalling and increased potential for DNA strand breaks (Bulgar et al., 2012; Weeks et al., 2014). Under these conditions, UNG2 activity counteracts the efficacy of thymidylate synthase inhibitors in cancer cells by removing uracil bases and initiating pathways to restore DNA integrity. Multiple studies have reported that genetic knockout of UNG2 results in cell lines that are hypersensitive to thymidylate synthase inhibitors, suggesting that UNG2 inhibitors might synergize with thymidylate synthase inhibitors (Bulgar et al., 2012; Christenson et al., 2021; Huehls et al., 2016;

Seiple et al., 2006; Showler & Weiser, 2020; Weeks et al., 2014; Yan et al., 2018). The molecular beacon substrate discussed in Section 2.2 is an example of a fluorescent probe that was already used in high-throughput screening to discover proof-of-principle inhibitors of UNG2 (Jiang et al., 2005; Jiang, Chung, Krosky, & Stivers, 2006; Nguyen et al., 2021). An important feature of this assay is the use of fluorescein as the fluorescent reporter for UNG2 activity, which has high excitation/emission wavelengths (495/520 nm) that do not typically overlap with the absorption wavelengths of high-throughput library compounds. Additional UNG2 assays have also been developed for potential use in high-throughput screening (Tao et al., 2015; Zhang, Li, Li, & Zhang, 2018; Zhu et al., 2018). Potent UNG2 inhibitors may also help us understand different mechanisms of toxicity that arise from different thymidylate synthase inhibitors in various cell lines (Bulgar et al., 2012; Christenson et al., 2021; Huehls et al., 2016; Seiple et al., 2006; Weeks et al., 2014; Yan et al., 2018).

There also remains much to explore regarding the influence of protein–protein interactions on the enzymatic activity of UNG2 during replication, in the presence and absence of thymidylate synthase inhibitors, and also during antibody diversification processes. In addition to binding PCNA and RPA, UNG2 has been found in cellular protein complexes with other base excision repair proteins and the scaffolding protein XRCC1 (Akbari et al., 2010). The architecture and stability of such complexes have not been reported, nor do we understand the influence of complex formation on uracil search and recognition mechanisms. The association of UNG2 with XRCC1 in base excision repair complexes likely complements its alternative association with PCNA or RPA to facilitate uracil excision in different DNA substrate contexts. Likewise, different proteins appear to tune the activity of UNG2 during class switch recombination and somatic hypermutation. In a recent example, there is evidence for a direct interaction between the N-terminal domain of UNG2 and FAM72A, a regulator of class switch recombination that antagonizes UNG2 activity on dsDNA (Feng et al., 2021; Rogier et al., 2021). The mechanism by which FAM72A modulates UNG2 activity, and the relevance of the DNA substrate structure, are currently unclear. Moreover, additional proteins including RPA are recruited to sites of programmed DNA mutagenesis and could influence FAM72A's interaction with UNG2 (Chaudhuri, Khuong, & Alt, 2004). Assays described in this chapter that elucidated many aspects of UNG2 function should provide the foundation for future experiments that examine these possibilities.

Acknowledgments

The authors thank Dr. James Stivers (Johns Hopkins University) for critical reading of the manuscript, and the authors thank NIH grant R01GM135152 for generous financial support.

References

Akbari, M., Otterlei, M., Peña-Diaz, J., & Krokan, H. E. (2007). Different organization of base excision repair of uracil in DNA in nuclei and mitochondria and selective upregulation of mitochondrial uracil-DNA glycosylase after oxidative stress. *Neuroscience*, *145*(4), 1201–1212. https://doi.org/10.1016/j.neuroscience.2006.10.010.

Akbari, M., Solvang-Garten, K., Hanssen-Bauer, A., Lieske, N. V., Pettersen, H. S., Pettersen, G. K., et al. (2010). Direct interaction between XRCC1 and UNG2 facilitates rapid repair of uracil in DNA by XRCC1 complexes. *DNA Repair*, *9*(7), 785–795. https://doi.org/10.1016/j.dnarep.2010.04.002.

Bellamy, S. R. W., Krusong, K., & Baldwin, G. S. (2007). A rapid reaction analysis of uracil DNA glycosylase indicates an active mechanism of base flipping. *Nucleic Acids Research*, *35*(5), 1478–1487. https://doi.org/10.1093/nar/gkm018.

Bharati, S., Krokan, H. E., Kristiansen, L., Otterlei, M., & Slupphaug, G. (1998). Human mitochondrial uracil-DNA glycosylase preform (UNG1) is processed to two forms one of which is resistant to inhibition by AP sites. *Nucleic Acids Research*, *26*(21), 4953–4959. https://doi.org/10.1093/nar/26.21.4953.

Buchinger, E., Wiik, S.Å., Kusnierczyk, A., Rabe, R., Aas, P. A., Kavli, B., et al. (2017). Backbone 1H, 13C and 15N chemical shift assignment of full-length human uracil DNA glycosylase UNG2. *Biomolecular NMR Assignments*, *12*, 15–22. https://doi.org/10.1007/s12104-017-9772-5.

Bulgar, A. D., Weeks, L. D., Miao, Y., Yang, S., Xu, Y., Guo, C., et al. (2012). Removal of uracil by uracil DNA glycosylase limits pemetrexed cytotoxicity: Overriding the limit with methoxyamine to inhibit base excision repair. *Cell Death & Disease*, *3*(1), e252. https://doi.org/10.1038/cddis.2011.135.

Burgers, P. M., & Yoder, B. L. (1993). ATP-independent loading of the proliferating cell nuclear antigen requires DNA ends. *The Journal of Biological Chemistry*, *268*(27), 19923–19926.

Chaudhuri, J., Khuong, C., & Alt, F. W. (2004). Replication protein a interacts with AID to promote deamination of somatic hypermutation targets. *Nature*, *430*(7003), 992–998. https://doi.org/10.1038/nature02821.

Chen, J., Le, S., Basu, A., Chazin, W. J., & Yan, J. (2015). Mechanochemical regulations of RPA's binding to ssDNA. *Scientific Reports*, *5*, 9296. https://doi.org/10.1038/srep09296.

Christenson, E. S., Gizzi, A., Cui, J., Egleston, M., Seamon, K. J., DePasquale, M., et al. (2021). Inhibition of human uracil DNA glycosylase sensitizes a large fraction of colorectal Cancer cells to 5-fluorodeoxyuridine and raltitrexed but not fluorouracil. *Molecular Pharmacology*, *99*(6), 412–425. https://doi.org/10.1124/molpharm.120.000191.

Cole, H. A., Tabor-Godwin, J. M., & Hayes, J. J. (2010). Uracil DNA glycosylase activity on nucleosomal DNA depends on rotational orientation of targets. *The Journal of Biological Chemistry*, *285*(4), 2876–2885. https://doi.org/10.1074/jbc.M109.073544.

Cravens, S. L., Hobson, M., & Stivers, J. T. (2014). Electrostatic properties of complexes along a DNA glycosylase damage search pathway. *Biochemistry*, *53*(48), 7680–7692. https://doi.org/10.1021/bi501011m.

Cravens, S. L., Schonhoft, J. D., Rowland, M. M., Rodriguez, A. A., Anderson, B. G., & Stivers, J. T. (2015). Molecular crowding enhances facilitated diffusion of two human DNA glycosylases. *Nucleic Acids Research*, *43*(8), 4087–4097. https://doi.org/10.1093/nar/gkv301.

Cravens, S. L., & Stivers, J. T. (2016). Comparative effects of ions, molecular crowding, and bulk DNA on the damage search mechanisms of hOGG1 and hUNG. *Biochemistry*, *55*(37), 5230–5242. https://doi.org/10.1021/acs.biochem.6b00482.

De Vlaminck, I., Vidic, I., van Loenhout, M. T. J., Kanaar, R., Lebbink, J. H. G., & Dekker, C. (2010). Torsional regulation of hRPA-induced unwinding of double-stranded DNA. *Nucleic Acids Research*, *38*(12), 4133–4142. https://doi.org/10.1093/nar/gkq067.

Doseth, B., Ekre, C., Slupphaug, G., Krokan, H. E., & Kavli, B. (2012). Strikingly different properties of uracil-DNA glycosylases UNG2 and SMUG1 may explain divergent roles in processing of genomic uracil. *DNA Repair*, *11*(6), 587–593. https://doi.org/10.1016/j.dnarep.2012.03.003.

Eckerich, C., Fackelmayer, F. O., & Knippers, R. (2001). Zinc affects the conformation of nucleoprotein filaments formed by replication protein a (RPA) and long natural DNA molecules. *Biochimica et Biophysica Acta (BBA) - molecular Cell Research*, *1538*(1), 67–75. https://doi.org/10.1016/S0167-4889(00)00138-5.

Eckstein, F. (2014). Phosphorothioates, essential components of therapeutic oligonucleotides. *Nucleic Acid Therapeutics*, *24*(6), 374–387. https://doi.org/10.1089/nat.2014.0506.

Eftedal, I., Guddal, P. H., Slupphaug, G., Volden, G., & Krokan, H. E. (1993). Consensus sequences for good and poor removal of uracil from double stranded DNA by uracil-DNA glycosylase. *Nucleic Acids Research*, *21*(9), 2095–2101.

Esadze, A., Rodriguez, G., Cravens, S. L., & Stivers, J. T. (2017). AP-endonuclease 1 accelerates turnover of human 8-Oxoguanine DNA glycosylase by preventing retrograde binding to the Abasic-site product. *Biochemistry*, *56*(14), 1974–1986. https://doi.org/10.1021/acs.biochem.7b00017.

Esadze, A., Rodriguez, G., Weiser, B. P., Cole, P. A., & Stivers, J. T. (2017). Measurement of nanoscale DNA translocation by uracil DNA glycosylase in human cells. *Nucleic Acids Research*, *45*(21), 12413–12424. https://doi.org/10.1093/nar/gkx848.

Feng, Y., Li, C., Stewart, J. A., Barbulescu, P., Seija Desivo, N., Álvarez-Quilón, A., et al. (2021). FAM72A antagonizes UNG2 to promote mutagenic repair during antibody maturation. *Nature*, *600*(7888), 324–328. https://doi.org/10.1038/s41586-021-04144-4.

Fischer, J. A., Muller-Weeks, S., & Caradonna, S. (2004). Proteolytic degradation of the nuclear isoform of uracil-DNA glycosylase occurs during the S phase of the cell cycle. *DNA Repair*, *3*(5), 505–513. https://doi.org/10.1016/j.dnarep.2004.01.012.

Grogan, B. C., Parker, J. B., Guminski, A. F., & Stivers, J. T. (2011). Effect of the thymidylate synthase inhibitors on dUTP and TTP pool levels and the activities of DNA repair glycosylases on uracil and 5-fluorouracil in DNA. *Biochemistry*, *50*(5), 618–627. https://doi.org/10.1021/bi102046h.

Guo, C., Zhang, X., Fink, S. P., Platzer, P., Wilson, K., Willson, J. K. V., et al. (2008). Ugene, a newly identified protein that is commonly overexpressed in Cancer and binds uracil DNA glycosylase. *Cancer Research*, *68*(15), 6118–6126. https://doi.org/10.1158/0008-5472.CAN-08-1259.

Hagen, L., Kavli, B., Sousa, M. M. L., Torseth, K., Liabakk, N. B., Sundheim, O., et al. (2008). Cell cycle-specific UNG2 phosphorylations regulate protein turnover, activity and association with RPA. *The EMBO Journal*, *27*(1), 51–61. https://doi.org/10.1038/sj.emboj.7601958.

Haldar, T., Jha, J. S., Yang, Z., Nel, C., Housh, K., Cassidy, O. J., et al. (2022). Unexpected complexity in the products arising from NaOH-, heat-, amine-, and glycosylase-induced Strand cleavage at an Abasic site in DNA. *Chemical Research in Toxicology*, *35*(2), 218–232. https://doi.org/10.1021/acs.chemrestox.1c00409.

Hedglin, M., Aitha, M., & Benkovic, S. J. (2017). Monitoring the retention of human proliferating cell nuclear antigen at primer/template junctions by proteins that bind single-stranded DNA. *Biochemistry*, *56*(27), 3415–3421. https://doi.org/10.1021/acs.biochem.7b00386.

Hinz, J. M., Rodriguez, Y., & Smerdon, M. J. (2010). Rotational dynamics of DNA on the nucleosome surface markedly impact accessibility to a DNA repair enzyme. *Proceedings of the National Academy of Sciences of the United States of America, 107*(10), 4646–4651. https://doi.org/10.1073/pnas.0914443107.

Holton, N. W., & Larson, E. D. (2016). G-quadruplex DNA structures can interfere with uracil glycosylase activity in vitro. *Mutagenesis, 31*(4), 385–392. https://doi.org/10.1093/mutage/gev083.

Huehls, A. M., Huntoon, C. J., Joshi, P. M., Baehr, C. A., Wagner, J. M., Wang, X., et al. (2016). Genomically incorporated 5-fluorouracil that escapes UNG-Initiated Base excision repair blocks DNA replication and activates homologous recombination. *Molecular Pharmacology, 89*(1), 53–62. https://doi.org/10.1124/mol.115.100164.

Imai, K., Slupphaug, G., Lee, W.-I., Revy, P., Nonoyama, S., Catalan, N., et al. (2003). Human uracil-DNA glycosylase deficiency associated with profoundly impaired immunoglobulin class-switch recombination. *Nature Immunology, 4*(10), 1023–1028. https://doi.org/10.1038/ni974.

Jiang, Y. L., Chung, S., Krosky, D. J., & Stivers, J. T. (2006). Synthesis and high-throughput evaluation of triskelion uracil libraries for inhibition of human dUTPase and UNG2. *Bioorganic & Medicinal Chemistry, 14*(16), 5666–5672. https://doi.org/10.1016/j.bmc.2006.04.022.

Jiang, Y. L., Krosky, D. J., Seiple, L., & Stivers, J. T. (2005). Uracil-directed ligand tethering: An efficient strategy for uracil DNA glycosylase (UNG) inhibitor development. *Journal of the American Chemical Society, 127*(49), 17412–17420. https://doi.org/10.1021/ja055846n.

Kaiser, S. M., & Emerman, M. (2006). Uracil DNA glycosylase is dispensable for human immunodeficiency virus type 1 replication and does not contribute to the antiviral effects of the cytidine deaminase Apobec3G. *Journal of Virology, 80*(2), 875–882. https://doi.org/10.1128/JVI.80.2.875-882.2006.

Kavli, B., Iveland, T. S., Buchinger, E., Hagen, L., Liabakk, N. B., Aas, P. A., et al. (2021). RPA2 winged-helix domain facilitates UNG-mediated removal of uracil from ssDNA; implications for repair of mutagenic uracil at the replication fork. *Nucleic Acids Research, 49*(7), 3948–3966. https://doi.org/10.1093/nar/gkab195.

Kavli, B., Sundheim, O., Akbari, M., Otterlei, M., Nilsen, H., Skorpen, F., et al. (2002). HUNG2 is the major repair enzyme for removal of Uracil from U:A Matches, U:G Mismatches, and U in single-stranded DNA, with hSMUG1 as a broad specificity backup. *Journal of Biological Chemistry, 277*(42), 39926–39936. https://doi.org/10.1074/jbc.M207107200.

Ko, R., & Bennett, S. E. (2005). Physical and functional interaction of human nuclear uracil-DNA glycosylase with proliferating cell nuclear antigen. *DNA Repair, 4*(12), 1421–1431. https://doi.org/10.1016/j.dnarep.2005.08.006.

Kozhukhar, N., Spadafora, D., Fayzulin, R., Shokolenko, I. N., & Alexeyev, M. (2016). The efficiency of the translesion synthesis across abasic sites by mitochondrial DNA polymerase is low in mitochondria of 3T3 cells. *Mitochondrial DNA. Part A, DNA Mapping, Sequencing, and Analysis, 27*(6), 4390–4396. https://doi.org/10.3109/19401736.2015.1089539.

Krokan, H., & Wittwer, C. U. (1981). Uracil DNa-glycosylase from HeLa cells: General properties, substrate specificity and effect of uracil analogs. *Nucleic Acids Research, 9*(11), 2599–2613. https://doi.org/10.1093/nar/9.11.2599.

Kumar, N. V., & Varshney, U. (1994). Inefficient excision of uracil from loop regions of DNA oligomers by E. coli uracil DNA glycosylase. *Nucleic Acids Research, 22*(18), 3737–3741. https://doi.org/10.1093/nar/22.18.3737.

Lao, Y., Lee, C. G., & Wold, M. S. (1999). Replication protein a interactions with DNA. 2. Characterization of double-stranded DNA-binding/Helix-destabilization activities and the role of the zinc-finger domain in DNA interactions. *Biochemistry, 38*(13), 3974–3984. https://doi.org/10.1021/bi982371m.

Liu, Z., Hu, Y., Gong, Y., Zhang, W., Liu, C., Wang, Q., et al. (2016). Hydrogen peroxide mediated mitochondrial UNG1-PRDX3 interaction and UNG1 degradation. *Free Radical Biology & Medicine, 99*, 54–62. https://doi.org/10.1016/j.freeradbiomed.2016.07.030.

Maksimenko, A., Ishchenko, A. A., Sanz, G., Laval, J., Elder, R. H., & Saparbaev, M. K. (2004). A molecular beacon assay for measuring base excision repair activities. *Biochemical and Biophysical Research Communications, 319*(1), 240–246. https://doi.org/10.1016/j.bbrc.2004.04.179.

Mehta, A., Raj, P., Sundriyal, S., Gopal, B., & Varshney, U. (2021). Use of a molecular beacon based fluorescent method for assaying uracil DNA glycosylase (Ung) activity and inhibitor screening. *Biochemistry and Biophysics Reports, 26*, 100954. https://doi.org/10.1016/j.bbrep.2021.100954.

Mol, C. D., Arvai, A. S., Slupphaug, G., Kavli, B., Alseth, I., Krokan, H. E., et al. (1995). Crystal structure and mutational analysis of human uracil-DNA glycosylase: Structural basis for specificity and catalysis. *Cell, 80*(6), 869–878. https://doi.org/10.1016/0092-8674(95)90290-2.

Nguyen, M. T., Moiani, D., Ahmed, Z., Arvai, A. S., Namjoshi, S., Shin, D. S., et al. (2021). An effective human uracil-DNA glycosylase inhibitor targets the open pre-catalytic active site conformation. *Progress in Biophysics and Molecular Biology, 163*, 143–159. https://doi.org/10.1016/j.pbiomolbio.2021.02.004.

Nilsen, H., Otterlei, M., Haug, T., Solum, K., Nagelhus, T. A., Skorpen, F., et al. (1997). Nuclear and mitochondrial uracil-DNA glycosylases are generated by alternative splicing and transcription from different positions in the UNG gene. *Nucleic Acids Research, 25*(4), 750–755.

Nilsen, H., Yazdankhah, S. P., Eftedal, I., & Krokan, H. E. (1995). Sequence specificity for removal of uracil from U.a pairs and U.G mismatches by uracil-DNA glycosylase from Escherichia coli, and correlation with mutational hotspots. *FEBS Letters, 362*(2), 205–209.

Olsen, L. C., Aasland, R., Wittwer, C. U., Krokan, H. E., & Helland, D. E. (1989). Molecular cloning of human uracil-DNA glycosylase, a highly conserved DNA repair enzyme. *The EMBO Journal, 8*(10), 3121–3125.

Otterlei, M., Warbrick, E., Nagelhus, T. A., Haug, T., Slupphaug, G., Akbari, M., et al. (1999). Post-replicative base excision repair in replication foci. *The EMBO Journal, 18*(13), 3834–3844. https://doi.org/10.1093/emboj/18.13.3834.

Parikh, S. S., Mol, C. D., Slupphaug, G., Bharati, S., Krokan, H. E., & Tainer, J. A. (1998). Base excision repair initiation revealed by crystal structures and binding kinetics of human uracil-DNA glycosylase with DNA. *The EMBO Journal, 17*(17), 5214–5226. https://doi.org/10.1093/emboj/17.17.5214.

Parikh, S. S., Walcher, G., Jones, G. D., Slupphaug, G., Krokan, H. E., Blackburn, G. M., et al. (2000). Uracil-DNA glycosylase-DNA substrate and product structures: Conformational strain promotes catalytic efficiency by coupled stereoelectronic effects. *Proceedings of the National Academy of Sciences of the United States of America, 97*(10), 5083–5088. https://doi.org/10.1073/pnas.97.10.5083.

Parker, J. B., Bianchet, M. A., Krosky, D. J., Friedman, J. I., Amzel, L. M., & Stivers, J. T. (2007). Enzymatic capture of an extrahelical thymine in the search for uracil in DNA. *Nature, 449*(7161), 433–437. https://doi.org/10.1038/nature06131.

Peled, J. U., Kuang, F. L., Iglesias-Ussel, M. D., Roa, S., Kalis, S. L., Goodman, M. F., et al. (2008). The biochemistry of somatic hypermutation. *Annual Review of Immunology, 26*, 481–511. https://doi.org/10.1146/annurev.immunol.26.021607.090236.

Pestryakov, P. E., Khlimankov, D. Y., Bochkareva, E., Bochkarev, A., & Lavrik, O. I. (2004). Human replication protein a (RPA) binds a primer–template junction in the absence of its major ssDNA-binding domains. *Nucleic Acids Research, 32*(6), 1894–1903. https://doi.org/10.1093/nar/gkh346.

Pettersen, H. S., Visnes, T., Vågbø, C. B., Svaasand, E. K., Doseth, B., Slupphaug, G., et al. (2011). UNG-initiated base excision repair is the major repair route for 5-fluorouracil in DNA, but 5-fluorouracil cytotoxicity depends mainly on RNA incorporation. *Nucleic Acids Research, 39*(19), 8430–8444. https://doi.org/10.1093/nar/gkr563.

Porecha, R. H., & Stivers, J. T. (2008). Uracil DNA glycosylase uses DNA hopping and short-range sliding to trap extrahelical uracils. *Proceedings of the National Academy of Sciences of the United States of America, 105*(31), 10791–10796. https://doi.org/10.1073/pnas.0801612105.

Rodriguez, G., Esadze, A., Weiser, B. P., Schonhoft, J. D., Cole, P. A., & Stivers, J. T. (2017). Disordered N-terminal domain of human uracil DNA glycosylase (hUNG2) enhances DNA translocation. *ACS Chemical Biology, 12*(9), 2260–2263. https://doi.org/10.1021/acschembio.7b00521.

Rodriguez, G., Orris, B., Majumdar, A., Bhat, S., & Stivers, J. T. (2020). Macromolecular crowding induces compaction and DNA binding in the disordered N-terminal domain of hUNG2. *DNA Repair, 86*, 102764. https://doi.org/10.1016/j.dnarep.2019.102764.

Rogier, M., Moritz, J., Robert, I., Lescale, C., Heyer, V., Abello, A., et al. (2021). Fam72a enforces error-prone DNA repair during antibody diversification. *Nature, 600*(7888), 329–333. https://doi.org/10.1038/s41586-021-04093-y.

Saha, T., Sundaravinayagam, D., & Di Virgilio, M. (2021). Charting a DNA repair roadmap for immunoglobulin class switch recombination. *Trends in Biochemical Sciences, 46*(3), 184–199. https://doi.org/10.1016/j.tibs.2020.10.005.

Sanderson, R. J., & Mosbaugh, D. W. (1996). Identification of specific carboxyl groups on uracil-DNA glycosylase inhibitor protein that are required for activity. *The Journal of Biological Chemistry, 271*(46), 29170–29181. https://doi.org/10.1074/jbc.271.46.29170.

Schonhoft, J. D., & Stivers, J. T. (2012). Timing facilitated site transfer of an enzyme on DNA. *Nature Chemical Biology, 8*(2), 205–210. https://doi.org/10.1038/nchembio.764.

Schonhoft, J. D., & Stivers, J. T. (2013). DNA translocation by human uracil DNA glycosylase: The case of ssDNA and clustered Uracils. *Biochemistry, 52*(15), 2536–2544. https://doi.org/10.1021/bi301562n.

Schormann, N., Ricciardi, R., & Chattopadhyay, D. (2014). Uracil-DNA glycosylases-structural and functional perspectives on an essential family of DNA repair enzymes. *Protein Science: A Publication of the Protein Society, 23*(12), 1667–1685. https://doi.org/10.1002/pro.2554.

Seiple, L., Jaruga, P., Dizdaroglu, M., & Stivers, J. T. (2006). Linking uracil base excision repair and 5-fluorouracil toxicity in yeast. *Nucleic Acids Research, 34*(1), 140–151. https://doi.org/10.1093/nar/gkj430.

Showler, M. S., & Weiser, B. P. (2020). A possible link to uracil DNA glycosylase in the synergistic action of HDAC inhibitors and thymidylate synthase inhibitors. *Journal of Translational Medicine, 18*(1), 377. https://doi.org/10.1186/s12967-020-02555-x.

Skjeldam, H. K., Kassahun, H., Fensgård, O., SenGupta, T., Babaie, E., Lindvall, J. M., et al. (2010). Loss of Caenorhabditis elegans UNG-1 uracil-DNA glycosylase affects apoptosis in response to DNA damaging agents. *DNA Repair, 9*(8), 861–870. https://doi.org/10.1016/j.dnarep.2010.04.009.

Sohail, A., Klapacz, J., Samaranayake, M., Ullah, A., & Bhagwat, A. S. (2003). Human activation-induced cytidine deaminase causes transcription-dependent, strand-biased C to U deaminations. *Nucleic Acids Research, 31*(12), 2990–2994. https://doi.org/10.1093/nar/gkg464.

Stivers, J. T. (1998). 2-aminopurine fluorescence studies of base stacking interactions at abasic sites in DNA: Metal-ion and base sequence effects. *Nucleic Acids Research, 26*(16), 3837–3844. https://doi.org/10.1093/nar/26.16.3837.

Talpaert-Borlé, M., Clerici, L., & Campagnari, F. (1979). Isolation and characterization of a uracil-DNA glycosylase from calf thymus. *The Journal of Biological Chemistry, 254*(14), 6387–6391.

Tao, J., Song, P., Sato, Y., Nishizawa, S., Teramae, N., Tong, A., et al. (2015). A label-free and sensitive fluorescent method for the detection of uracil-DNA glycosylase activity. *Chemical Communications (Cambridge, England)*, *51*(5), 929–932. https://doi.org/10.1039/c4cc06170e.

Wang, Z., & Mosbaugh, D. W. (1989). Uracil-DNA glycosylase inhibitor gene of bacteriophage PBS2 encodes a binding protein specific for uracil-DNA glycosylase. *The Journal of Biological Chemistry*, *264*(2), 1163–1171.

Weeks, L. D., Zentner, G. E., Scacheri, P. C., & Gerson, S. L. (2014). Uracil DNA glycosylase (UNG) loss enhances DNA double strand break formation in human cancer cells exposed to pemetrexed. *Cell Death & Disease*, *5*(2), e1045. https://doi.org/10.1038/cddis.2013.477.

Weil, A. F., Ghosh, D., Zhou, Y., Seiple, L., McMahon, M. A., Spivak, A. M., et al. (2013). Uracil DNA glycosylase initiates degradation of HIV-1 cDNA containing misincorporated dUTP and prevents viral integration. *Proceedings of the National Academy of Sciences of the United States of America*, *110*(6), E448–E457. https://doi.org/10.1073/pnas.1219702110.

Weiser, B. P. (2020). Analysis of uracil DNA glycosylase (UNG2) stimulation by replication protein a (RPA) at ssDNA-dsDNA junctions. *Biochimica et Biophysica Acta (BBA) - Proteins and Proteomics*, *1868*(3), 140347. https://doi.org/10.1016/j.bbapap.2019.140347.

Weiser, B. P., Rodriguez, G., Cole, P. A., & Stivers, J. T. (2018). N-terminal domain of human uracil DNA glycosylase (hUNG2) promotes targeting to uracil sites adjacent to ssDNA-dsDNA junctions. *Nucleic Acids Research*, *46*(14), 7169–7178. https://doi.org/10.1093/nar/gky525.

Weiser, B. P., Stivers, J. T., & Cole, P. A. (2017). Investigation of N-terminal Phospho-regulation of uracil DNA glycosylase using protein Semisynthesis. *Biophysical Journal*, *113*(2), 393–401. https://doi.org/10.1016/j.bpj.2017.06.016.

Wollen Steen, K., Doseth, B., Westbye, P., Akbari, M., Kang, D., Falkenberg, M., et al. (2012). MtSSB may sequester UNG1 at mitochondrial ssDNA and delay uracil processing until the dsDNA conformation is restored. *DNA Repair*, *11*(1), 82–91. https://doi.org/10.1016/j.dnarep.2011.10.026.

Xiao, G., Tordova, M., Jagadeesh, J., Drohat, A. C., Stivers, J. T., & Gilliland, G. L. (1999). Crystal structure of Escherichia coli uracil DNA glycosylase and its complexes with uracil and glycerol: Structure and glycosylase mechanism revisited. *Proteins: Structure, Function, and Bioinformatics*, *35*(1), 13–24. https://doi.org/10.1002/(SICI)1097-0134(19990401)35:1<13::AID-PROT2>3.0.CO;2-2.

Yan, Y., Qing, Y., Pink, J. J., & Gerson, S. L. (2018). Loss of uracil DNA glycosylase selectively Resensitizes p53-mutant and -deficient cells to 5-FdU. *Molecular Cancer Research: MCR*, *16*(2), 212–221. https://doi.org/10.1158/1541-7786.MCR-17-0215.

Ye, Y., Stahley, M. R., Xu, J., Friedman, J. I., Sun, Y., McKnight, J. N., et al. (2012). Enzymatic excision of uracil residues in nucleosomes depends on the local DNA structure and dynamics. *Biochemistry*, *51*(30), 6028–6038. https://doi.org/10.1021/bi3006412.

Zan, H., White, C. A., Thomas, L. M., Mai, T., Li, G., Xu, Z., et al. (2012). Rev1 recruits Ung to switch regions and enhances dU glycosylation for immunoglobulin class switch DNA recombination. *Cell Reports*, *2*(5), 1220–1232. https://doi.org/10.1016/j.celrep.2012.09.029.

Zhang, Y., Li, Q.-N., Li, C.-C., & Zhang, C.-Y. (2018). Label-free and high-throughput bioluminescence detection of uracil-DNA glycosylase in cancer cells through tricyclic cascade signal amplification. *Chemical Communications (Cambridge, England)*, *54*(51), 6991–6994. https://doi.org/10.1039/c8cc03769h.

Zhu, J., Hao, Q., Liu, Y., Guo, Z., Rustam, B., & Jiang, W. (2018). Integrating DNA structure switch with branched hairpins for the detection of uracil-DNA glycosylase activity and inhibitor screening. *Talanta*, *179*, 51–56. https://doi.org/10.1016/j.talanta.2017.10.052.

CHAPTER FIFTEEN

Measurement of kinetic isotope effects on peptide hydroxylation using MALDI-MS

Michael A. Mingroni, Vanessa Chaplin Momaney, Alexandra N. Barlow, Isabella Jaen Maisonet, and Michael J. Knapp*

Department of Chemistry, University of Massachusetts, Amherst, MA, United States
*Corresponding author: e-mail address: mjknapp@umass.edu

Contents

Abstract

Primary kinetic isotope effects (KIEs) provide unique insight into enzymatic reactions, as they can reveal rate-limiting steps and detailed chemical mechanisms. HIF hydroxylases, part of a family of 2-oxoglutarate (2OG) oxygenases are central to the regulation of many crucial biological processes through O_2-sensing, but present a challenge to monitor due to the large size of the protein substrate and the similarity between native and hydroxylated substrate. MALDI-TOF MS is a convenient tool to measure peptide masses, which can also be used to measure the discontinuous kinetics of peptide hydroxylation for Factor Inhibiting HIF (FIH). Using this technique, rate data can be observed from the mole-fraction of CTAD and CTAD-OH in small volumes, allowing noncompetitive H/D KIEs to be measured. Slow dCTAD substrate leads to extensive uncoupling of O_2 consumption from peptide hydroxylation, leading to enzyme autohydroxylation, which is observed using UV–vis spectroscopy. Simultaneously measuring both the normal product, CTAD-OH, and the uncoupled product, autohydroxylated enzyme, the KIE on the microscopic step of hydrogen atom transfer (HAT) can be estimated. MALDI-MS analysis is a strong method for monitoring reactions that hydroxylate peptides, and can be generalized to other similar reactions, and simultaneous kinetic detection of branched products can provide valuable insight on microscopic KIEs at intermediate mechanistic steps.

Abbreviations

2OG	2-oxoglutarate
CTAD	C-terminal transactivation domain
FIH	factor inhibiting HIF
HAT	hydrogen atom transfer
HIF	hypoxia inducible factor
KIE	kinetic isotope effect
LC-ESI	liquid chromatography-electrospray ionization
MALDI-TOF	matrix assisted laser desorption ionization-time of flight
UV–Vis	ultraviolet–visible spectroscopy
α-CHCA	α-cyano-4-hydroxycinnamic acid

1. Introduction

Primary kinetic isotope effects (KIEs) provide unique insight into enzymatic reactions, as they can reveal rate-limiting steps and detailed chemical mechanisms (Kohen & Limbach 2005; Karandashev, Xu, Meuwly, Vaníček, & Richardson, 2017). This is particularly important for 2-oxoglutarate (2OG) oxygenases as the enzymatic reactions serve to regulate many crucial biological processes such as transcription, metabolism, and proliferation (Islam, Leissing, Chowdhury, Hopkinson, & Schofield, 2018; Schofield & Ratcliffe, 2004; Semenza, 2011). HIF hydroxylases, in particular, are central to these processes through O_2-sensing, but present a challenge to monitor due to the large size of the protein substrate and the similarity between native and hydroxylated protein (Semenza, 2014). In order to screen for enzyme inhibitors and interrogate the chemical mechanisms of these enzymes, a strong method for monitoring reactions that hydroxylate peptides is necessary.

Our research group studies the chemistry of an enzyme called Factor Inhibiting HIF-1 (FIH), which hydroxylates the Asn803 residue within the C-terminal transactivation domain (CTAD) domain of HIF-1α (Iyer, Chaplin, Knapp, & Solomon, 2018; Lando, Peet, Whelan, Gorman, & Whitelaw, 2002). The chemical mechanism of FIH proceeds through two sequential half-reactions, in which O_2 reacts with 2OG to form succinate and the ferryl active oxidant (Eq. 1a); then the ferryl reacts with the methylene position of Asn803 within CTAD, to form the hydroxylated product (CTAD-OH, Eq. 1b).

$$O_2 + 2OG + Fe^{2+} \rightarrow CO_2 + \text{succinate} + (FeO)^{2+} \qquad (1a)$$

$$(FeO)^{2+} + CTAD \rightarrow Fe^{2+} + CTAD - OH \qquad (1b)$$

Several approaches can be used to measure the activity of FIH, as for other 2OG hydroxylases. Quantifying O_2 uptake, CO_2 release, or succinate production are accessible methods, but they suffer from measuring only the oxidative reaction (Eq. 1a) which can be uncoupled from the subsequent hydroxylation (Eq. 1b). This uncoupling of O_2 consumption from product formation occurs when the active oxidant in many 2OG oxygenases, an Fe(IV)-oxo, fails to react with the prime substrate (McCusker & Klinman, 2009).

Although the reaction in FIH is tightly coupled when using wild-type enzyme and abundant CTAD peptide, under other conditions FIH will autohydroxylate the FIH-Trp296 residue to inactivate the enzyme (Chen et al., 2008). This internal competition (Scheme 1) between the sidechains of the prime substrate (CTAD-Asn803) and the enzyme (FIH-Trp296) leads to variable amounts of prime product and autohydroxylated enzyme depending on the enzyme variant. Using CTAD containing the 2,3,3-d_3-Asn803 residue (dCTAD), we found autohydroxylation was a common outcome when FIH was confronted with a slow substrate (Mingroni & Knapp, 2021). As the internal competition was dependent on the isotopic composition of CTAD, the KIE interpretations were obscured by variable extents of autohydroxylation.

The KIE of FIH can reveal much about its mechanistic chemistry. Given the second half-reaction (Eq. 1b) involves the cleavage of the

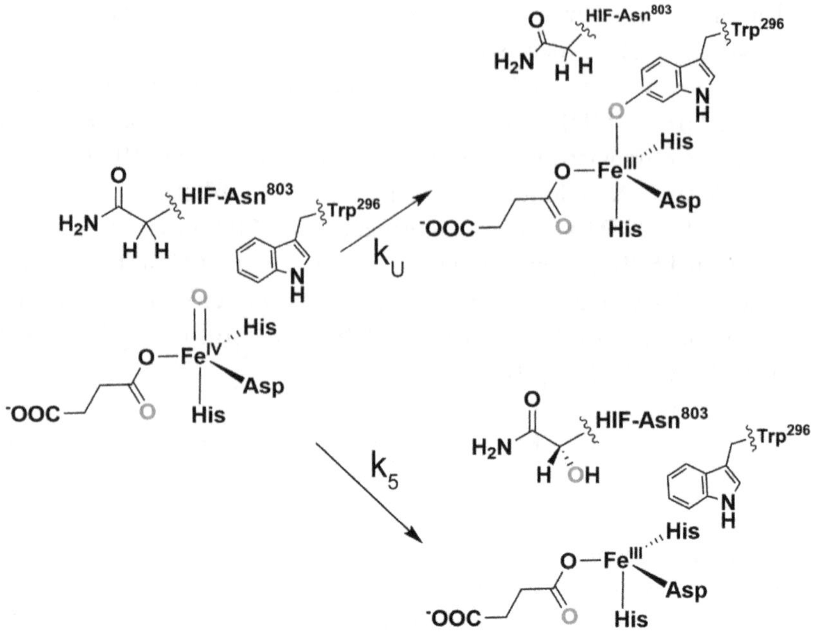

Scheme 1 Internal competition between hydroxylation at the CTAD (HIF-Asn803) and autohydroxylation at FIH-Trp296.

C_β-H bond of Asn803, the size of the H/D KIE will reveal the extent to which the C—H cleavage limits overall turnover of FIH, as in the case of taurine dioxygenase (Price, Barr, Glass, Krebs, & Bollinger, 2003). Furthermore, there is a potential for hydrogen atom tunneling, leading to large KIEs which may reveal a richer reaction coordinate involving protein dynamics (Klinman & Kohen, 2013). By simultaneously measuring both the prime product formation and the autohydroxylated enzyme, it is possible to estimate the KIE on the C—H cleavage step. MALDI provides an excellent method to measure discontinuous enzyme activity that circumvents the problem with uncoupled O_2 consumption. By using a high-resolution MALDI, both substrate and product can be observed with small sample volumes, thereby avoiding any need for chromatographic separation.

1.1 MALDI for discontinuous kinetics

There are numerous advantages to using MALDI to measure protein hydroxylation, and a few disadvantages. As MALDI sample preparation involves drying peptide solutions with a crystallization matrix, some of the pros and cons are defined by this. MALDI sampling is discontinuous, which means that each sample or time-point must be prepared separately and therefore requires more handling than continuous methods such as ultraviolet–visible spectroscopy (UV–vis). Sample sizes can be quite small, with our method easily using $1 \mu L$ of peptide solution per data point. MALDI tolerates biological buffers well, which is a significant advantage over methods requiring LC-based separations such as electrospray mass spectrometry (LC-ESI-MS) (Dreisewerd, 2014).

Some of the complications arise from the need for the MALDI matrix to crystallize; however, peptide solutions typically work well for MALDI. As buffer components remain on the matrix spots, any buffer components can interfere, including trivial issues such as the appearance of both Na^+ and K^+ adducts of peptides (Lou, Miley, & Van Dongen, 2021). Wide isotopic envelopes for each peptide can lead to overlap between product and substrate (Anderson et al., 2012). Absolute quantitation is not possible with conventional MALDI, as the ionization efficiency of different peptides can vary widely. Perhaps the largest challenge in using MALDI for activity assays is the variation in signal intensity due to subtle changes in conditions of crystallization, drying rate, laser intensity, or even variation within each spot (Albrethsen, 2007). Consequently, there needs to be an internal standard, so that the signal intensity can be normalized.

Our method uses the unreacted substrate as an internal standard for the hydroxylated product, and relies on the assumption that both substrate and product have the same ionization efficiency. In order to obtain reaction progress curves it is necessary to measure the intensity of both the product and the substrate peaks for each of our time-points and from that we can then calculate the mole-fraction of the product as a function of time. An example spectrum for CTAD hydroxylation (Fig. 1) shows the mass envelope for the CTAD substrate and the CTAD-OH product (top); an example spectrum using the dCTAD substrate (bottom) similarly shows substrate and product mass envelopes. This works remarkably well, providing linear progress curves which we have benchmarked through comparison with rates obtained from O_2 uptake using WT-FIH and CTAD. Initial steady-state rates for product formation are obtained by collecting several time-points under conditions of varied substrate concentration, which are then fit to the nonlinear Michaelis Menten equation.

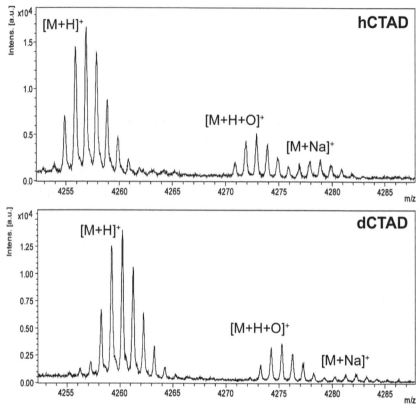

Fig. 1 MALDI-MS spectra following 2 min. Reaction with FIH. (top) CTAD with natural isotopic abundance (CTAD, $[MH]^+ = 4254$ calc; CTAD-OH, $[MOH]^+ = 4270$); (bottom) dCTAD with d3-Asn803 residue (dCTAD, $[MH]^+ = 4257$ calc; dCTAD-OH, $[MOH]^+ = 4273$). In each panel the $[MNa]^+$ ion is also observed.

2. Equipment, buffers and reagents, procedures

2.1 Equipment

- Bruker Daltonics UltraFlextreme MALDI-TOF Mass Spectrometer

2.2 Procedure—FIH activity assay and MALDI target preparation

1. Prepare a saturated solution of α-cyano-4-hydroxycinnamic acid (α-CHCA) in 75% acetonitrile/0.2% trifluoroacetic acid.
2. Add 9 μL of saturated α-CHCA to a small tube for each time point.

3. Thaw prepared aliquots of 100 mM ascorbate, 5 mM 2-oxoglutarate, 2 mM purified CTAD, and 20 μM FIH.

4. Prepare a fresh solution of 2.5 mM FeSO$_4$ in ultrapure H$_2$O.

5. To a 0.5 mL Eppendorf tube, prepare a reaction solution by adding 40 μL of 50 mM HEPES, pH 7.00, and, in the following sequential order, 1 μL ascorbate (2 mM), 1 μL 2OG (100 μM), 1 μL FeSO$_4$ (50 μM), and 2 μL CTAD (80 μM).

6. Incubate the reaction solution at 37 °C for 3 min.

7. Initiate the reaction by adding 5 μL of 20 μM FIH to reaction solution (final concentration, 2 μM, final volume, 50 μL).

8. Remove 3 μL of progressing reaction every 20 s, for 140 s, and quench it by adding it to 9 μL of saturated α-CHCA solution.

9. Repeat steps 5 through 8 with 1 μL (40 μM), 1.5 μL (60 μM), 2 μL (80 μM), 3 μL (120 μM), 4.5 μL (180 μM), 6 μL (240 μM), and 8 μL (320 μM) CTAD, with reaction volumes kept at constant 50 μL by adjusting final HEPES concentrations to compensate for additional volume contributions of CTAD.

10. Prepare the MALDI samples. Vortex the quenched sample in α-CHCA solution, and immediately add 0.5 μL to a single spot on an MTP 384 ground steel MALDI target plate. Repeat for each timepoint. Allow the sample to air dry.

11. Mount the prepared target in Bruker Daltonics UltraFlextreme MALDI-TOF.

12. Shoot samples at 50% laser intensity at 2000 Hz collection rate, 5000 GS/s, on random walk sample collection mode within 100 μm raster area. Spectra are collected over a mass range of 500–5000 m/z ([M+H]$_{CTAD}$ = 4254, [M+H+O]$_{CTAD}$ = 4270, [M+H]$_{dCTAD}$ = 4257, [M+H+O]$_{dCTAD}$ = 4272).

13. MALDI-TOF spectra are analyzed using FlexAnalysis software by performing a baseline subtraction, and collecting the peak intensity of the most abundant isotope of the isotopic envelope for both the unreacted peptide substrate, and the hydroxylated product.

14. Peak intensities recorded in a spreadsheet are used to calculated mole fraction of the hydroxylated product ($\chi_{CTAD-OH}$) using the following equation:

$$\chi_{CTAD-OH} = \frac{I_{CTAD-OH}}{I_{CTAD} + I_{CTAD-OH}} * [CTAD]_{initial} \qquad (2)$$

where I_{CTAD} and $I_{CTAD-OH}$ are the intensities of unreacted peptide and hydroxylated product, respectively, and $[CTAD]_{initial}$ is the initial concentration of the peptide added to the reaction.

15. Linear progress curves are generated by plotting $\chi_{CTAD-OH}$ vs time, and rate of reaction is determined from the slope/[FIH].

16. Michaelis-Menten kinetic constants are ascertained by plotting rate of reaction vs [CTAD], and fit in OriginPro to derive K_M and k_{cat}.

Notes

1. 50 mM HEPES buffer must be prepared in a glass container and used within 7 to 10 days of preparation, as trace amounts of decomposition products interfere with the assay.

2. The order of addition is important, as the reducing nature of ascorbate prevents the rapid oxidation of $FeSO_4$ at neutral pH.

3. Assay CTAD concentration is at the K_M of CTAD for FIH.

3. Measurement of KIEs

The ability to collect initial rate data using MALDI at varied peptide concentration allowed us to plot the steady state kinetics data using initial rates. This provided data that was directly comparable to other datasets monitoring substrate consumption or product formation, and showed that peptide hydroxylation followed typical saturation behavior. Fitting the initial rate data to the Michaelis Menten equation provided kinetic parameters, k_{cat} and k_{cat}/K_M, for the appropriate substrate peptide.

Of note is the dynamic range limits on conventional MALDI instruments. Suitable signal to noise ratios were easily obtained for peptide concentrations above 2 μM, however typical enzyme concentrations were in the range of 0.5–2 μM, which can make low peptide concentrations (<20 μM) deviate from the steady state approximation. This became somewhat limiting for the datasets using dCTAD, as we were unable to use peptide concentrations lower than the estimated K_M. Although introducing appreciable uncertainty to the k_{cat}/K_M parameter, we were most interested in the k_{cat} parameter because of the irreversible decarboxylation step which preceded the reaction with peptide. Due to this irreversible chemical step, k_{cat}/K_M cannot exhibit a KIE from the hydrogen atom transfer (HAT) step.

Fitting the initial rate data using CTAD or dCTAD showed kinetic parameters (Table 1). Noncompetitive KIEs were simply the ratios of the

Table 1 Steady state parameters for FIH at 37.0 °C, monitoring peptide hydroxylation with MALDI-TOF.[a]

	hCTAD	dCTAD
k_{cat} (min^{-1})	27 ± 1	2.7 ± 0.2
K_M (µM)	130 ± 12	8 ± 4
k_{cat}/K_M (min^{-1} µM^{-1})	0.21 ± 0.03	0.4 ± 0.2

[a]Assays were conducted at 37.0 °C and contained ascorbate (2 mM), 2OG (100 µM), FeSO$_4$ (50 µM), CTAD (20–320 µM), and FIH (2 µM) in 50 mM HEPES pH 7.00.

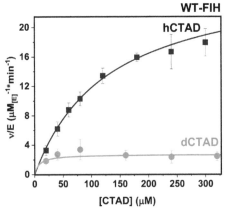

Fig. 2 Initial rate data for FIH at 37.0 °C, monitoring peptide hydroxylation with MALDI-TOF. Assays were conducted at 37.0 °C and contained ascorbate (2 mM), 2OG (100 µM), FeSO$_4$ (50 µM), CTAD (squares) or dCTAD (circles) (20–320 µM), and FIH (2 µM) in 50 mM HEPES pH 7.00.

kinetic parameter for CTAD and dCTAD, along with propagated uncertainties. Example data are shown, Fig. 2, with the $^{D}k_{cat} = 10 \pm 1$. The magnitude of k_{cat} indicated that turnover under saturated conditions was appreciably determined by the microscopic HAT step, k_5 (Scheme 1). However, uncoupled turnover in which the ferryl non-productively forms and dissipates without hydroxylating the peptide is known to happen for slow variants of FIH, as for other 2OG oxygenases, raising the potential for this uncoupling to interfere with the KIE measurement (Iyer et al., 2018; Koehntop, Marimanikkuppam, Ryle, Hausinger, & Que, 2006; Saban et al., 2011; Taabazuing, Fermann, Garman, & Knapp, 2016).

4. Uncoupling of O_2 consumption from CTAD hydroxylation

Uncoupling of O_2 consumption from CTAD hydroxylation can happen for WT-FIH when CTAD is absent, or for FIH variants even in the presence of CTAD. In the course of using dCTAD, we suspected that this slow substrate might also lead to uncoupling for WT-FIH. If the uncoupling were more extensive for dCTAD, then this would create numerous artifacts in the KIE measurements by some combination of enzyme inactivation and uncoupled O_2 consumption.

To demonstrate uncoupling in FIH generally, we present a dataset showing O_2 consumption without CTAD-OH formation. A Clark sensor was used to measure initial rates in terms of substrate, or $v_0 = d[O_2]/dt$, under conditions in which we varied the initial concentration of O_2 in each reaction. Clark sensors, also called O_2-electrodes, are well established tools to directly measure the dissolved O_2 concentration.

O_2 consumption was measured at 37.0 °C with an Oxygraph Plus instrument (Hansatech, Norfolk, UK), using a water-jacketed reaction vessel to control temperature. An oxygen electrode disk consisting of a central platinum cathode and a concentric silver anode resides at the bottom of the reaction vessel. An oxygen-permeable PTFE membrane combined with a paper wick allows for the signal to be conducted across the electrode disc through a KCl bridge. The initial $[O_2]$ was controlled by a variable ratio of compressed gasses, N_2 and O_2, which were mixed inline and equilibrated over the reaction vessel. Reactions were initiated by adding FIH with a gas-tight syringe, then the reaction progress measured by monitoring the $[O_2]$ was as a function of time. O_2 consumption was corrected for the baseline drift of the Clark electrode by subtracting the background drift from a plot of the $-d[O_2]/dt$ trace.

The high level of uncoupling for some FIH variants is demonstrated by initial-rate measurements varying initial $[O_2]$ in the absence of CTAD. As shown in Fig. 3, the D201E variant consumed O_2 even in the absence of CTAD, demonstrating uncoupling for this variant. In light of the extensive uncoupling of O_2 consumption from CTAD hydroxylation for this and other FIH variants, it was clear that any H/D KIE measurements would need to account for the potential uncoupling for WT-FIH with deuterated CTAD substrate.

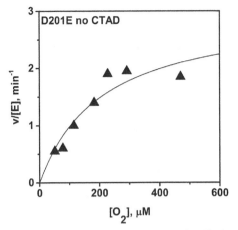

Fig. 3 Initial rate data for the D201E FIH variant using the Clark sensor, fitted to the Michaelis-Menten equation. Steady-state kinetics varying O_2 using an oxygen sensor monitoring O_2 consumption. Assays were conducted at 37.0 °C and contained ascorbate (50 µM), 2OG (100 µM), $FeSO_4$ (50 µM), and FIH (1–5 µM) in 50 mM HEPES pH 7.00.

4.1 Equipment

- Bruker Daltonics UltraFlextreme MALDI-TOF Mass Spectrometer
- Hansatech Oxygraph+ Oxygen Electrode System
- Agilent HP 8453 UV–visible Spectrophotometer

4.2 Procedure—O_2 consumption assay

1. Prepare the electrode membrane as per the manufacturer's instructions
2. Calibrate the electrode using air saturated water and dithionite
3. Set the targeted O_2 concentration by adjusting the ratio of O_2 and N_2 gasses using a flow meter
4. Add the 400 µL reaction mixture (50 µM ascorbate, 100 µM 2OG, 50 µM $FeSO_4$, and 80 µM CTAD in 50 mM HEPES pH 7.00) to the electrode chamber
   a. Note: Typical activity assay conditions include 2 mM ascorbate, however ascorbate consumes O_2. Ascorbate level was reduced for this assay to minimize baseline O_2 contribution while keeping the Fe^{2+} reduced
5. Incubate the reaction mixture at 37.0 °C until a stable O_2 baseline reading is achieved

6. Initiate the reaction with 10 μM of cold FIH using a gas-tight Hamilton syringe

7. Record the O_2 consumption over time until the rate of O_2 consumption resembles the baseline slope

8. Upon reaching baseline O_2 consumption, remove a 5 μL aliquot of reaction mixture and quench in α-CHCA matrix

9. Analyze the sample for CTAD-OH formation using MALDI as described in the previous section

5. Autohydroxylation assay and simultaneous detection of hydroxylated CTAD product

As the active oxidant in FIH can partition between two reaction paths, only one of which is the HAT of interest, determining the KIE on that HAT path must account for both coupled and uncoupled reactions. The uncoupled reaction in FIH leads to autohydroxylation of the FIH residue Trp296, which is observable through formation of a Fe^{3+}-O-Trp296 chromophore. Measuring both the coupled product, CTAD-OH, and the uncoupled product, Fe^{3+}-O-Trp296, makes it possible to back out the KIE on the coupled reaction. This is a matter of thinking of the reaction as an internal competition between the FIH-Trp296 and the protiated or deuterated CTAD peptide.

These internal competition assays were performed under single turnover conditions. The purpose of the assays was to measure the partitioning of the ferryl oxidant between two possible fates: either catalyzing CTAD hydroxylation to form product, or autohydroxylation of the FIH residue Trp296. The hydroxylated CTAD was measured using MALDI, but autohydroxylation was most effectively measured by the appearance of the chromophore formed by coordination of the HO-Trp296 residue to the Fe cofactor (Chen, Comeaux, Eyles, et al., 2008).

Competition assays were initiated by adding 500 μM 2OG to the other reaction components in an optical cuvette, with spectra collected to monitor the appearance of the Fe^{3+}-O-Trp296 chromophore at 583 nm. CTAD hydroxylation was measured by withdrawing 3 μL of the reaction mixture and quenching it in 10 μL of MALDI matrix at defined time points. The yield of HO-Trp296 (U, for uncoupled enzyme) and CTAD-OH (P, for product) were plotted as a function of time (Fig. 4), but it was apparent that the CTAD hydroxylation was quite rapid when using the CTAD peptide (0.5 min) or the dCTAD peptide (5 min).

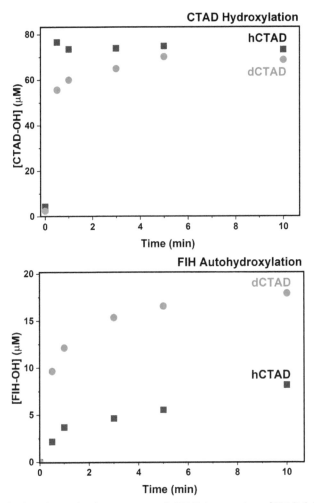

Fig. 4 Autohydroxylation in the presence of CTAD (squares) or dCTAD (circles). Top: CTAD hydroxylation measured using MALDI-TOF. Bottom: Autohydroxylation measured by UV–Vis. Reactions were conducted at 37.0 °C, and contained FIH (100 μM), $FeSO_4$ (100 μM), 2OG (500 μM), and CTAD or dCTAD (100 μM) in 50 mM HEPES, pH 7.00.

Following the work of Fitzpatrick, the KIE on the HAT step ($^D k_5$) can be experimentally determined from the observed KIE on the uncoupling reaction (Fitzpatrick, 2006). This is simply a matter of calculating the observed KIE ($^D k_{obs}$) on the ratio of hydroxylated CTAD product (P) and autohydroxylated enzyme (U) (Eq. 3) while using the CTAD peptide which was protiated or deuterated. The first ten minutes of this data showed that $^D k_{obs} = 1.147 \pm 0.005$. Although this is a small observed KIE, it was supported by the Clark electrode data showing excess O_2 consumption in the presence of dCTAD.

Using the rate law of parallel reactions, $^D k_5$ is able to be derived from the acquired $^D k_{obs}$, and the ratio of rate constants between HAT with CTAD (k_5) and the uncoupling (k_U) (Eq. 4). As there was not a direct measurement of k_5 nor of k_U, we estimated each rate constant in order to obtain a lower bound for the microscopic KIE on the HAT step. To estimate k_5, steady state and pre-steady state kinetic data were considered, as true HAT kinetics are masked by the prior rate limiting decarboxylation step (Hangasky, Gandhi, Valliere, Ostrom, & Knapp, 2014). Pre-steady state stopped-flow studies conducted by Tarhonskaya et al. observed a rate of product formation ($k_{obs} = 0.33\,s^{-1}$), without observing accumulation of the Fe(IV)-oxo species, further indicating that the HAT occurs faster than formation of the ferryl intermediate (Tarhonskaya et al., 2015). As such, the k_{cat} of $30\,min^{-1}$, is a representation of the rate limiting step, and we estimate the rate of k_5 to occur approximately 4 times faster at $120\,min^{-1}$.

The value for k_U was determined experimentally, wherein the range of potential values was considered from enzyme autohydroxylation in both the presence and absence of substrate. Enzyme autohydroxylation occurs as a result of an uncoupled reaction with activated O_2, and its kinetics in the absence of peptide substrate is dependent on the passive release of the axial aquo ligand. The appearance of autohydroxylated FIH as determined from absorbance at 435 nm provided a lower bound k_U of $0.055 \pm 0.003\,min^{-1}$. When autohydroxylation was monitored in the presence of $100\,\mu M$ CTAD, a small initial spike in absorbance occurred, which correlated with the rapid hydroxylation of CTAD peptide product within the first 30 s, followed by slow autohydroxylation as the reaction was dominated by passive O_2 binding once the availability of peptide substrate was depleted. Autohydroxylation occurring during the initial spike is likely substrate driven uncoupling, wherein the Fe(IV)-oxo intermediate reacted with Trp296 in small amounts, despite the availability of peptide, and the k_U with substrate bound was found to be $1.1 \pm 0.3\,min^{-1}$. Because substrate bound k_U was representative of intermediate competition, $^D k_5$ was calculated from this value, with a $k_5/k_U = 86 \pm 18$, and a $^D k_5 > 14$.

$$^D k_{obs} = \frac{\left(\frac{P}{U + P}\right)_H}{\left(\frac{P}{U + P}\right)_D} \tag{3}$$

$$^D k_5 = \left[{}^D k_{obs}\left(1 + \frac{k_5}{k_U}\right)\right] - \frac{k_5}{k_U} \tag{4}$$

5.1 Equipment

- Bruker Daltonics Ultraflextreme MALDI-TOF Mass Spectrometer
- Agilent HP 8453 UV–visible Spectrophotometer

5.2 Procedure—Autohydroxylation assay and simultaneous detection of hydroxylated product

1. Preheat UV–vis cuvette water jacketed holder to $37.0\,^{\circ}$C.
2. Prepare a saturated solution of α-CHCA in 75% acetonitrile/0.2% trifluoroacetic acid.
3. Add 6 µL of saturated α-CHCA to a small tube for each time point.
4. Thaw prepared aliquots of 5 mM 2-oxoglutarate, 2 mM purified CTAD, and 1.5 mM FIH.
5. Prepare a fresh solution of 5 mM $FeSO_4$ in ultrapure H_2O.
6. Blank the UV–vis spectrometer with 50 mM HEPES, pH 7.00
7. To a quartz cuvette, prepare the reaction solution by adding 91.6 µL 50 mM HEPES, pH 7.00, 8 µL 1.5 mM FIH (100 µM), 2.4 µL 5 mM $FeSO_4$ (100 µM), and 6 µL 2 mM CTAD (100 µM).
8. Incubate cuvette at $37.0\,^{\circ}$C in the water jacket for 3 min
9. Remove 2 µL of reaction mix and transfer to a tube containing saturated α-CHCA as a t_0 sample. Additionally, collect a UV–vis spectra.
10. Initiate the reaction with 12 µL of 5 mM 2OG (500 µM).
11. To collect a time course of simultaneous samples, remove 2 µL of reaction mixture, and immediate collect a UV–vis spectra as sample mix is quenched in a tube of saturated α-CHCA, at 0.25, 0.5, 1, 1.5, 2, 3, 5, 10, 15, 20, 30, 45, and 60 min.
12. Monitor autohydroxylated enzyme absorption at $\lambda = 583$ nm
13. For hydroxylated peptide analysis, prepare the MALDI samples by vortexing the quenched sample, and immediately adding 0.5 µL to a single spot on an MTP 384 ground steel MALDI target plate, and allow sample to air dry
14. Mount the prepared target in Bruker Daltonics UltraFlextreme MALDI-TOF.
15. Shoot samples at 50% laser intensity at 2000 Hz collection rate, 5000 GS/s, on random walk sample collection mode within 100 µm raster area. Spectra was collected over a mass range of 500–5000 m/z ($[M+H]_{CTAD} = 4254$, $[M+H+O]_{CTAD} = 4270$, $[M+H]_{dCTAD} = 4257$, $[M+H+O]_{dCTAD} = 4272$).

Notes

1. Autohydroxylation is prepared as a single turnover experiment, with a stoichiometric excess of 2OG

6. Conclusion

MALDI is a convenient tool to measure peptide masses, which can also be used to measure the discontinuous kinetics of peptide hydroxylation as shown for FIH. The mole-fraction of CTAD and CTAD-OH can be observed using small sample volumes without need for chromatography. Noncompetitive H/D KIEs can also be measured, however the slow dCTAD substrate leads to extensive uncoupling of O_2 consumption from peptide hydroxylation. Simultaneously measuring both the normal product, CTAD-OH, and the uncoupled product, autohydroxylated enzyme, the KIE on peptide hydroxylation can be estimated. These approaches can be generalized to other hydroxylation reactions.

Acknowledgments

We thank the U.S. National Institutes of Health for funding (1R01-GM077413 to MJK), the UMass Commonwealth Honors College for student support; mass spectral data were obtained at the University of Massachusetts Mass Spectrometry Core Facility, RRID: SCR_019063.

References

Albrethsen, J. (2007). Reproducibility in protein profiling by MALDI-TOF mass spectrometry. *Clinical Chemistry*, *53*(5), 852–858. https://doi.org/10.1373/clinchem.2006.082644.

Anderson, N. L., Razavi, M., Pearson, T. W., Kruppa, G., Paape, R., & Suckau, D. (2012). Precision of heavy–light peptide ratios measured by MALDI-TOF mass spectrometry. *Journal of Proteome Research*, *11*(3), 1868–1878. https://doi.org/10.1021/pr201092v.

Chen, Y. H., Comeaux, L. M., Eyles, S. J., & Knapp, M. J. (2008). Auto-hydroxylation of FIH-1: An Fe(II), α-ketoglutarate-dependent human hypoxia sensor. *Chemical Communications*, *39*, 4768–4770. https://doi.org/10.1039/b809099h.

Chen, Y. H., Comeaux, L. M., Herbst, R. W., Saban, E., Kennedy, D. C., Maroney, M. J., et al. (2008). Coordination changes and auto-hydroxylation of FIH-1: Uncoupled O2-activation in a human hypoxia sensor. *Journal of Inorganic Biochemistry*, *102*(12), 2120–2129. https://doi.org/10.1016/j.jinorgbio.2008.07.018.

Dreisewerd, K. (2014). Recent methodological advances in MALDI mass spectrometry. *Analytical and Bioanalytical Chemistry*, *406*(9–10), 2261–2278. https://doi.org/10.1007/s00216-014-7646-6.

Fitzpatrick, P. F. (2006). Isotope effects from partitioning of intermediates in enzyme-catalyzed hydroxylation reactions. In A. Kohen, & H.-H. Limbach (Eds.), *Isotope effects in chemistry and biology* (1st ed., pp. 861–874). CRC Press.

Hangasky, J. A., Gandhi, H., Valliere, M. A., Ostrom, N. E., & Knapp, M. J. (2014). The rate-limiting step of O2 activation in the α-ketoglutarate oxygenase factor inhibiting hypoxia inducible factor. *Biochemistry*, *53*(51), 8077–8084. https://doi.org/10.1021/bi501246v.

Islam, M. S., Leissing, T. M., Chowdhury, R., Hopkinson, R. J., & Schofield, C. J. (2018). 2-Oxoglutarate-dependent oxygenases. *Annual Review of Biochemistry*, *87*(1), 585–620. https://doi.org/10.1146/annurev-biochem-061516-044724.

Iyer, S. R., Chaplin, V. D., Knapp, M. J., & Solomon, E. I. (2018). O2 activation by Nonheme FeII α-ketoglutarate-dependent enzyme variants: Elucidating the role of the facial triad carboxylate in FIH. *Journal of the American Chemical Society*, *140*(37), 11777–11783. https://doi.org/10.1021/jacs.8b07277.

Karandashev, K., Xu, Z.-H., Meuwly, M., Vaníček, J., & Richardson, J. O. (2017). Kinetic isotope effects and how to describe them. *Structural Dynamics*, *4*(6), 061501. https://doi.org/10.1063/1.4996339.

Klinman, J. P., & Kohen, A. (2013). Hydrogen tunneling links protein dynamics to enzyme catalysis. *Annual Review of Biochemistry*, *82*(1), 471–496. https://doi.org/10.1146/annurev-biochem-051710-133623.

Koehntop, K. D., Marimanikkuppam, S., Ryle, M. J., Hausinger, R. P., & Que, L. (2006). Self-hydroxylation of taurine/α-ketoglutarate dioxygenase: Evidence for more than one oxygen activation mechanism. *Journal of Biological Inorganic Chemistry*, *11*(1), 63–72. https://doi.org/10.1007/s00775-005-0059-4.

Kohen, A., & Limbach, H.-H. (2005). *Isotope effects in chemistry and biology* (1st ed.). CRC Press. https://doi.org/10.1201/9781420028027.

Lando, D., Peet, D. J., Whelan, D. A., Gorman, J. J., & Whitelaw, M. L. (2002). Asparagine hydroxylation of the HIF transactivation domain a hypoxic switch. *Science*, *295*(5556), 858–861. https://doi.org/10.1126/science.1068592.

Lou, X., Miley, G., & Van Dongen, J. L. J. (2021). Dual roles of [CHCA + Na/K/Cs] + as a cation adduct or a protonated salt for analyte ionization in matrix-assisted laser desorption/ionization mass spectrometry. *Rapid Communications in Mass Spectrometry*, *35*(13), 1–4. https://doi.org/10.1002/rcm.9111.

McCusker, K. P., & Klinman, J. P. (2009). Modular behavior of tauD provides insight into the origin of specificity in α-ketoglutarate-dependent nonheme iron oxygenases. *Proceedings of the National Academy of Sciences of the United States of America*, *106*(47), 19791–19795. https://doi.org/10.1073/pnas.0910660106.

Mingroni, M. A., & Knapp, M. J. (2021). Kinetic studies of the hydrogen atom transfer in a hypoxia-sensing enzyme, FIH-1: KIE and O 2 reactivity. *Biochemistry*, *60*(44), 3315–3322. https://doi.org/10.1021/acs.biochem.1c00476.

Price, J. C., Barr, E. W., Glass, T. E., Krebs, C., & Bollinger, J. M. (2003). Evidence for hydrogen abstraction from C1 of taurine by the high-spin Fe(IV) intermediate detected during oxygen activation by taurine:α-ketoglutarate dioxygenase (TauD). *Journal of the American Chemical Society*, *125*(43), 13008–13009. https://doi.org/10.1021/ja037400h.

Saban, E., Chen, Y. H., Hangasky, J. A., Taabazuing, C. Y., Holmes, B. E., & Knapp, M. J. (2011). The second coordination sphere of FIH controls hydroxylation. *Biochemistry*, *50*(21), 4733–4740. https://doi.org/10.1021/bi102042t.

Schofield, C. J., & Ratcliffe, P. J. (2004). Oxygen sensing by HIF hydroxylases. *Nature Reviews Molecular Cell Biology*, *5*(5), 343–354. https://doi.org/10.1038/nrm1366.

Semenza, G. L. (2011). Oxygen sensing, homeostasis, and disease. *New England Journal of Medicine*, *365*(6), 537–547. https://doi.org/10.1056/NEJMra1011165.

Semenza, G. L. (2014). Oxygen sensing, hypoxia-inducible factors, and disease pathophysiology. *Annual Review of Pathology: Mechanisms of Disease*, *9*, 47–71. https://doi.org/10.1146/annurev-pathol-012513-104720.

Taabazuing, C. Y., Fermann, J., Garman, S., & Knapp, M. J. (2016). Substrate promotes productive gas binding in the α-ketoglutarate-dependent oxygenase FIH. *Biochemistry*, *55*(2), 277–286. https://doi.org/10.1021/acs.biochem.5b01003.

Tarhonskaya, H., Hardy, A. P., Howe, E. A., Loik, N. D., Kramer, H. B., McCullagh, J. S. O., et al. (2015). Kinetic investigations of the role of factor inhibiting hypoxia-inducible factor (FIH) as an oxygen sensor. *Journal of Biological Chemistry*, *290*(32), 19726–19742. https://doi.org/10.1074/jbc.M115.653014.

CHAPTER SIXTEEN

Continuous photometric activity assays for lytic polysaccharide monooxygenase—Critical assessment and practical considerations

Lorenz Schwaiger (iD), **Alice Zenone** (iD), **Florian Csarman*** (iD), **and Roland Ludwig** (iD)

Department of Food Science and Technology, Institute of Food Technology, University of Natural Resources and Life Sciences, Vienna, Austria
*Corresponding author: e-mail address: florian.csarman@boku.ac.at

Contents

Abstract

Lytic polysaccharide monooxygenase (LPMO) is a monocopper-dependent enzyme that cleaves glycosidic bonds by using an oxidative mechanism. In nature, they act in concert with cellobiohydrolases to facilitate the efficient degradation of lignocellulosic biomass. After more than a decade of LPMO research, it has become evident that LPMOs are abundant in all domains of life and fulfill a diverse range of biological functions. Independent of their biological function and the preferred polysaccharide substrate, studying and characterizing LPMOs is tedious and so far mostly relied on the discontinuous analysis of the solubilized reaction products by HPLC/MS-based methods. In the absence of appropriate substrates, LPMOs can engage in two off-pathway reactions, i.e., an oxidase and a peroxidase-like activity. These futile reactions have been exploited to set up easy-to-use continuous spectroscopic assays.

Methods in Enzymology, Volume 679
ISSN 0076-6879
https://doi.org/10.1016/bs.mie.2022.08.054

As the natural substrates of newly discovered LPMOs are often unknown, widely applicable, simple, reliable, and robust spectroscopic assays are required to monitor LPMO expression and to perform initial biochemical characterizations, e.g., thermal stability measurements. Here we provide detailed descriptions and practical protocols to perform continuous photometric assays using either 2,6-dimethoxyphenol (2,6-DMP) or hydrocoerulignone as colorimetric substrates as a broadly applicable assay for a range of LPMOs. In addition, a turbidimetric measurement is described as the currently only method available to continuously monitor LPMOs acting on amorphous cellulose.

1. Introduction

The discovery that the hydrolytic depolymerization of recalcitrant carbohydrates is greatly facilitated by the action of oxidative enzymes led to a paradigm shift in our understanding of lignocellulose degradation (Bissaro, Várnai, Røhr, & Eijsink, 2018; Johansen, 2016; Müller, Chylenski, Bissaro, Eijsink, & Horn, 2018). In 2010, Vaaje-Kolstad and co-workers reported an explanation for an earlier observed boosting effect on the activity of hydrolytic enzymes (Merino & Cherry, 2007; Vaaje-Kolstad, Horn, van Aalten, Synstad, & Eijsink, 2005) by the currently named lytic polysaccharide monooxygenases (LPMOs) (Horn, Vaaje-Kolstad, Westereng, & Eijsink, 2012), which oxidatively cleave glycosidic bonds (Vaaje-Kolstad et al., 2010). So far, LPMO has been found in different organisms with very different lifestyles and substrate specificities ranging from cellulose, chitin, and hemicelluloses to starch (CAZy auxiliary families AA9–11, AA13–17 (Drula et al., 2022)). Besides facilitating glycosidic bond cleavage, recent findings identified alternative biological functions for LPMOs (Vandhana et al., 2022). For instance, LPMOs play a role in pathogenesis, as shown recently in mammals (Askarian et al., 2021) as well as plants (Sabbadin et al., 2021). Independent of their origin and substrate specificities, they all share a strictly conserved active site architecture consisting of a Cu(II) (Quinlan et al., 2011) coordinated via two histidine residues (Aachmann, Sørlie, Skjåk-Bræk, Eijsink, & Vaaje-Kolstad, 2012), referred to as the histidine brace motif (Aachmann et al., 2012; Hemsworth et al., 2013; Hemsworth, Davies, & Walton, 2013; Quinlan et al., 2011). The cleavage of the glycosidic bond proceeds via an oxidative mechanism depending on suitable reductants (Kracher et al., 2016) and an oxygen-containing co-substrate. Depending on the substrate specificity of the LPMO, the hydroxylation of the scissile bond can result in oligosaccharides oxidized at the C1 (Vaaje-Kolstad et al., 2010) and/or C4 position of the sugar moiety (Beeson, Phillips, Cate, & Marletta, 2012; Phillips, Beeson, Cate, & Marletta, 2011). Initially, molecular O_2 was suggested to function as co-substrate (Beeson et al., 2012; Vaaje-Kolstad et al., 2010), thus the name monooxygenase. However, today there is ample evidence from several kinetic studies that catalytic efficiencies of the peroxygenase reaction are several orders of magnitude higher, indicating that H_2O_2 is the preferred co-substrate (Bissaro et al., 2017; Bissaro, Kommedal, Røhr, & Eijsink, 2020; Hedison et al., 2021; Jones, Transue, Meier, Kelemen, & Solomon, 2020; Kont, Bissaro, Eijsink, & Väljamäe, 2020; Kuusk et al., 2018, 2019; Kuusk & Väljamäe, 2021; Müller et al., 2018; Rieder, Petrović, Väljamäe, Eijsink, & Sørlie, 2021; Rieder, Stepnov, Sørlie, & Eijsink, 2021).

Since the discovery of LPMO, analytical methods have been developed for the detection and quantification of products released by LPMO reactions and these methodologies have been summarized and reviewed comprehensively elsewhere (Eijsink et al., 2019; Westereng et al., 2018). Despite years of research, measuring LPMO activities on their natural heterogeneous substrates remains challenging. Most approaches rely on stopped reactions followed by retrospective analysis of the solubilized products (Eijsink et al., 2019). To determine the amount of total oxidized sites in the soluble or insoluble fraction an additional hydrolytic conversion step is required (Courtade, Forsberg, Heggset, Eijsink, & Aachmann, 2018). In any case, advanced analytical equipment is a prerequisite to perform these measurements. For example, reaction products of cellulose-active LPMOs have been analyzed using HPEAC-PAD and MALDI-TOF (Forsberg et al., 2011; Westereng et al., 2018) and similarly hemicelluloses-derived degradation products using HPEAC-PAD and ESI-MS (Isaksen et al., 2014) and HPAEC-MALDI-TOF (Agger et al., 2014). The activity of chitin-active LPMOs has been assessed using a combination of UHPLC-HILIC and MALDI-TOF (Loose, Forsberg, Fraaije, Eijsink, & Vaaje-Kolstad, 2014; Vaaje-Kolstad et al., 2010) and RP-UHPLC-MS (Frommhagen et al., 2017). Recently, HPAEC-PAD was used to determine catalytic rates of *Neurospora crassa* LPMO9C acting on cellopentaose by stopping reactions in the range of 10–60 s and subsequently analyzing the formed products offline (Rieder, Stepnov, et al., 2021). Even more demanding is to determine catalytic rates on more complex substrates like cellulose or chitin (Kont et al., 2020). One approach is based on stopped substrate conversion experiments followed by measuring the increase in radioactivity from reactions with *Hj*AA9A acting on ^{14}C-labeled cellulose (Kuusk & Väljamäe, 2021) or

*Sm*LPMO10A (formerly CBP21) acting on ^{14}C-labeled chitin nano-whiskers (Kuusk et al., 2018, 2019).

To date, there is no methodology available that can measure the peroxygenase activity of LPMOs continuously except turbidimetric measurement. This technique was successfully used to continuously track the LPMO-catalyzed reaction on amorphous cellulose by following the change in light scattering at either 620 nm (PASC) (Breslmayr et al., 2022; Filandr et al., 2020) or 420 nm (Hansson et al., 2017). However, this method has several practical limitations that are further discussed later.

Due to the lack of practicable and reliable continuous assays, detailed kinetic characterizations of LPMOs on natural carbohydrate substrates are still scarce. Monitoring the progress of reactions continuously is superior to discontinuous measurements. There have been attempts to continuously monitor the progress of LPMO reactions, with limited applicability. For example, Kojima and co-workers used the decrease in dynamic viscosity to determine the degree of LPMO activity on hemicellulose substrates (Kojima et al., 2016). This study was performed before the discovery that H_2O_2 is utilized much more efficiently as co-substrate than O_2, but might also be applicable with the better co-substrate. This method is limited to LPMO reactions on substrates which lead to a significant change in dynamic viscosity. In another approach, the stoichiometric oxidation of ascorbic acid under apparent monooxygenase conditions can be used to monitor the progress of the LPMO reaction. Note that the consumption of ascorbic acid can only be monitored continuously in the presence of fully soluble substrates in the range of 250–290 nm (Karayannis, Samios, & Gousetis, 1977; Stepnov et al., 2022).

Several discontinuous photometric assays have been developed to monitor the progress of LPMOs acting on different substrates with standard UV/Vis-spectrophotometers. The colorimetric assay developed by Wang and co-workers (Wang, Li, Wong, Aachmann, & Hsieh, 2018) uses the increase in adsorption of Ni^{2+} ions to aldonic acids, the product of C1-oxidizing LPMOs, to estimate their activities in the presence of carbohydrate substrates. Pyrocatechol violet is used as a complexometric indicator of the Ni^{2+} concentration to follow the formation of C1-oxidized products by LPMOs over time. However, the low change in absorption (0.1 AU) for a calibration function in the range of 1–8 mM carboxylic acids and the low sensitivity limits the application of this assay when considering that the amount of oxidized products formed in LPMO assays is often below 1 mM (Hegnar et al., 2021; Kracher et al., 2020). Another assay specifically aimed at quantifying C1-oxidized products uses β-glucosidase to convert the soluble products and to some extent C1-oxidized sites on the insoluble fraction to glucose and gluconic acid. The amount of gluconic acid obtained corresponds to the number of oxidized sites and is quantified using a commercially available D-gluconic acid/D-glucono-δ-lactone assay kit. Although limited to C1-oxidized products, the proposed methodology can be used in a more reasonable concentration range with a limit of detection of around $2\,\mu M$ (Keller et al., 2020). For the indirect quantification of C4-oxidized products, the TTC assay (2,3,5-triphenyl-2H tetrazolium chloride) can be used to quantify the increasing number of reducing-end sugars. The cleavage of cellulose chains by C4-oxidizing LPMOs generates a ketoaldose group at the non-reducing end, while on the opposite chain a new reducing-end is generated (Hemsworth, Johnston, Davies, & Walton, 2015). The reaction of TTC with aldehyde groups leads to reduction of TTC to the red-colored formazan, which can be quantified by UV/Vis-spectrometry at 546 nm. Initially introduced by Obolenskaya, Yelnitskaya, and Leonovitch (1991), this assay was recently applied to study synergistic effects of LPMO9A and Cel45A from *Trichoderma reseei* (Ceccherini et al., 2021), as well as to investigate the efficiency of oxidative modifications on cotton and softwood pulp fibers by *Hj*AA9A (Marjamaa et al., 2022). The major advantage of this assay is the possibility to study modifications on solid material without the need for additional hydrolyzation steps. Another advantage is that this assay allows direct quantification of C4-oxidized products, which is traditionally quite difficult using HPEAC-PAD because they undergo partial on-column degradation due to the high pH used (Westereng et al., 2016). However, the time-dependent instability of formazan might influence reproducibility (Ceccherini et al., 2021).

Another method to spectroscopically follow the release of soluble cellooligosaccharides was introduced by Wu and co-workers based on gluco-oligosaccharide oxidase used to convert the solubilized products from LPMOs acting on amorphous cellulose (PASC) to H_2O_2 (Wu et al., 2022). The formation of H_2O_2 is further quantified using a horse-radish peroxidase (HRP) dependent colorimetric assay (4-amino-antipyrine and 3,5-dichloro-2-hydroxybenzene-sulfonic acid, measured at 515 nm) (Wu et al., 2022). Prior to the quantification of reaction products, ascorbic acid used as reductant for LPMO has to be removed due to possible interferences with the HRP colorimetric assay (Mehlhorn, Lelandais, Korth, & Foyer, 1996; Rodrigues & Gomes, 2010), using ascorbate oxidase. The advantage of this assay is that it is possible to detect products of C1- and C4-oxidizing LPMOs. However, the assay is limited to soluble products.

Besides introducing oxidative chain breaks by their peroxygenase activity (Fig. 1A–C), substrate unbound LPMOs engage in two main off-pathway

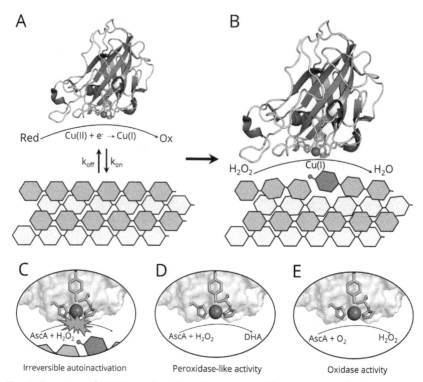

Fig. 1 Overview of LPMO-catalyzed reactions (exemplified structure of LPMO9C from *Neurospora crassa*, PDB ID: 4D7U). (A) The catalytic cycle of LPMOs involves a priming reduction of the active site Cu(II) to Cu(I), the concomitant oxidation of the used reductant (Red → Ox) (Kracher et al., 2016) and binding to the carbohydrate substrate (Kracher, Andlar, Furtmüller, & Ludwig, 2018). (B) Reduced LPMOs then utilize H_2O_2 as co-substrate and induce glycosidic bond cleavage by hydroxylating their polysaccharide substrates either at the C1 or C4 carbon and release of H_2O (Bissaro et al., 2017; Wang, Johnston, et al., 2018). After completing one cycle of the peroxygenase reaction, LPMOs can retain the Cu(I) state and could potentially enter another catalytic cycle (Bissaro et al., 2020; Müller et al., 2018). (C) However, even in the presence of appropriate carbohydrate substrates, but too high H_2O_2 concentrations, they easily engage in non-productive reactions which lead to severe oxidative damages of the active site and subsequent irreversible inactivation (Bissaro et al., 2017). (D) In the absence of polysaccharide substrates, LPMOs are able to oxidize low-molecular-weight reductants like ascorbic acid (AscA) in a H_2O_2 dependent manner. This reaction has been termed a peroxidase-like activity (Breslmayr et al., 2018; Kuusk & Väljamäe, 2021). (E) Similarly, unbound reduced LPMOs show oxidase activity with the concomitant production of H_2O_2 from O_2 (Kittl, Kracher, Burgstaller, Haltrich, & Ludwig, 2012).

reactions (Fig. 1D and E). These futile side-reactivities have been exploited to set up easy-to-use and fast UV/Vis-spectroscopic assays for the assessment of enzyme activity.

The first of these assays based on LPMO's side reaction was developed by Kittl and co-workers (Kittl et al., 2012) using the Amplex Red/HRP system to quantify the oxidase activity of LPMOs by monitoring the production of H_2O_2 under aerobic conditions in the presence of reductant and absence of cellulosic substrates. Common pitfalls and limitations of the Amplex Red/HRP are reviewed in the chapter "Looking at LPMO reactions through the lens of the HRP/Amplex Red assay" by Stepnov and Eijsink of this issue.

Another off-pathway reaction involves oxidation of the employed reductant in the presence of H_2O_2. This reaction has been termed a peroxidase-like reaction, and leads, in the case of ascorbic acid, to the formation of dehydroascorbic acid (DHA) (Kuusk & Väljamäe, 2021) followed by its decomposition to a multitude of different species (Stepnov et al., 2022). This off-pathway reaction was also exploited by Breslmayr et al. (2018, 2019) by using suitable phenolic compounds (2,6-DMP, hydrocoerulignone) that can serve as reductants of LPMOs and form detectable colored products during the LPMO-catalyzed H_2O_2-dependent oxidation reaction.

The difficulties related to measuring LPMO activities in day-to-day laboratory practice highlight the importance of easy-to-use continuous UV–Vis spectroscopic assays. This work aims to give an overview of currently available assays that can be used to determine LPMO activities and to provide practical measurement protocols. LPMO assays based on turbidity measurements using amorphous cellulose will be discussed as an example of a continuous spectroscopic method to monitor the real peroxygenase reaction of LPMOs as well as two assays based on artificial phenolic substrates (2,6-DMP and hydrocoerulignone).

2. Continuous photometric assays

The activity of LPMOs can be measured continuously either by the colorimetric assays based on 2,6-dimethoxyphenol (2,6-DMP) or hydrocoerulignone oxidation developed by Breslmayr and co-workers (Breslmayr et al., 2018, 2019). Both colorimetric assays make use of an off-pathway reaction of LPMOs, the H_2O_2-dependent oxidation of the used reductant. 2,6-DMP reduces the LPMO-Cu(II) active site and by consuming one molecule of H_2O_2, 2,6-DMP is oxidized to the corresponding phenoxy

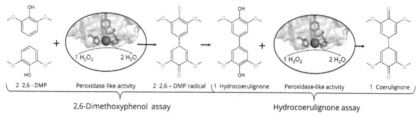

Fig. 2 Reaction scheme of the LPMO-catalyzed oxidation of 2,6-DMP and hydro-coerulignone (exemplary active site structure of *Neurospora crassa* LPMO9C, PDB ID: 4D7U). LPMOs catalyze the H_2O_2-dependent oxidation of two molecules of 2,6-DMP to the corresponding phenoxy radicals, followed by formation of one molecule of hydrocoerulignone, which is converted to the orange-colored product coerulignone consuming another molecule of H_2O_2. In summary, using 2,6-DMP as artificial substrate the LPMO catalyzed-oxidation reaction proceeds via the hydrocoerulignone intermediate, while this compound can also be used directly as artificial substrate in the hydrocoerulignone assay.

radical. In the subsequent reaction step, dimerization of phenoxy radicals results in the formation of hydrocoerulignone, which is also used as reductant by LPMOs. Similarly, hydrocoerulignone is oxidized by LPMO using one molecule of H_2O_2 to the orange-colored coerulignone, which can be detected at 469 nm. Both peroxidase-like reactions follow a 1:1 stoichiometry. The overall reaction scheme is summarized in Fig. 2 and Eqs. (1)–(3).

$$2\ 2,6\text{-DMP} + 1\ H_2O_2 \rightarrow 1\ \text{hydrocoerulignone} + 2\ H_2O \qquad (1)$$

$$1\ \text{hydrocoerulignone} + 1\ H_2O_2 \rightarrow 1\ \text{coerulignone} + 2\ H_2O \qquad (2)$$

Overall reaction scheme.

$$2\ 2,6\text{-DMP} + 2\ H_2O_2 \rightarrow 1\ \text{coerulignone} + 4\ H_2O \qquad (3)$$

The obtained enzyme activities are lower in the 2,6-DMP assay because the oxidation of 2,6-DMP is the rate-limiting step. By using hydrocoerulignone, the intermediate of the reaction, the obtained enzyme activities increase fivefold, which also increases the sensitivity of the hydrocoerulignone assay. It is important to note that some LPMOs might be able to react more readily with hydrocoerulignone. Thus, a comparison of different LPMOs might not be informative as the employed artificial substrate does not reflect catalytic turnovers on natural substrates. However, it is possible to compare and investigate trends, for example optimizing reaction conditions or studying buffer and pH dependencies. In addition, the obtained rates strongly depend on the chosen reaction conditions, such as buffer species and ionic strength. Thus, caution must be taken when setting up these assays. Recording a blank that is representative of the used buffer system and the ionic species present is recommended. By using these colorimetric assays various aspects of different LPMOs have been studied and reported in literature including pH-dependent activities of homologously expressed LPMO9A from *Penicillium verruculosum* (Semenova et al., 2020), characterizing a new AA9 LPMO from compost samples (Ma et al., 2022), studying pH and temperature effects of an AA10 LPMO from *Bacillus amyloliquefaciens* (Guo et al., 2022), characterizing several novel AA9 LPMOs from *Thermothielavioides terrestris* (Tõlgo et al., 2022) or performing kinetic characterizations of LPMO9A and its variants from *Talaromyces amestolkiae* (Méndez-Líter et al., 2021). In conclusion, especially the 2,6-DMP assay has been used extensively to perform basic biochemical characterizations of novel LPMOs. A detailed description of the 2,6-DMP assay can be found in Section 2.1, for the hydrocoerulignone assay in Section 2.2.

2.1 2,6-DMP assay

2.1.1 Equipment
- Perkin Elmer Lambda 35 UV/Vis Spectrophotometer equipped with 8+1 water-thermostatted cell changer
- Julabo SE-12 heating circulator for temperature control
- Semi-micro cuvettes (Greiner Bio-One, PS, 1 cm pathlength)
- Quartz cuvette (Hellma, semi-micro 1 cm pathlength)

2.1.2 Preparation of stock solutions
All aqueous solutions, reagents and buffers have to be prepared in highly purified water.

2.1.2.1 H_2O_2 stock solution
30% (w/w, ~9.8 M) is a commonly available formulation of H_2O_2. To obtain a concentration of ~500 mM, 50 μL of the H_2O_2 (30%) is diluted in 950 μL highly pure water, which can be stored for up to 3 months at 4 °C. The H_2O_2 solution is further diluted 1:100 to the final concentration of 5 mM, which will be used as assay stock solution. The H_2O_2 assay stock solution should be prepared fresh every day.

Note: As the H_2O_2 concentration can change due to a disproportionation reaction, it is highly recommended to verify the concentration of purchased H_2O_2 solutions by measuring the absorbance at 240 nm using the extinction coefficient of 43.6 M^{-1} cm^{-1}. The absorbance at 240 nm of the H_2O_2 solution is measured in a quartz cuvette (1 cm pathlength) blanked against water. A 5 mM H_2O_2 solution will result in an absorbance of 0.218.

2.1.2.2 2,6-DMP stock solution
For the preparation of 1 mL (sufficient for 10 assays) 15.5 mg of 2,6-DMP (CAS Number: 91-10-1; orange/brown crystals) is dissolved in highly pure water resulting in a 100 mM stock solution. The solution must be stored on ice or at 4 °C in the dark and is stable for a maximum of 1 week.

Note: Use a sonicated water bath to facilitate the dissolution of 2,6-DMP in water.

2.1.2.3 Assay buffer
The buffering species, as well as buffer strength, have considerable effects on the results of the activity measurement. For measurements at pH 6.0 and 7.5, an acetate/imidazole buffer is recommended with a stock concentration of 116 mM resulting in a final concentration of 100 mM (50 mM imidazole + 50 mM acetate) within the assay.

For the preparation of 100 mL assay buffer use 0.348 g (0.332 mL) acetic acid (100%) and 0.399 g imidazole and dissolve in 90 mL of highly pure water. Adjust the pH value to 6.0 or 7.5 with 1 M sodium hydroxide and fill to a final volume of 100 mL with highly pure water. The buffer should be stored at room temperature and is stable for a maximum of 2 weeks.

Note: Other recommended buffers are 100 mM pyridine/imidazole (for pH 6.0–7.5), 100 mM sodium succinate/phosphate (for pH 6.0–7.5) or 50 mM imidazole (for pH 7.5).

2.1.2.4 LPMO preparation
This assay can be used for the determination of LPMO activity of purified enzyme samples (recommended) or of culture supernatants. Samples should be clarified by centrifugation (5 min at $10,000 \times g$ in Eppendorf tube) and diluted in assay buffer or highly pure water, respectively. The sample should be stored on ice or at 4 °C.

Note: For measurements at pH 6.0 an LPMO protein concentration between 1.0 and 3.0 μM is recommended. For assays at pH 7.5 a concentration between 0.1 and 1.0 μM. Please note that these values depend on the enzyme used and the quality of the enzyme preparation.

2.1.3 Measurement protocol
All measurements should be performed at 30 °C.
For measurements in sodium acetate/imidazole buffer pH 6.0
1. Prewarm cuvettes (1 cm pathlength) to 30 °C
2. Add 860 μL of sodium acetate/imidazole buffer (116 mM, pH 6.0)
3. Add 100 μL of 2,6-DMP stock (100 mM)

4. Add 20 µL of H_2O_2 solution (5 mM)
5. Incubate for 15 min at 30 °C
6. Transfer the cuvette with the assay mix into a spectrophotometer equipped with a temperature-controlled cuvette holder set to 30 °C
7. Start the reaction by the addition of 20 µL of diluted sample
8. Immediately start the measurement and record the increase in absorbance at 469 nm for 300 s

For measurements in sodium acetate/imidazole pH 7.5
1. Prewarm cuvettes (1 cm pathlength) to 30 °C
2. Add 860 µL of sodium acetate/imidazole buffer (116 mM, pH 7.5)
3. Dilute the DMP stock solution 1:10 to a concentration of 10 mM
4. Add 100 µL of diluted 2,6-DMP stock (10 mM)
5. Add 20 µL of H_2O_2 solution (5 mM)
6. Incubate for 15 min at 30 °C
7. Transfer the cuvette with the assay mix into a spectrophotometer equipped with a temperature-controlled cuvette holder set to 30 °C
8. Start the reaction by the addition of 20 µL of diluted sample
9. Immediately start the measurement and record the increase in absorbance at 469 nm for 300 s

2.1.4 Activity calculation

The slope of the initial linear increase in absorbance at 469 nm ($\Delta Abs\,min^{-1}$, Fig. 3) is used to calculate the enzyme activity [$U\,L^{-1}$] by multiplication with the enzyme factor (EF), Eq. (4):

$$
\begin{aligned}
volumetric\ activity&\left[UL^{-1}\right] \\
&= slope\left[\Delta Abs\ min^{-1}\right] * EF \\
&= \frac{slope[\Delta Abs\ min^{-1}] * final\ volume[\mu L] * volume\ factor}{pathlength[cm] * \varepsilon_{469}[\mu M^{-1} cm^{-1}] * sample\ volume[\mu L]} \\
&= \frac{slope[\Delta Abs\ min^{-1}] * 1000\mu L}{1\ cm * 0.0532\,\mu M^{-1} cm^{-1} * 20\mu L} \\
&= slope\left[\Delta Abs\ min^{-1}\right] * 939.85
\end{aligned}
\tag{4}
$$

2.2 Hydrocoerolignone assay

2.2.1 Equipment

- Perkin Elmer Lambda 35 UV/Vis Spectrophotometer equipped with 8+1 water-thermostatted cell changer
- Julabo SE-12 heating circulator for temperature control

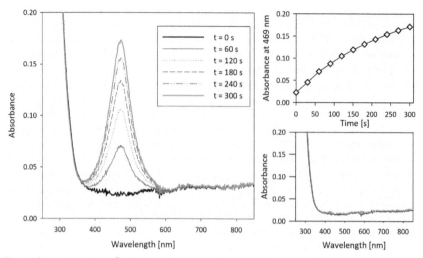

Fig. 3 Determination of LPMO activity using the 2,6-DMP assay. The measurement was performed in acetate/imidazole buffer at pH 6.0 using a final concentration of 0.4 μM *Neurospora crassa* LPMO9C following the described protocol. Spectra were recorded every 60 s for over a total assay time of 300 s (left panel). The activity is calculated using equation described above from the initial slope of the absorbances measured at 469 nm (top right). A blank reaction was performed without LPMO applying the same volume of water instead of enzyme (bottom right).

- Semi-micro cuvettes (Greiner Bio-One, PS, 1 cm pathlength)
- Quartz cuvette (Hellma, semi-micro 1 cm pathlength)

2.2.2 Preparation of stock solutions

All aqueous solutions, reagents and buffers have to be prepared in highly pure water.

2.2.2.1 H_2O_2 stock solution

A commonly available formulation contains 30% (w/w, ~9.8 M) H_2O_2. To obtain a concentration of ~500 mM, 50 μL of the H_2O_2 (30%) is diluted in 950 μL highly pure water, which can be stored for up to 3 months at 4 °C. The H_2O_2 solution is further diluted 1:100 to the final concentration of 5 mM, which will be used as assay stock solution.

The H_2O_2 assay stock solution should be prepared fresh every day.

Note: As H_2O_2 concentrations of the commercially available stocks could change due to disproportionation reactions, it is highly recommended to verify the concentration by measuring the absorbance at 240 nm using the extinction coefficient of $43.6 \, M^{-1} \, cm^{-1}$. The absorbance at 240 nm of the H_2O_2 solution is therefore measured in a quartz cuvette (1 cm pathlength) blanked against water. A 5 mM H_2O_2 solution will result in an absorbance of 0.218.

2.2.2.2 Hydrocoerulignone stock

For the preparation of 0.25 mL sock solution with a concentration of 50 mM (sufficient for 12 assays) 3.8 mg of hydrocoerulignone (CAS 612-69-1; brown powder) is dissolved in dimethyl sulfoxide (DMSO) for 60 min under constant mixing using a rotary shaker. The solution is clarified by centrifugation for 5 min at $10,000 \times g$ and the slightly brown to orange supernatant is transferred to a fresh tube. The solution is stored for at least 60 min at room temperature prior to use. The stock solution can be stored at room temperature for several days.

Note: The use of transparent tubes or glass vials for the preparation of hydrocoerulignone stocks is recommended to avoid the contamination with DMSO-soluble extractables from plastic tubes.

Note: Due to variabilities between different hydrocoerulignone preparation and to verify the correct concentration of dissolved hydrocoerulignone, we recommend checking the concentration by UV/Vis-spectroscopy using $\varepsilon_{280} = 16,260 \, M^{-1} \, cm^{-1}$ as molar extinction coefficient for hydrocoerulignone (Breslmayr et al., 2019). For the determination of the absorbance the hydrocoerulignone stock is diluted 1:20 in DMSO and 20 µL is transferred into a quartz cuvette with a pathlength of 1.0 cm and mixed with 980 µL of highly pure water. A 50 mM stock solution will result in an absorbance of 0.813 using this procedure (with 2% (*v*/v) DMSO in water used as blank).

Note: The concentration of hydrocoerulignone to be used for the assay is 25 mM resulting in a final concentration of 0.5 mM in the assay mix. To optimize the reaction conditions for different LPMO variants, hydrocoerulignone stock concentrations between 10 and 50 mM can be used corresponding to a final concentration range from 0.2 to 1.0 mM in the assay mix.

Note: Dilutions of the hydrocoerulignone stock should be prepared in DMSO as the stability of hydrocoerulignone is reduced in aqueous solutions.

2.2.2.3 Assay buffer

As already described for the 2,6-DMP assay, buffer composition and buffer strength have a strong influence on LPMO activity and assay blank reactions.

Due to the generally higher reaction rates with hydrocoerulignone compared to the rates determined with 2,6-DMP as substrate, but also because of the increase in the rates of the background reactions with increasing pH, the hydrocoerulignone assay is only recommended for pH 6.0. An assay buffer stock solution containing 106 mM sodium acetate buffer pH 6.0 which will result in a final concentration of 100 mM in the assay reaction is therefore highly recommended. For the preparation of 100 mL assay buffer stock use 0.637 g (0.606 mL) acetic acid (100%) and dilute in 90 mL of highly pure water. Adjust the pH value to 6.0 with 1 M sodium hydroxide and fill to a final volume of 100 mL with highly pure water. The buffer should be stored at room temperature and is stable for a maximum of 2 weeks.

Note: A 100 mM pyridine buffer or a 100 mM imidazole buffer titrated to pH 6.0 with HCl can be used as an alternative to sodium acetate.

2.2.2.4 LPMO preparation

LPMO activity of clear culture supernatants or purified enzymes can be determined using the hydrocoerulignone assay. All samples need to be clarified by centrifugation (5 min at 16,000 rcf in Eppendorf tube) and diluted in assay buffer or highly pure water, respectively. The sample should be stored on ice or at 4 °C.

Note: The recommended LPMO concentration in the assay is between 0.1 and 0.3 μM.

2.2.3 Measurement protocol

All measurements should be performed at 30 °C!
1. Prewarm cuvettes (1 cm pathlength) to 30 °C
2. Add 940 μL of sodium acetate buffer (106 mM, pH 6.0)
3. Add 20 μL of hydrocoerulignone stock (25 mM)
4. Add 20 μL of H_2O_2 solution (5 mM)
5. Incubate for 15 min at 30 °C
6. Transfer the cuvette with the assay mix into a spectrophotometer equipped with a temperature-controlled cuvette holder set to 30 °C
7. Start the reaction by the addition of 20 μL of diluted sample
8. Immediately start the measurement and record the increase in absorbance at 469 nm for 200 s

Due to the autoxidation of hydrocoerulignone to coerulignone, a blank reaction containing the same reaction mix but highly pure water instead of LPMO has to be performed and the rate of the blank has to be subtracted.

2.2.4 Data evaluation

The slope of the initial linear increase in absorbance at 469 nm (ΔAbs min^{-1}, Fig. 3) is used to calculate the enzyme activity [$U L^{-1}$] by multiplication with the enzyme factor (EF), Eq. (5):

$$
\begin{aligned}
volumetric\ activity & \left[UL^{-1}\right] \\
&= slope\left[\Delta Abs\ min^{-1}\right] * EF \\
&= \frac{slope\left[\Delta Abs\ min^{-1}\right] * final\ volume[\mu L] * volume\ factor}{pathlength[cm] * \varepsilon_{469}\left[\mu M^{-1} cm^{-1}\right] * sample\ volume[\mu L]} \qquad (5) \\
&= \frac{slope\left[\Delta Abs\ min^{-1}\right] * 1000\mu L}{1\ cm * 0.0532\ \mu M^{-1} cm^{-1} * 20\mu L} \\
&= slope\left[\Delta Abs\ min^{-1}\right] * 939.85
\end{aligned}
$$

3. Continuous turbidimetric assay

Turbidimetric measurements are currently the only tool to continuously monitor the peroxygenase activity of LPMOs acting on cellulose. However, they can only be used to study LPMOs acting on phosphoric acid swollen cellulose (PASC), an amorphous cellulose preparation. Initially, developed to determine the activities of cellulases acting on PASC (Nummi, Fox, Niku-Paavola, & Enari, 1981; Wood & Bhat, 1988), it was adapted in several studies to continuously study LPMOs (Breslmayr et al., 2022; Filandr et al., 2020; Hansson et al., 2017). Due to the insolubility of PASC in aqueous solutions, light is scattered when passing through a PASC suspension. The amount of scattered light changes, when an LPMO is acting on PASC eventually solubilizing a fraction of the amorphous cellulose preparation leading to less light scattering and consequently a decrease in measured absorbance over time. A common limitation of turbidimetric measurements is that only certain substrate concentrations can be used, otherwise absorbance values will be too high due to intense light scattering. A suggested assay setup is outlined in Section 3.3.

3.1 Equipment

- Hitachi U-3000 spectrophotometer equipped with a thermoelectric cell holder and an SPR-10 temperature and stirrer controller
- Julabo SE-12 heating circulator
- 6 mm cross-shaped magnetic stirring bar for 1 cm cuvette
- Quartz cuvette (Hellma, macro, 3.5 mL, 1 cm pathlength)

3.2 Preparation of stock solutions

All aqueous solutions, reagents and buffers have to be prepared in highly pure water.

3.2.1 Phosphoric acid swollen cellulose (PASC)

For the preparation of highly amorphous PASC from microcrystalline cellulose (Avicel, Sigma Aldrich), 4 g of cellulose powder are suspended in 100 mL ice-cold phosphoric acid (85% w/w) and stirred at 4 °C. After 18 h 1.9 L of cold deionized water is added and the dissolved cellulose is precipitated forming a white cellulose gel. Swollen cellulose is recovered by vacuum filtration using a paper filter and washed with 2 L of ice-cold deionized water. The resulting material is further washed with 2 L of sodium bicarbonate (2.0% w/v) to neutralize residual phosphoric acid and 1 L of sodium phosphate buffer (50 mM, pH 6.0). PASC is further homogenized using an Ultra Turrax dispenser. The preparation can be stored at 4 °C and stable for 4 weeks.

The dry weight of the PASC is determined by transferring aliquots (~1 g) into pre-weighed glass tubes and drying at 105 °C to a constant weight. The dry weight is calculated as $g g^{-1}$ and for the standard assay a PASC stock with a concentration of $4 g L^{-1}$ dry weight is prepared.

Note: Previous studies showed that the activity of LPMOs and cellulolytic enzymes, in general, is strongly affected by the crystallinity as well as by the degree of polymerization of their cellulosic substrates (Hall, Bansal, Lee, Realff, & Bommarius, 2010; Valenzuela et al., 2019). As different PASC preparations may show certain differences regarding those characteristics as a result of variations in the production procedure and reagents used, we recommend using the same PASC stock for comparative characterizations (Percival Zhang, Cui, Lynd, & Kuang, 2006).

Note: Measuring the optical density at 620 nm a linear relation was determined for PASC concentrations between 0 and $1.4 g L^{-1}$. A final concentration of $0.8 g L^{-1}$ is recommended for the assay (Filandr et al., 2020).

3.2.2 Ascorbate stock solution

A 10 mM stock of ascorbate is prepared by dissolving 17.6 mg of ascorbic acid in 10 mL of highly pure water. The solution should be stored at 4 °C or on ice and has to be prepared fresh every day.

Note: The use of highly pure water for the preparation of stocks of ascorbic acid is highly recommended as traces of metal ions promote autooxidation reactions and the formation of reactive oxygen species (Eijsink et al., 2019).

3.2.3 Assay buffer

To obtain a final concentration of 50 mM sodium phosphate pH 6.0 in the assay, an assay buffer stock with a concentration of 83 mM is used. For the preparation of 100 mL, 0.996 g of anhydrous sodium dihydrogen phosphate is dissolved in 90 mL of deionized water. The pH is adjusted to 6.0 using 1 M sodium hydroxide and the volume is adjusted to 100 mL with deionized water. The buffer is stored at room temperature for a maximum of 2 weeks.

3.2.4 LPMO preparation

For the turbidimetric determination of LPMO activity purified LPMO preparations should be used. A final concentration of 3.0 µM LPMO in the assay is recommended to obtain the best results. Dilutions should be prepared in deionized water or assay buffer. The sample should be stored on 4 °C or on ice.

Note: The turbidimetric LPMO assay should only be used for highly pure LPMO preparations as hydrolytic cellulose-degrading enzymes would significantly influence the results of the measurements.

3.3 Turbidimetric measurement protocol

1. A spectrophotometer equipped with a temperature-controlled cuvette holder set to 30 °C and a built-in stirrer is necessary for the measurement
2. Use a 3 mL quartz cuvette (1 cm pathlength) with a 6 mm cross-shopped stirrer bar for the measurement
3. Prewarm the PASC stock solution and the assay buffer stock solution to 30 °C
4. Transfer 1.8 mL of buffer stock solution (83 mM, pH 6.0) into the quartz cuvette
5. Start the stirrer and set stirring rate to 400 rpm
6. Add 0.6 mL PASC ($4\,g\,L^{-1}$)
7. Add 0.3 mL LPMO stock (30 µM)
8. Start recording the optical density at 620 nm and equilibrate for at least 30 s
9. Start the reaction by the addition of 0.3 mL ascorbate stock (10 mM)
10. Record the decrease of the optical density at 620 nm for 900 s

Control reaction without LPMO and without ascorbate should be performed to validate the experimental setup. These control reactions should result in stable baselines without significant slopes of the optical density.

3.4 Activity calculation

The initial decrease in absorbance is fitted using a linear function and the rate is expressed as ΔOD_{620} min^{-1} (Fig. 4). Following the LPMO-catalyzed reaction on PASC by turbidimetry allows to directly investigate LPMO activity on a natural amorphous substrate. Nevertheless, as the method is based on light scattering and PASC represents a highly polydisperse suspension of cellulose particles, no authentic turnover numbers can be derived from the absorbance values. The assay is therefore limited and can be used for qualitative or comparative studies.

The sensitivity of the turbidimetric assay can be increased by the addition of H_2O_2 to enhance the reaction rate of LPMO on PASC. Filandr and co-workers reported two different approaches either by the titration of H_2O_2 aliquots or by the in-situ production of H_2O_2 using glucose oxidase as an auxiliary enzyme (Filandr et al., 2020). However, a too high concentration of H_2O_2 has to be avoided since it leads to fast inactivation of LPMO (Bissaro et al., 2017; Kuusk et al., 2018).

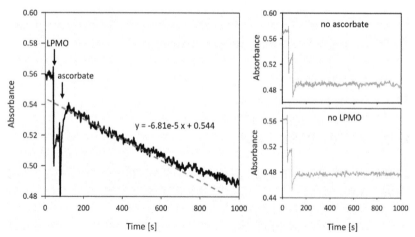

Fig. 4 Turbidimetric determination of LPMO activity using 0.8 mg mL^{-1} PASC as substrate. The assay was performed as described in the protocol with a final concentration of 3 µM NcLPMO9C. In the control experiments (right panels) water was added instead of LPMO or ascorbate, respectively. Data are shown as raw values measures at 620 nm using a Hitachi U-3000 spectrophotometer at a measuring speed of 2 s^{-1} and no data correction was applied.

4. Summary and concluding remarks

Within a short period of time, a multitude of methods has been developed to characterize LPMOs acting on their natural complex carbohydrate substrates (Eijsink et al., 2019). The need for easy-to-use assays to detect LPMOs led to the development of several spectroscopic assays. Continuous colorimetric spectroscopic assays have proven to be useful to monitor expression levels of LPMOs, perform basic characterizations of newly expressed LPMOs or test the effect of different reaction conditions such as pH and buffer species on LPMO reactivity (Breslmayr et al., 2019; Kittl et al., 2012). Although useful, the measured activities of these assays do not correspond to the catalytic turnover of polysaccharide substrates. Turbidimetric measurements are currently the only continuous assay available to monitor the real peroxygenase reaction of LPMOs acting on amorphous cellulose preparations (PASC) (Filandr et al., 2020). However, this approach is limited to a narrow concentration range of the substrate and it is challenging to obtain enzyme activities in μmol/min.

In conclusion, continuous assays are an important tool to characterize LPMO-catalyzed reactions. The wide application of the 2,6-DMP assay highlights the need for reliable continuous assays. It is therefore of highest interest to focus future research on developing a continuous LPMO assay that allows the measurement of LPMO activity on any kind of soluble or solid carbohydrate substrate.

References

Aachmann, F. L., Sørlie, M., Skjåk-Bræk, G., Eijsink, V. G. H., & Vaaje-Kolstad, G. (2012). NMR structure of a lytic polysaccharide monooxygenase provides insight into copper binding, protein dynamics, and substrate interactions. *Proceedings of the National Academy of Sciences of the United States of America*, *109*(46), 18779–18784. https://doi.org/10.1073/PNAS.1208822109.

Agger, J. W., Isaksen, T., Várnai, A., Vidal-Melgosa, S., Willats, W. G. T., Ludwig, R., et al. (2014). Discovery of LPMO activity on hemicelluloses shows the importance of oxidative processes in plant cell wall degradation. *Proceedings of the National Academy of Sciences of the United States of America*, *111*(17), 6287–6292. https://doi.org/10.1073/PNAS.1323629111.

Askarian, F., Uchiyama, S., Masson, H., Sørensen, H. V., Golten, O., Bunæs, A. C., et al. (2021). The lytic polysaccharide monooxygenase CbpD promotes Pseudomonas aeruginosa virulence in systemic infection. *Nature Communications*, *12*(1). https://doi.org/10.1038/s41467-021-21473-0.

Beeson, W. T., Phillips, C. M., Cate, J. H. D., & Marletta, M. A. (2012). Oxidative cleavage of cellulose by fungal copper-dependent polysaccharide monooxygenases. *Journal of the American Chemical Society*, *134*(2), 890–892. https://doi.org/10.1021/ja210657t.

Bissaro, B., Kommedal, E., Røhr, Å. K., & Eijsink, V. G. H. (2020). Controlled depolymerization of cellulose by light-driven lytic polysaccharide oxygenases. *Nature Communications*, *11*(1). https://doi.org/10.1038/s41467-020-14744-9.

Bissaro, B., Røhr, Å. K., Müller, G., Chylenski, P., Skaugen, M., Forsberg, Z., et al. (2017). Oxidative cleavage of polysaccharides by monocopper enzymes depends on H2O2. *Nature Chemical Biology*, *13*, 1123–1128. https://doi.org/10.1038/nchembio.2470.

Bissaro, B., Streit, B., Isaksen, I., Eijsink, V. G. H., Beckham, G. T., DuBois, J. L., et al. (2020). Molecular mechanism of the chitinolytic peroxygenase reaction. *Proceedings of the National Academy of Sciences of the United States of America*, *117*(3), 1504–1513. https://doi.org/10.1073/pnas.1904889117.

Bissaro, B., Várnai, A., Røhr, Å. K., & Eijsink, V. G. H. (2018). Oxidoreductases and reactive oxygen species in conversion of lignocellulosic biomass. *Microbiology and Molecular Biology Reviews*, *82*(4). https://doi.org/10.1128/MMBR.00029-18.

Breslmayr, E., Daly, S., Požgajčić, A., Chang, H., Rezić, T., Oostenbrink, C., et al. (2019). Improved spectrophotometric assay for lytic polysaccharide monooxygenase. *Biotechnology for Biofuels*, *12*, 283. https://doi.org/10.1186/s13068-019-1624-3.

Breslmayr, E., Hanžek, M., Hanrahan, A., Leitner, C., Kittl, R., Šantek, B., et al. (2018). A fast and sensitive activity assay for lytic polysaccharide monooxygenase. *Biotechnology for Biofuels*, *11*, 79. https://doi.org/10.1186/s13068-018-1063-6.

Breslmayr, E., Poliak, P., Požgajči, A., Schindler, R., Kracher, D., Oostenbrink, C., et al. (2022). Inhibition of the peroxygenase lytic polysaccharide monooxygenase by carboxylic acids and amino acids. *Antioxidants*, *11*(6), 1096. https://doi.org/10.3390/ANTIOX11061096.

Ceccherini, S., Rahikainen, J., Marjamaa, K., Sawada, D., Grönqvist, S., & Maloney, T. (2021). Activation of softwood Kraft pulp at high solids content by endoglucanase and lytic polysaccharide monooxygenase. *Industrial Crops and Products*, *166*, 113463. https://doi.org/10.1016/J.INDCROP.2021.113463.

Courtade, G., Forsberg, Z., Heggset, E. B., Eijsink, V. G. H., & Aachmann, F. L. (2018). The carbohydrate-binding module and linker of a modular lytic polysaccharide monooxygenase promote localized cellulose oxidation. *Journal of Biological Chemistry*, *293*(34), 13006–13015. https://doi.org/10.1074/JBC.RA118.004269.

Drula, E., Garron, M. L., Dogan, S., Lombard, V., Henrissat, B., & Terrapon, N. (2022). The carbohydrate-active enzyme database: Functions and literature. *Nucleic Acids Research*, *50*(D1), D571–D577. https://doi.org/10.1093/NAR/GKAB1045.

Eijsink, V. G. H., Petrovic, D., Forsberg, Z., Mekasha, S., Røhr, Å. K., Várnai, A., et al. (2019). On the functional characterization of lytic polysaccharide monooxygenases (LPMOs). *Biotechnology for biofuels*, *1*(58). https://doi.org/10.1186/s13068-019-1392-0.

Filandr, F., Man, P., Halada, P., Chang, H., Ludwig, R., & Kracher, D. (2020). The H2O2-dependent activity of a fungal lytic polysaccharide monooxygenase investigated with a turbidimetric assay. *Biotechnology for Biofuels*, *13*, 37. https://doi.org/10.1186/s13068-020-01673-4.

Forsberg, Z., Vaaje-kolstad, G., Westereng, B., Bunsæ, A. C., Stenstrøm, Y., Mackenzie, A., et al. (2011). Cleavage of cellulose by a CBM33 protein. *Protein Science*, *20*(9), 1479–1483. https://doi.org/10.1002/PRO.689.

Frommhagen, M., Mutte, S. K., Westphal, A. H., Koetsier, M. J., Hinz, S. W. A., Visser, J., et al. (2017). Boosting LPMO-driven lignocellulose degradation by polyphenol oxidase-activated lignin building blocks. *Biotechnology for Biofuels*, *10*, 121. https://doi.org/10.1186/s13068-017-0810-4.

Guo, X., An, Y., Jiang, L., Zhang, J., Lu, F., & Liu, F. (2022). The discovery and enzymatic characterization of a novel AA10 LPMO from bacillus amyloliquefaciens with dual substrate specificity. *International Journal of Biological Macromolecules*, *203*, 457–465. https://doi.org/10.1016/J.IJBIOMAC.2022.01.110.

Hall, M., Bansal, P., Lee, J. H., Realff, M. J., & Bommarius, A. S. (2010). Cellulose crystallinity—A key predictor of the enzymatic hydrolysis rate. *FEBS Journal*, *277*(6), 1571–1582. https://doi.org/10.1111/j.1742-4658.2010.07585.x.

Hansson, H., Karkehabadi, S., Mikkelsen, N., Douglas, N. R., Kim, S., Lam, A., et al. (2017). High-resolution structure of a lytic polysaccharide monooxygenase from Hypocrea jecorina reveals a predicted linker as an integral part of the catalytic domain. *The Journal of Biological Chemistry*, *292*(46), 19099–19109. https://doi.org/10.1074/JBC.M117.799767.

Hedison, T. M., Breslmayr, E., Shanmugam, M., Karnpakdee, K., Heyes, D. J., Green, A. P., et al. (2021). Insights into the H2O2-driven catalytic mechanism of fungal lytic polysaccharide monooxygenases. *The FEBS Journal*, *288*(13), 4115–4128. https://doi.org/10.1111/FEBS.15704.

Hegnar, O. A., Østby, H., Petrovi, D. M., Olsson, L., Várnai, A., & Eijsink, V. G. H. (2021). Quantifying oxidation of cellulose-associated glucuronoxylan by two lytic polysaccharide monooxygenases from Neurospora crassa. *Applied and Environmental Microbiology*, *87*(24), e01652-21. https://doi.org/10.1128/AEM.01652-21.

Hemsworth, G. R., Davies, G. J., & Walton, P. H. (2013). Recent insights into copper-containing lytic polysaccharide mono-oxygenases. *Current Opinion in Structural Biology*, *23*(5), 660–668. https://doi.org/10.1016/J.SBI.2013.05.006.

Hemsworth, G. R., Johnston, E. M., Davies, G. J., & Walton, P. H. (2015). Lytic polysaccharide monooxygenases in biomass conversion. *Trends in Biotechnology*, *33*(12), 747–761. https://doi.org/10.1016/J.TIBTECH.2015.09.006.

Hemsworth, G. R., Taylor, E. J., Kim, R. Q., Gregory, R. C., Lewis, S. J., Turkenburg, J. P., et al. (2013). The copper active site of CBM33 polysaccharide oxygenases. *Journal of the American Chemical Society*, *135*(16), 6069–6077. https://doi.org/10.1021/ja402106e.

Horn, S. J., Vaaje-Kolstad, G., Westereng, B., & Eijsink, V. G. H. (2012). Novel enzymes for the degradation of cellulose. *Biotechnology for Biofuels*, *5*(1). https://doi.org/10.1186/1754-6834-5-45.

Isaksen, T., Westereng, B., Aachmann, F. L., Agger, J. W., Kracher, D., Kittl, R., et al. (2014). A C4-oxidizing lytic polysaccharide monooxygenase cleaving both cellulose and cello-oligosaccharides. *The Journal of Biological Chemistry*, *289*(5), 2632–2642. https://doi.org/10.1074/JBC.M113.530196.

Johansen, K. S. (2016). Discovery and industrial applications of lytic polysaccharide monooxygenases. *Biochemical Society Transactions*, *44*(1), 143–149. https://doi.org/10.1042/BST20150204.

Jones, S. M., Transue, W. J., Meier, K. K., Kelemen, B., & Solomon, E. I. (2020). Kinetic analysis of amino acid radicals formed in H$_2$O$_2$-driven Cu^I LPMO reoxidation implicates dominant homolytic reactivity. *Proceedings of the National Academy of Sciences*, *117*(22), 11916–11922. https://doi.org/10.1073/PNAS.1922499117.

Karayannis, M. I., Samios, D. N., & Gousetis, C. H. P. (1977). A study of the molar absorptivity of ascorbic acid at different wavelengths and pH values. *Analytica Chimica Acta*, *93*, 275–279. https://doi.org/10.1016/0003-2670(77)80032-9.

Keller, M. B., Felby, C., Labate, C. A., Pellegrini, V. O. A., Higasi, P., Singh, R. K., et al. (2020). A simple enzymatic assay for the quantification of C1-specific cellulose oxidation by lytic polysaccharide monooxygenases. *Biotechnology Letters*, *42*, 93–102. https://doi.org/10.1007/S10529-019-02760-9.

Kittl, R., Kracher, D., Burgstaller, D., Haltrich, D., & Ludwig, R. (2012). Production of four Neurospora crassa lytic polysaccharide monooxygenases in Pichia pastoris monitored by a fluorimetric assay. *Biotechnology for Biofuels*, *5*(1), 79. https://doi.org/10.1186/1754-6834-5-79.

Kojima, Y., Várnai, A., Ishida, T., Sunagawa, N., Petrovic, D. M., Igarashi, K., et al. (2016). A lytic polysaccharide monooxygenase with broad xyloglucan specificity from the brown-rot fungus Gloeophyllum trabeum and its action on cellulose-xyloglucan complexes. *Applied and Environmental Microbiology*, *82*(22), 6557–6572. https://doi.org/10.1128/AEM.01768-16.

Kont, R., Bissaro, B., Eijsink, V. G. H., & Väljamäe, P. (2020). Kinetic insights into the peroxygenase activity of cellulose-active lytic polysaccharide monooxygenases (LPMOs). *Nature Communications*, *11*, 5786. https://doi.org/10.1038/S41467-020-19561-8.

Kracher, D., Andlar, M., Furtmüller, P. G., & Ludwig, R. (2018). Active-site copper reduction promotes substrate binding of fungal lytic polysaccharide monooxygenase and reduces stability. *The Journal of Biological Chemistry*, *293*(5), 1676–1687. https://doi.org/10.1074/JBC.RA117.000109.

Kracher, D., Forsberg, Z., Bissaro, B., Gangl, S., Preims, M., Sygmund, C., et al. (2020). Polysaccharide oxidation by lytic polysaccharide monooxygenase is enhanced by engineered cellobiose dehydrogenase. *FEBS Journal*, *287*(5), 897–908. https://doi.org/10.1111/febs.15067.

Kracher, D., Scheiblbrandner, S., Felice, A. K. G., Breslmayr, E., Preims, M., Ludwicka, K., et al. (2016). Extracellular electron transfer systems fuel cellulose oxidative degradation. *Science*, *352*(6289), 1098–1101. https://doi.org/10.1126/science.aaf3165.

Kuusk, S., Bissaro, B., Kuusk, P., Forsberg, Z., Eijsink, V. G. H., Sørlie, M., et al. (2018). Kinetics of H2O2-driven degradation of chitin by a bacterial lytic polysaccharide monooxygenase. *Journal of Biological Chemistry*, *293*(2), 523–531. https://doi.org/10.1074/jbc.M117.817593.

Kuusk, S., Kont, R., Kuusk, P., Heering, A., Sørlie, M., Bissaro, B., et al. (2019). Kinetic insights into the role of the reductant in H2O2-driven degradation of chitin by a bacterial lytic polysaccharide monooxygenase. *Journal of Biological Chemistry*, *294*(5), 1516–1528. https://doi.org/10.1074/jbc.RA118.006196.

Kuusk, S., & Väljamäe, P. (2021). Kinetics of H2O2-driven catalysis by a lytic polysaccharide monooxygenase from the fungus Trichoderma reesei. *Journal of Biological Chemistry*, *297*(5), 101256. https://doi.org/10.1016/J.JBC.2021.101256.

Loose, J. S. M., Forsberg, Z., Fraaije, M. W., Eijsink, V. G. H., & Vaaje-Kolstad, G. (2014). A rapid quantitative activity assay shows that the Vibrio cholerae colonization factor GbpA is an active lytic polysaccharide monooxygenase. *FEBS Letters*, *588*(18), 3435–3440. https://doi.org/10.1016/J.FEBSLET.2014.07.036.

Ma, L., Li, G., Xu, H., Liu, Z., Wan, Q., Liu, D., et al. (2022). Structural and functional study of a novel lytic polysaccharide monooxygenase cPMO2 from compost sample in the oxidative degradation of cellulose. *Chemical Engineering Journal*, *433*(1), 134509. https://doi.org/10.1016/J.CEJ.2022.134509.

Marjamaa, K., Rahikainen, J., Karjalainen, M., Maiorova, N., Holopainen-Mantila, U., Molinier, M., et al. (2022). Oxidative modification of cellulosic fibres by lytic polysaccharide monooxygenase AA9A from Trichoderma reesei. *Cellulose*, *29*, 6021–6038. https://doi.org/10.1007/s10570-022-04648-w.

Mehlhorn, H., Lelandais, M., Korth, H. G., & Foyer, C. H. (1996). Ascorbate is the natural substrate for plant peroxidases. *FEBS Letters*, *378*(3), 203–206. https://doi.org/10.1016/0014-5793(95)01448-9.

Méndez-Líter, J. A., Ayuso-Fernández, I., Csarman, F., de Eugenio, L. I., Míguez, N., Plou, F. J., et al. (2021). Lytic polysaccharide monooxygenase from Talaromyces amestolkiae with an enigmatic linker-like region: The role of this enzyme on cellulose saccharification. *International Journal of Molecular Sciences*, *22*(24), 13611. https://doi.org/10.3390/IJMS222413611.

Merino, S. T., & Cherry, J. (2007). Progress and challenges in enzyme development for biomass utilization. *Advances in Biochemical Engineering/Biotechnology*, *108*, 95–120. https://doi.org/10.1007/10_2007_066.

Müller, G., Chylenski, P., Bissaro, B., Eijsink, V. G. H., & Horn, S. J. (2018). The impact of hydrogen peroxide supply on LPMO activity and overall saccharification efficiency of a commercial cellulase cocktail. *Biotechnology for Biofuels*, *11*, 209. https://doi.org/10.1186/s13068-018-1199-4.

Nummi, M., Fox, P. C., Niku-Paavola, M. L., & Enari, T. M. (1981). Nephelometric and turbidometric assays of cellulase activity. *Analytical Biochemistry*, *116*(1), 133–136. https://doi.org/10.1016/0003-2697(81)90334-1.

Obolenskaya, A. V., Yelnitskaya, Z. P., & Leonovitch, A. A. (1991). *Laboratory studies of wood and cellulose chemistry*. Moscow: Ekologiya.

Percival Zhang, Y. H., Cui, J., Lynd, L. R., & Kuang, L. R. (2006). A transition from cellulose swelling to cellulose dissolution by o-phosphoric acid: Evidence from enzymatic hydrolysis and supramolecular structure. *Biomacromolecules*, *7*(2), 644–648. https://doi.org/10.1021/bm050799c.

Phillips, C. M., Beeson, W. T., Cate, J. H., & Marletta, M. A. (2011). Cellobiose dehydrogenase and a copper-dependent polysaccharide monooxygenase potentiate cellulose degradation by Neurospora crassa. *ACS Chemical Biology*, *6*(12), 1399–1406. https://doi.org/10.1021/cb200351y.

Quinlan, R. J., Sweeney, M. D., Lo Leggio, L., Otten, H., Poulsen, J. C. N., Johansen, K. S., et al. (2011). Insights into the oxidative degradation of cellulose by a copper metalloenzyme that exploits biomass components. *Proceedings of the National Academy of Sciences of the United States of America*, *108*(37), 15079–15084. https://doi.org/10.1073/pnas.1105776108.

Rieder, L., Petrović, D., Väljamäe, P., Eijsink, V. G. H., & Sørlie, M. (2021). Kinetic characterization of a putatively chitin-active LPMO reveals a preference for soluble substrates and absence of monooxygenase activity. *ACS Catalysis*, *11*(18), 11685–11695. https://doi.org/10.1021/acscatal.1c03344.

Rieder, L., Stepnov, A. A., Sørlie, M., & Eijsink, V. G. H. (2021). Fast and specific peroxygenase reactions catalyzed by fungal mono-copper enzymes. *Biochemistry*, *60*(47), 3633–3643. https://doi.org/10.1021/acs.biochem.1c00407.

Rodrigues, J.v., & Gomes, C. M. (2010). Enhanced superoxide and hydrogen peroxide detection in biological assays. *Free Radical Biology & Medicine*, *49*(1), 61–66. https://doi.org/10.1016/J.FREERADBIOMED.2010.03.014.

Sabbadin, F., Urresti, S., Henrissat, B., Avrova, A. O., Welsh, L. R. J., Lindley, P. J., et al. (2021). Secreted pectin monooxygenases drive plant infection by pathogenic oomycetes. *Science*, *373*(6556), 774–779. https://doi.org/10.1126/science.abj1342.

Semenova, M. V., Gusakov, A. V., Telitsin, V. D., Rozhkova, A. M., Kondratyeva, E. G., & Sinitsyn, A. P. (2020). Purification and characterization of two forms of the homologously expressed lytic polysaccharide monooxygenase (PvLPMO9A) from Penicillium verruculosum. *Biochimica et Biophysica Acta (BBA)—Proteins and Proteomics*, *1868*(1), 140297. https://doi.org/10.1016/J.BBAPAP.2019.140297.

Stepnov, A. A., Christensen, I. A., Forsberg, Z., Aachmann, F. L., Courtade, G., & Eijsink, V. G. H. (2022). The impact of reductants on the catalytic efficiency of a lytic polysaccharide monooxygenase and the special role of dehydroascorbic acid. *FEBS Letters*, *596*(1), 53–70. https://doi.org/10.1002/1873-3468.14246.

Tõlgo, M., Hegnar, O. A., Østby, H., Várnai, A., Vilaplana, F., Eijsink, V. G. H., et al. (2022). Comparison of six lytic polysaccharide monooxygenases from Thermothielavioides terrestris shows that functional variation underlies the multiplicity of LPMO genes in filamentous Fungi. *Applied and Environmental Microbiology*, *88*(6). https://doi.org/10.1128/AEM.00096-22.

Vaaje-Kolstad, G., Horn, S. J., van Aalten, D. M. F., Synstad, B., & Eijsink, V. G. H. (2005). The non-catalytic chitin-binding protein CBP21 from Serratia marcescens is essential for chitin degradation. *Journal of Biological Chemistry*, *280*(31), 28492–28497. https://doi.org/10.1074/JBC.M504468200.

Vaaje-Kolstad, G., Westereng, B., Horn, S. J., Liu, Z., Zhai, H., Sørlie, M., et al. (2010). An oxidative enzyme boosting the enzymatic conversion of recalcitrant polysaccharides. *Science*, *330*(6001), 219–222. https://doi.org/10.1126/science.119223.

Valenzuela, S.v., Valls, C., Schink, V., Sánchez, D., Roncero, M. B., Diaz, P., et al. (2019). Differential activity of lytic polysaccharide monooxygenases on celluloses of different crystallinity. Effectiveness in the sustainable production of cellulose nanofibrils. *Carbohydrate Polymers*, *207*, 59–67. https://doi.org/10.1016/J.CARBPOL.2018.11.076.

Vandhana, T. M., Reyre, J. L., Sushmaa, D., Berrin, J. G., Bissaro, B., & Madhuprakash, J. (2022). On the expansion of biological functions of lytic polysaccharide monooxygenases. *New Phytologist*, *233*(6), 2380–2396. https://doi.org/10.1111/NPH.17921.

Wang, B., Johnston, E. M., Li, P., Shaik, S., Davies, G. J., Walton, P. H., et al. (2018). QM/MM studies into the H2O2-dependent activity of lytic polysaccharide monooxygenases: Evidence for the formation of a caged hydroxyl radical intermediate. *ACS Catalysis*, *8*(2), 1346–1351. https://doi.org/10.1021/acscatal.7b03888.

Wang, D., Li, J., Wong, A. C. Y., Aachmann, F. L., & Hsieh, Y. S. Y. (2018). A colorimetric assay to rapidly determine the activities of lytic polysaccharide monooxygenases. *Biotechnology for Biofuels*, *11*, 215. https://doi.org/10.1186/S13068-018-1211-Z/TABLES/1.

Westereng, B., Arntzen, M. T., Aachmann, F. L., Várnai, A., Eijsink, V. G. H., & Agger, J. W. (2016). Simultaneous analysis of C1 and C4 oxidized oligosaccharides, the products of lytic polysaccharide monooxygenases acting on cellulose. *Journal of Chromatography A*, *1445*, 46–54. https://doi.org/10.1016/J.CHROMA.2016.03.064.

Westereng, B., Loose, J. S. M., Vaaje-Kolstad, G., Aachmann, F. L., Sørlie, M., & Eijsink, V. G. H. (2018). Analytical tools for characterizing cellulose-active lytic polysaccharide monooxygenases (LPMOs). *Methods in Molecular Biology*, *1796*, 219–246. https://doi.org/10.1007/978-1-4939-7877-9_16.

Wood, T. M., & Bhat, K. M. (1988). Methods for measuring cellulase activities. *Methods in Enzymology*, *160*, 87–112. https://doi.org/10.1016/0076-6879(88)60109-1.

Wu, S., Tian, J., Xie, N., Adnan, M., Wang, J., & Liu, G. (2022). A sensitive, accurate, and high-throughput gluco-oligosaccharide oxidase-based HRP colorimetric method for assaying lytic polysaccharide monooxygenase activity. *Biotechnology for Biofuels and Bioproducts*, *15*, 15. https://doi.org/10.1186/s13068-022-02112-2.